A Pictorial Approach to Molecular Bonding and Vibrations

Second Edition

Springer
New York
Berlin
Heidelberg
Barcelona
Budapest
Hong Kong
London
Milan
Paris
Santa Clara
Singapore
Tokyo

John G. Verkade

A Pictorial Approach to Molecular Bonding and Vibrations

Second Edition

With 322 Illustrations

Springer

John G. Verkade
Department of Chemistry
Iowa State University of
Science and Technology
Ames, IA 50011
USA

The interactive instructional software program Node Game is freely available for downloading from http://www.public.iastate.edu/~jverkade/nodegame.html. Windows®, Macintosh®, and Unix® versions are available.

Library of Congress Cataloging-in-Publication Data
Verkade, John G., 1935–
A pictorial approach to molecular bonding and vibrations / John G. Verkade. — 2nd ed.
p. cm.
Rev. ed. of: A pictorial approach to molecular bonding. c1986.
Includes bibliographical references and index.
ISBN 0-387-94811-2 (hbk.: alk. paper)
1. Chemical bonds. I. Verkade, John G., 1935– Pictorial approach to molecular bonding. II. Title.
QD461.V45 1996
541.2′24—dc20 96-19135

Printed on acid-free paper.

Production coordinated by Chernow Editorial Services, Inc., and managed by Bill Imbornoni; manufacturing supervised by Jacqui Ashri.
Typeset by Asco Trade Typesetting Ltd., Hong Kong.
Printed and bound by Maple-Vail Book Manufacturing Group, York, PA.
Printed in the United States of America.

9 8 7 6 5 4 3 2 1

ISBN 0-387-94811-2 Springer-Verlag New York Berlin Heidelberg SPIN 10524022

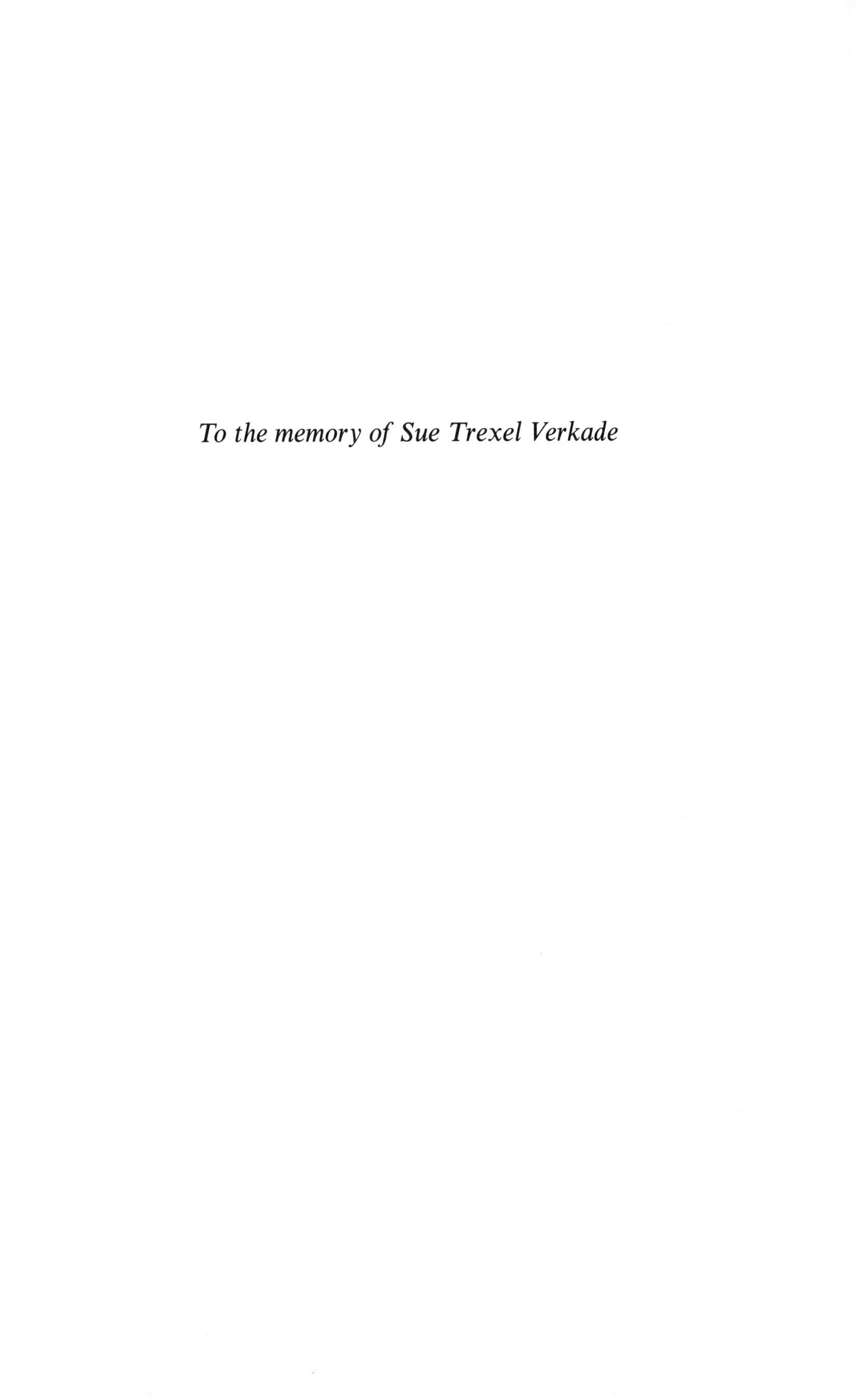

To the memory of Sue Trexel Verkade

Preface to the Second Edition

Many chemistry faculties are currently developing undergraduate and graduate chemistry curricula that include pedagogical reforms as well as student exposure to emerging technologies in chemically related areas. To create room for additional material in a "constant volume" curriculum, something must be dropped. It is the author's belief, however, that by providing students with a simple (but not simplistic) learning tool for pictorializing the bonding in a very wide variety of chemical systems in a highly unified manner, the total time spent teaching bonding concepts in inorganic, organic and physical chemistry courses could actually be significantly reduced, creating room for new topics without the need to eliminate material. To integrate the learning approach to bonding developed in this book across the chemistry curriculum, the following teaching sequence could be considered: Chapters 1–6 in the freshman chemistry year; Chapters 7, 9 (sections 1–3), 11 (sections 1 and 2), and 13 in organic chemistry; and Chapters 8, 9 (sections 2, 4, and 5), 10, 11 (section 3), 12, and 14 in inorganic chemistry. Chapters 1–3 and, to some extent, 4–6 could be briefly reviewed in the physical chemistry sequence as part of the quantum mechanics and bonding covered there. Although organic chemists do not usually include Te_4^{2+} (Chapter 7.1) or P_4 (Chapter 9.2) as organic species, Te_4^{2+} is a good example for demonstrating resonance forms and aromaticity, and the bonding in P_4 is analogous to that in tetrahedrane. This suggested manner in which to cover the material would also serve to demonstrate to the student the applicability of the delocalized and localized generator orbital (GO) bonding approach across chemical disciplines.

At the graduate level, the pictorial approach developed in this book could be taught prior to the inorganic or physical chemistry group theory course. Students who take this route generally find the more formalistic aspects of group theory considerably easier to apply. The generator orbital approach to bonding has also been found to be more attractive to most organic and analytical graduate students than a traditional group theory course, which they tend to avoid in any case.

Should science-oriented high-school students be exposed to the generator approach to bonding? To the extent that these students are learning about atomic orbitals and molecular bonding in their high-school courses, it has been the author's experience that they are easily able to apply the generator orbital concept to diatomic molecules, H_2O, NH_3, and CH_4. Using a computer to do some of the interactive exercises in Node Game would also be a particularly attractive feature of the bonding portion of a high-school course. Admittedly, this book is not written for high-school students, but an enterprising teacher could adapt appropriate portions of this material for class presentation.

An indication of the growing usefulness of the generator orbital approach to molecular bonding is its citation in an undergraduate inorganic text,[a] its inclusion as the basis of a bonding chapter in a forthcoming revision of another,[b] and its citation in two research publications.[c,d]

The present edition differs from the first in four important respects. First, to take advantage of the increasing computer literacy of students and their growing access to computers, I was fortunate to be able to establish a collaboration with Dr. Peter Boysen, a systems analyst in the Iowa State University Computer Science Department, in creating the Node Game computer software with which students can learn the generator approach to drawing delocalized MOs through a set of practice exercises. With the Homework Drawing Board in Node Game, students are able to provide pictorial answers in printed form for homework and "take-home" tests. Peter and I are grateful for the able help of Carolyn Pitcher, an Honors Program chemical engineering major in my research group, and Mark Van Gorp, a graduate student in his group, in creating Node Game. Second, the coverage of the present edition has been expanded to include a chapter on delocalized MOs for one-, two-, and three-dimensional polymers, a chapter on delocalized MOs of spheroidal molecules such as fullerenes, and a chapter on solid-state structures. I am grateful to my departmental colleague, Professor Gordon Miller, for his useful insights in these areas and for coauthoring the first two of these chapters. Third, the problem sets at the end of each chapter have been revised to include molecules that have been discovered since the publication of the first edition in 1986. Fourth, most of the mathematical considerations that are found in the first three chapters of the first edition are now found in appendices to accommodate students who desire a less formal introduction to the very pictorially oriented later chapters. High school students and most undergraduates have found this strategy appealing in the author's experience. Students who are more mathematically inclined will find the

---

[a] G.L. Miessler and D.A. Tarr "Inorganic Chemistry", Prentice-Hall: Englewood Cliffs, New Jersey, 1991.

[b] G. Wulfsberg "Principles of Descriptive Inorganic Chemistry", Brooks/Cole Publishing Co.: Monterey, California, 1987.

[c] D.M.P. Mingos and J.C. Hawes, "Complementary Spherical Electron Density Model", Structure and Bonding, *63*, 3 (1985).

[d] A.J. Stone, Molecular Physics, A new Approach to Bonding in Transition Metal Clusters *41*, 1339 (1980).

appendices helpful. It should be noted that end-of-chapter exercises that are based on material found in the appendices are numbered in bold face type.

Comments and criticisms of this book and of Node Game are welcome. Please feel free to contact me by letter, phone (515-294-5023), fax (515-294-0105) or email (jverkade@iastate.edu).

My heartfelt thanks go to my late wife Sue, whose unfailing support and patience were truly wonderful in helping me finish this project and to whose gentle memory this book is lovingly dedicated. I am also grateful to Denise Miller for helping me prepare the manuscript for this revision and to Deborah Schlagel for corrections and helpful suggestions.

John G. Verkade
Ames, Iowa

Preface to the First Edition

With the development of accurate molecular calculations in recent years, useful predictions of molecular electronic properties are currently being made. It is therefore becoming increasingly important for the nontheoretically oriented chemist to appreciate the underlying principles governing molecular orbital formation and to distinguish them from the quantitative details associated with particular molecules. It seems highly desirable then that the nontheoretician be able to deduce results of general validity without esoteric mathematics. In this context, pictorial reasoning is particularly useful. Such an approach is virtually indispensable if bonding concepts are to be taught to chemistry students early in their careers.

Undergraduate chemistry majors typically find it difficult to formulate molecular orbital schemes, especially delocalized ones, for molecules more complicated than diatomics. The major reason for this regrettable situation is the general impracticability of teaching group theory before students take organic and inorganic courses, wherein the applications of these concepts are most beneficial. Consequently many students graduate with the misconception that the ground rules governing bonding in molecules such as NH_3 are somehow different from those which apply to aromatic systems such as C_6H_6. Conversely, seniors and many graduate students are usually only vaguely, if at all, aware that sigma bonding (like extended pi bonding) can profitably be described in a delocalized manner when discussing the UV-photoelectron spectrum of CH_4, for example. Moreover, many graduate students who have had group theory find it difficult to visualize pictorially the linear combinations of AOs which make up MOs and to picture the relative movements of the atoms in the normal vibrational modes of even very symmetrical polyatomic systems.

In 1968, Professor Klaus Ruedenberg and the author became aware of this dilemma, and to remedy the situation we jointly developed a new course designed to teach the basic elements of chemical bonding to undergraduates. In order to do justice to the point of view of the theoretical as well as the experimental chemist, we team-taught the course for several years and, as a result of many hours of teaching and many more hours of discussion, a set of printed class notes came into existence. Using the nodal symmetries of atomic orbitals, Klaus outlined during one of these warmly acknowledged conversations how the intuitive reasoning associated with the "united atom model," originally introduced by R.S. Mulliken in connection with correlation diagrams, can provide a tool for students to learn how to construct MOs in very simple systems of high symmetry. Further reflection along these lines led the author to determine with simple sketches whether the nodal properties of atomic orbitals placed at the center of more complicated and less symmetrical molecules could be effectively used as a device to generate their MOs as LCAOs. This "generator orbital" method appeared to be widely applicable and, after Klaus justified its generality on group theoretical grounds in 1973, we used this approach successfully ever since in a course on bonding taught mainly to undergraduate majors and interested graduate students. In 1975, Professor David K. Hoffman pointed out that the generator orbital approach can also be used for a pictorial deduction of localized MOs and of normal vibrational modes in molecules. Dave's contributions of these ideas are also most warmly acknowledged and their development is included here.

Although the generator orbital concept can be presented in a mathematical and group theoretical framework, the concept and its applications lend themselves exceedingly well to a pictorial approach. The only prerequisites are high school level chemistry, geometry, physics, and trigonometry. Although integrals, vectors, and matrices are *briefly* touched upon, a working knowledge of these subjects is not necessary for understanding the generator orbital concept and its applications.

Some of the results of the fruitful collaborations with Klaus and David were published in article form.[a] We believe that more widespread pedagogical benefit can be realized by developing a textbook containing the ideas and applications contained in these papers, as well as in the class notes which by 1977 had become quite voluminous. After the appearance of the articles in that year, the three of us completed a preliminary manuscript for such a book and made several serious attempts to bring it into publishable form. Because of the press of other commitments as well as philosophical differences (stemming from our respective scientific backgrounds) concerning the manner of presentation of the material, it devolved in the author, by mutual agreement, to finish the project. The author, therefore, takes responsibility for errors and ambiguities which will undoubtedly be found, and it is hoped that these will be brought to his attention. The contents of essentially all of the first three chapters were adapted from more extensive class notes prepared by Klaus, and substantial portions of Chapters 4, 5, and 6 were developed in more

[a] See D.K. Hoffman, K. Ruedenberg, and J.G. Verkade, "A Novel Pictorial Approach to Teaching MO Bonding Concepts in Polyatomic Molecules," J. Chem. Ed., *54*, 590 (1977); and D.K. Hoffman, K. Ruedenberg, and J.G. Verkade, "Molecular Orbital Bonding Concepts in Polyatomic Molecules: A Novel Pictorial Approach," Structure and Bonding, *33*, 59 (1977).

extensive form by Klaus and David. David also made many helpful suggestions for the remaining chapters.

Many teachers of inorganic chemistry are currently seeking ways to reintroduce more descriptive chemistry at both the undergraduate and graduate levels. Part of the motivation in writing this book was to provide students with some simple tools for rationalizing the bonding in a very wide variety of molecules in a highly unified theoretically sound manner. By utilizing the generator orbital approach, the total time spent in teaching bonding concepts in inorganic, organic, and physical chemistry courses can actually be significantly reduced.

My loving thanks go to my wife, Sue, whose support and patience were truly wonderful in helping me finish this project. I also thank Mrs. Joyce Gilbert and Mrs. Peggy Biskner for their excellent deciphering and typing skills in bringing my handwritten manuscript into readable form. My thanks also go to Professor Walter Struve and the members of my 1984 and 1985 Structure and Bonding classes for reading the final manuscript and making many helpful suggestions.

John G. Verkade
Ames, Iowa
July 1986

Contents

CHAPTER 1

The Orbital Picture for Bound Electrons

The primary purpose of this book is to equip the nontheoretically oriented chemical scientist with simple and effective tools for pictorializing molecular bonding and motions. To accomplish this goal we consider in this chapter and Chapter 2 some fundamental properties of electrons and orbitals, which are required for the introduction of "generator orbitals." Generator orbitals (which will be defined later) are the tools with which the three-dimensional visualization of molecular orbitals is achieved in subsequent chapters. We will see that the vibrational modes of molecules are also easily visualized with this approach. After we apply the generator orbital concept in Chapter 3 to bonding and to vibrational motion in diatomics, we will apply this concept for the same purposes to more complex polyatomic molecules possessing a variety of geometrical shapes.

1.1. Traveling and Standing Waves

Because electrons can behave as either traveling or standing waves, it is important to understand some properties of waves. When a guitar string is plucked, the magnitude of the string displacement is called its *amplitude*. The louder the instrument is played, the greater the amplitude. It is interesting, however, that the amplitude of a plucked string varies over its length, and it also changes with time t. Thus *the amplitude for such a one-dimensional wave is a function of both position* (e.g., along x) *and time*: $f(x, t)$. As we will see shortly, the standing wave pattern in a plucked string fixed at both ends is composed of two traveling waves. In a traveling wave (e.g., as you would see by shaking one end of a rope in an up and down motion) you would observe a sinusoidal pattern moving with constant velocity along the x direction (i.e., the length) of the rope. For more about waves, see appendix I.A.

To see the meaning of λ, suppose the wave is frozen at a fixed time $t = t_0$. The amplitude is now a sinusoidal function of space only (see Figure 1.1). *The distance at which the entire pattern repeats itself is called the wavelength,* λ.

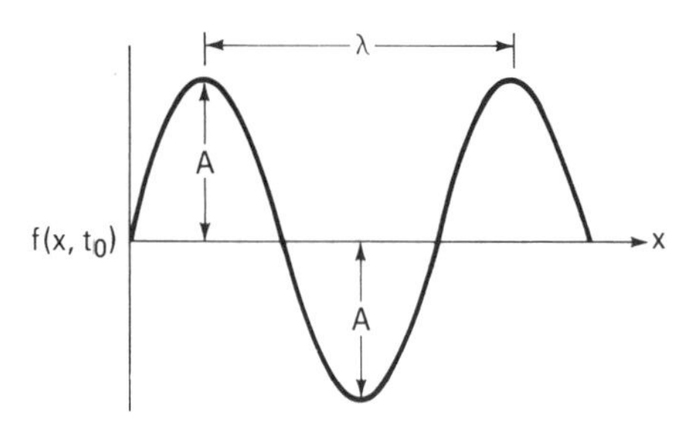

FIGURE 1.1 A "snapshot" of a sinusoidal wave pattern traveling in the x direction with wavelength λ. A is the value of the maximum or minimum wave amplitude. The vertical coordinate is called the *amplitude function.*

The reciprocal of λ is called the *wave number*, $\bar{\nu}$, and it represents the number of full waves per centimeter.

Suppose you view the wave through a vertical slit at a fixed point anywhere along x (say, $x = x_0$) in Figure 1.1. You will then see an amplitude that oscillates from a positive value to an equal but negative value with time (Figure 1.2). The time it takes for the amplitude to return to its initial value is called the *period*, τ. The number of times this phenomenon occurs per second is given by $1/\tau$, and this quantity is called the *frequency*, ν. The speed v of an advancing crest in a traveling wave is given by $v = \lambda\nu$.

The surface of a wave coming in on a beach is a two-dimensional traveling wave (Figure 1.3) since it has troughs and crests possessing length along y. To visualize a three-dimensional traveling wave, consider what happens when a sound wave is produced. The air, which was of constant density before the sound, now possesses regions of higher and lower density which are depicted in Figure 1.4, wherein only the x and z dimensions are shown for clarity. In this figure, the air density in any given plane perpendicular to x is represented by a square lattice work of dots. Although this density in any given yz plane is constant, it decreases to a minimum as you travel in the x direction, and

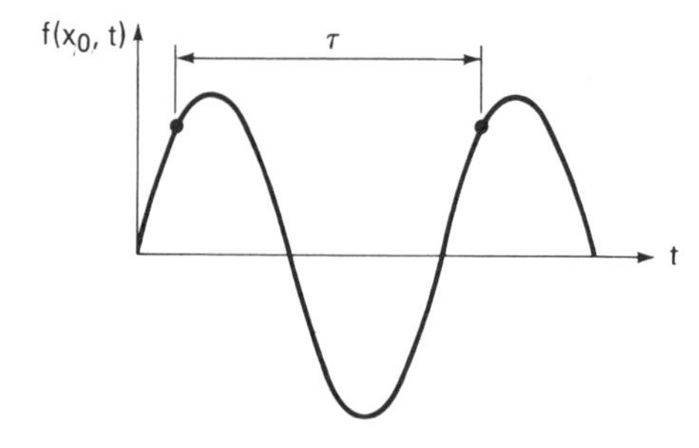

FIGURE 1.2 The behavior of the amplitude of a traveling wave as viewed at a fixed point $x = x_0$, where τ is the period.

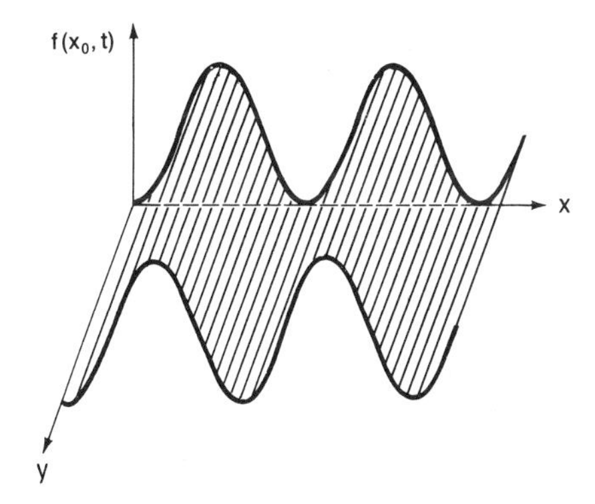

FIGURE 1.3 A two-dimensional traveling wave represented by the surface of an ocean.

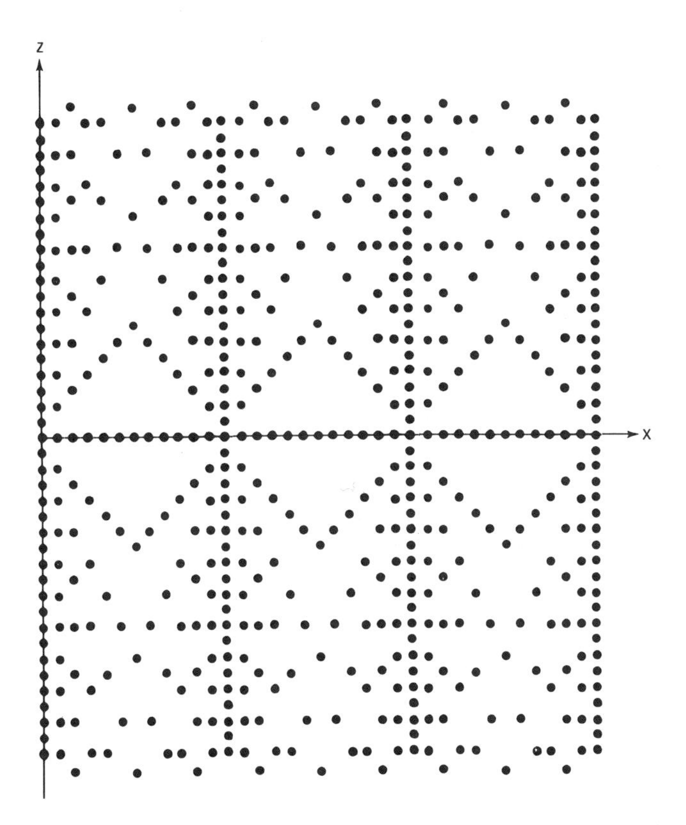

FIGURE 1.4 A snapshot of a two-dimensional cross section of a three-dimensional traveling sound wave caused by a solid yz surface vibrating back and forth across $x = 0$, causing compressions and rarefactions among the air molecules to move along the x direction.

then it increases to a maximum. Thus an air "compression" has formed at $x = 0$ owing to movement to the right of the yz surface during its vibratory motion. Meanwhile, an air "rarefaction" (minimum air density) is formed on the left side of the yz plane during this movement. The propagation of such a plane wave caused by the vibration of the yz surface back and forth in the x direction is characterized by positive and negative deviations from the average density of air. These deviations are called *compressions* and *rarefactions*, respectively. The magnitude of these deviations is the amplitude.

To understand *standing* waves, it helps to examine the traveling waves we have considered for points, lines, or planes where the amplitude is zero and where it is maximum. Zero-amplitude regions are called *nodes*, and *antinodes* are places of maximum amplitude. In a traveling wave, nodes and antinodes move at the same velocity as the whole pattern. In a standing wave, the nodes and antinodes are fixed in space.

There is an important difference between nodes and antinodes in a standing wave. Nodes are time-independent points, lines, or planes, while antinodes oscillate with time. In a one-dimensional wave, the amplitude points oscillate between positive and negative values of z (Figure 1.5) and in a two-dimensional wave the amplitude line does the same. In a three-dimensional plane wave, a density oscillates from a maximum to a minimum in any yz plane along x.

Standing waves occur when *boundary conditions* are imposed. Clamps at both ends of a vibrating string impose the important boundary condition that the string is stationary at these positions and so two nodes appear at these endpoints. The wave, therefore, *must* be a standing wave and in fact the length (L) of the domain between the endpoints must be a multiple of the length between any *two* nodes ($\frac{1}{2}\lambda$). Because of this, the only wavelengths permitted are $\lambda_n = 2L/n$, with $n = 1, 2, 3, 4$, etc. In other words, λ is *quantized* to particular values and the λ_n are the *characteristic values*, or *eigenvalues*, of the wavelength.

It is an interesting observation that two identical traveling waves propagating in opposite directions form a standing wave when superposed. Thus, for example, even though neither one of two superposed traveling one-

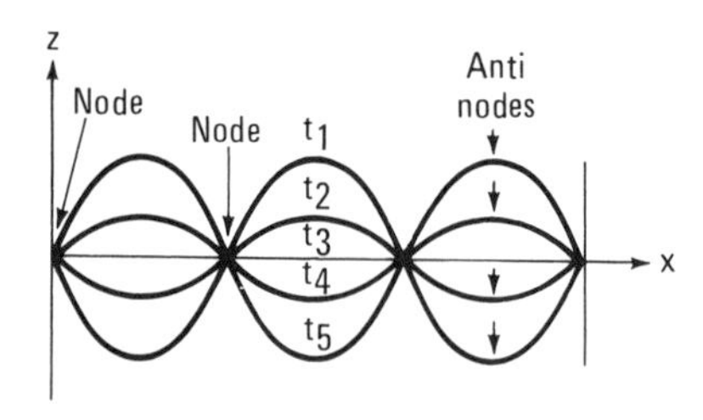

FIGURE 1.5 A standing wave bounded at the extreme ends, showing the time-independent nodes and time-dependent antinodes.

dimensional waves contains points at which the amplitude vanishes *at all times*, such points (the nodes) do exist in the standing wave.

1.2. Wave Energy and Interference

All waves contain energy. Such energy can appear as kinetic and potential energy of matter (as in water waves or sound waves) or it can be electromagnetic (as in light waves). In all cases, the *amount of wave energy per unit volume* (i.e., the energy density or intensity) *is proportional to the square of the amplitude* (i.e., f^2). In a traveling wave, energy flows along the direction of propagation. In a standing wave the energy is *localized* in the regions between the nodes, and the nodes always have zero energy density.

If two or more sound waves having different frequencies encounter each other, their amplitudes superpose to form a new wave $f(x, y, z, t) = f_1(x, y, z, t) + f_2(x, y, z, t) + \cdots + f_n(x, y, z, t)$. Because the amplitudes of the constituent waves enhance one another at some points and cancel each other at others, a new set of oscillations is created. This has important consequences for the composite wave. For example, the energy density at most points in the new wave [i.e., $f^2 = (f_1 + f_2 + f_3 + \cdots + f_n)^2$] is not the same as the sum of the energy densities of the constituent waves [i.e., $f_1^2 + f_2^2 + \cdots + f_n^2$]. The difference of these two sums plotted along the direction of propagation gives an *interference* pattern. Such interference of two frequencies (Figure 1.6a) gives rise to the "beat" (Figure 1.6b) of fluctuating sound intensity heard when two identical or different instruments are playing slightly different pitches (frequencies) at the same intensity.

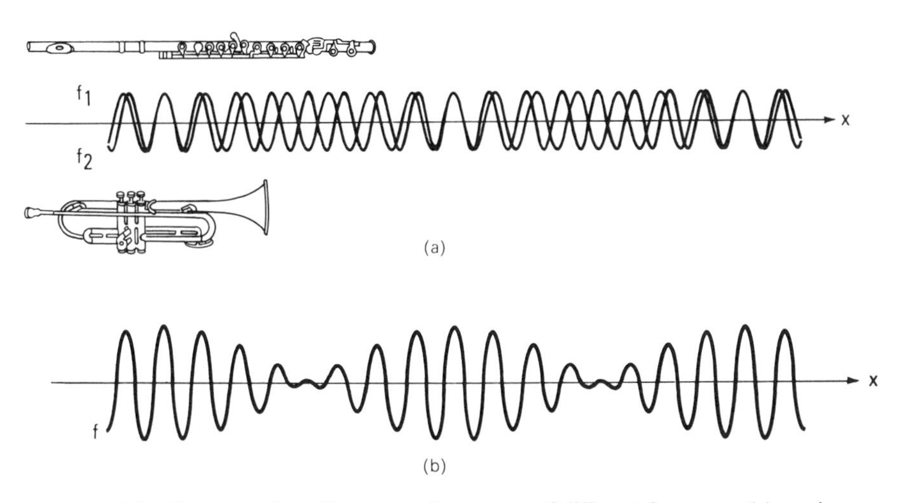

FIGURE 1.6 Superposition of two traveling waves of different frequency (a) to give a composite traveling wave (b) having "beats" of intensity. The closer the two frequencies are to each other, the further apart the beats are heard.

1.3. Electron Orbitals

Under many conditions, electrons behave as waves while under others they possess the characteristics of particles. This should not be disconcerting. The fundamental mathematical theory actually encompasses both the classical (particle) and the quantum mechanical (wave) descriptions harmoniously. It is only because of our limited ability of visualization that these approaches appear to be so different. Free electrons (i.e., in the absence of external forces and thus moving with constant velocity in a straight line) are found to act like three-dimensional traveling plane waves. In certain types of planar molecules such as butadiene ($CH_2{=}CH{-}CH{=}CH_2$) electrons behave like plane waves enclosed in a rectangular box. In a molecule in which double bonds alternate with single bonds (i.e., *conjugated* π bond systems), the π electrons (of which there are four in butadiene) move across *all the carbon atoms and also spend part of their time between the central two carbons*, even though the way we normally write the formula does not indicate this. In many respects, the behavior and properties of these π electrons can be described in terms of a free electron wave constrained to move along the carbon–carbon bond skeleton. Although we will not detail such a description here, these boundary conditions (as in the case of the string clamped at both ends) lead to quantized wavelengths $\lambda_k = 2L/k$, where L is related to the length of the carbon skeleton and $k = 1, 2, 3$, etc. The kinetic energy in ergs of a free electron (which is the only kind of energy it has, even when constrained to remain in a confined space) is given by Equation 1.1 (which we do not derive), where m = the mass of the electron and h is Planck's constant (6.6×10^{-27} erg-sec). This energy quantization is

$$\varepsilon_k = \left(\frac{h^2}{8mL^2}\right)k^2 \tag{1.1}$$

characteristic of *all* electrons in atoms as well as molecules. Such electrons are all restricted to certain regions of space by electrostatic nuclear attractions and are said to be *bound* electrons. *All bound electrons (atomic and molecular) have wave character and possess quantized energy levels.* Although unbound electrons also have wave properties, their energies are not quantized. Quantization is the result of imposing boundaries on a wave. The energy levels of bound electron waves are spaced differently from one atom or molecule to another.

What does an amplitude function of an electron wave signify? For a string it is displacement of a point along z, for a water wave it is a similar displacement of a line parallel to y, and for a sound wave it is a deviation of air density in the xz plane from the average. For electron waves the amplitude function is related to the spatial *distribution of the electron density* and is denoted by $\psi(x, y, z, t)$. How we reconcile the ideas of electrons as particles and as clouds having density will be made more clear shortly. For a free electron, ψ has the mathematical features described earlier for traveling waves, but for a bound electron (i.e., a localized electron wave with boundary conditions on its length) the properties of ψ are those of standing waves. Any *three-dimensional* standing wave is a product of a space function ϕ_n and a time function. For a three-dimensional standing electron wave we can then write:

$$\psi_n(x, y, z, t) = \phi_n(x, y, z)f_n(t). \tag{1.2}$$

where ϕ_n represents the spatial part and f_n the time part of the amplitude function ψ_n. For a one-dimensional standing wave this space factor has the sinusoidal form

$$\phi_n(x) = A \sin \frac{2\pi x}{\lambda_n}. \tag{1.3}$$

It contains the quantized wavelength λ_n discussed earlier which, in turn, is related to the orbital energy ε_n given by Equation 1.1. The *space* amplitudes ϕ_n associated with the first four ε_n values (i.e., $n = 1, 2, 3, 4$) are shown in Figure 1.7. The mathematical form for the time factors in Equation 1.2 is thus the same for all states, and it is only the value of the energy ε_n that differentiates between different states and different systems. *However, the spatial amplitudes for bound electrons differ markedly, and it is these functions that characterize the individual natures of various electronic states.* In general (i.e., for electrons under the influence of forces) these space amplitudes are three-dimensional in nature $[\phi_n(x, y, z)]$ and their shapes are not sinusoidal. One important feature of the electron waves shown in Figure 1.7 that carries over

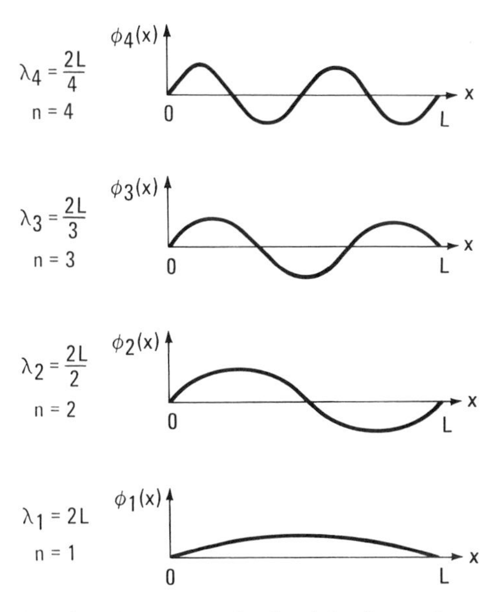

FIGURE 1.7 A series of space amplitudes ϕ_n for four values of ε_n ($n = 1$ to 4).

into *all* standing electron waves is that they, too, possess *nodes*, *positive lobes* (amplitudes), and *negative lobes*.

Single-electron standing waves are called *orbitals*, and this term applies to the total wave function $\psi_n(x, y, z, t)$ as well as to only its spatial amplitude $\phi_n(x, y, z)$. An electron described by ψ_n is said to occupy the orbital ψ_n or ϕ_n.

1.4. Normalization and Orthogonality

These formidable words denote rather simple concepts, which can be developed after we consider how the electron density is related to the electron wave function. Earlier it was stated that the amplitude function ψ "describes the distribution of an electron." How can this be if ψ can have negative values? The answer is that the fraction of electron density in, or the probability of finding an electron in a particular small volume element in space, is related to ψ^2, which must always have a positive value. (If you would like to know more about this, see Appendix I.B.) By adding up all the volume elements, the total fraction of the electron found (or the probability of finding the electron in the total volume) must equal 1. For a one-dimensional standing electron wave (Figure 1.7), the fraction of an electron lying along any small length element along x or the probability of finding it there is related to ϕ^2 as a function of x, which is plotted in Figure 1.8. Because ϕ^2 tells us the essential characteristics of a given system (since f_n^2 does not influence ψ^2) we will deal mainly with the *spatial orbitals* $\phi(x, y, z)$ rather than total functions $\psi(x, y, z, t)$.

To help us understand orbital occupancies of ground and excited states of a system, it is very helpful to relate the geometrical shapes of different orbitals to their energies. Much information of this kind is contained in the nodes and antinodes (lobes) of orbitals. For example, for our one-dimensional electron wave, the energies of the orbitals ϕ_n, as given by Equation 1.1 rise monotonically with n. Note that the number of nodes *inside* the confined space is $n - 1$.

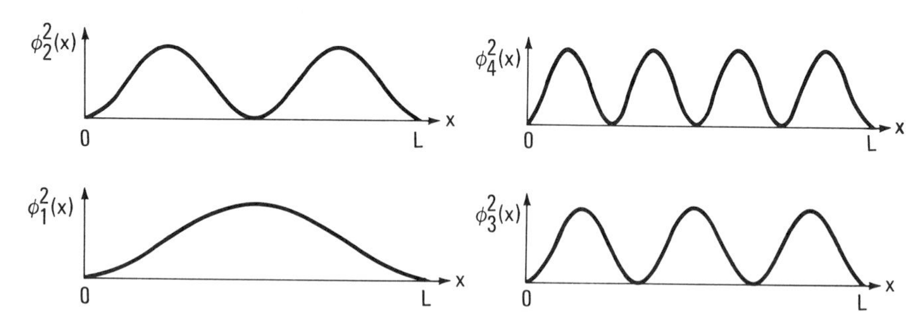

FIGURE 1.8 The ϕ_n of Figure 1.7 are squared to show their probability densities $\phi_n^2(x)$.

To help us generalize this observation to other systems, the concept of *orbital orthogonality* is important. To appreciate this concept consider what happens when we generalize the normalization concept to *two* different spatial orbitals ϕ and ψ. (Since the time factor is no longer necessary for us, ψ will be used to denote another spatial orbital.)

It turns out that the net overlap for any pair of normalized orbitals on an atom is zero. This means that even though orbitals interpenetrate one another, their amplitudes in regions of overlap cancel to zero. In other words, the product of ϕ and ψ will be zero as long as their individual total probability densities (which are related to ϕ^2 and ψ^2) = 1. For more on overlap integrals, see Appendix I.B.

The relation between orbital orthogonality and orbital nodal character can be seen by examining the one-dimensional electron wave. For example, are ϕ_1 and ϕ_2 orthogonal? The product of the two orbitals is plotted in Figure 1.9 and the contributions of the positive and negative lobes cancel by symmetry. *The pivotal point here is that one of the orbitals* (ϕ_2) *has a node where the other* (ϕ_1) *does not.* Similar reasoning applies to three-dimensional orbitals. Each orbital of an atom or molecule has a characteristic nodal structure that renders all the nondegenerate orbitals orthogonal to each other in the same way that ϕ_1 and ϕ_2 in Figure 1.9 are. *As a general rule, the number of nodes of an orbital in a given system increases with the orbital energy.*

It is important to note that observations on mass or charge density of the electron tell us only about ψ^2 but not directly about ψ. In fact, there is no observable property of the electron that allows us to observe ψ directly. To calculate any electronic property, however, we must in general have the amplitude ψ. Thus an important area of research is the development of more accurate wave functions which describe electron behavior in atoms and molecules. It should be noted that the calculation of any observable property always involves the wave function ψ multiplied either by itself or by some derivative of ψ. Consequently, the wave function ψ yields the same value for any property as the wave function $-\psi$. The sign of ψ is therefore arbitrary. In other words, ψ and $-\psi$ are two equivalent alternative mathematical descriptions of the same physical orbital.

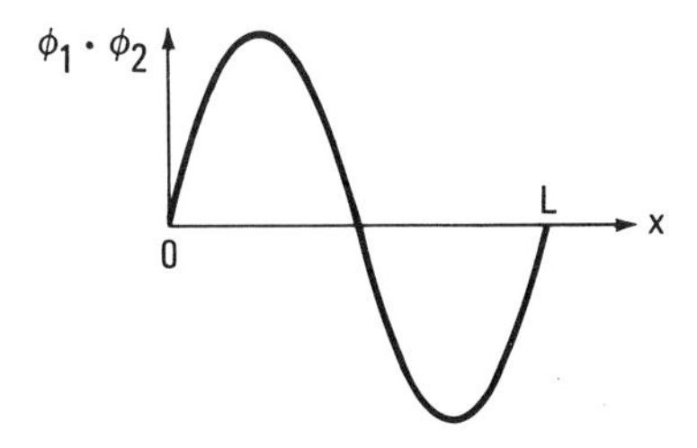

FIGURE 1.9 Plot of $\phi_1 \cdot \phi_2$ showing the total area under the curve to be zero.

1.5. Systems with Many Electrons

In atoms and molecules with many electrons, electrons occupy various orbitals. Also, quantum states of the total system differ by their occupation patterns. The possible occupation patterns are restricted by the *Pauli exclusion principle*, which specifies that each orbital cannot be occupied by more than two electrons. Moreover, two electrons occupying an orbital must have opposite spin rotations. The concept of an electron as a particle spinning on its axis seems odd when we have been discussing the wave nature of an electron in a confined space. Recall that particle and wave descriptions are only mathematical models that help us understand submicroscopic phenomena.

The total energy of a system of electrons in a molecule depends upon the energies of the quantized orbitals that are occupied by electrons. This overall energy can often be approximated by the sum of the energies of all the electrons. In butadiene, for example, we can sketch the *orbital energy level diagram* for the π system as shown in Figure 1.10a. Here, E_0 represents the lowest energy, or *ground state*, of the π system with two electrons in $k = 1$ and two electrons in $k = 2$ (Equation 1.1). Two other possible occupation patterns are shown in Figures 1.10b and c. These patterns are called *excited states* since their total energies are larger than E_0. Thus $E_0 = 2\varepsilon_1 + 2\varepsilon_2$ and $E_1 = 2\varepsilon_1 + \varepsilon_2 + \varepsilon_3$. The *total* energies of the ground and excited states relative to ε_1, and ε_2, and ε_3 are shown in Figure 1.10d. Excited states can be achieved by pumping energy into the ground state. The frequency of electromagnetic radiation required for such *absorption* of energy is given, for example, by $\Delta E = E_1 - E_0 = h\nu$ in Figure 1.11. This frequency can be radiated by

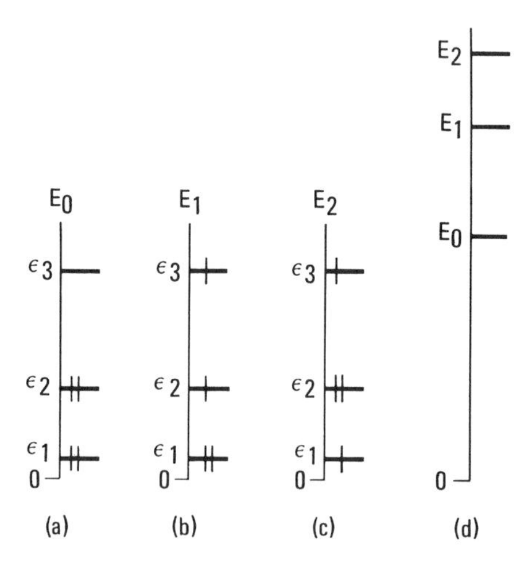

FIGURE 1.10 Orbital energy level diagrams for a ground (a) and two excited [(b), (c)] states and an energy level diagram (d) for the states represented in (a)–(c).

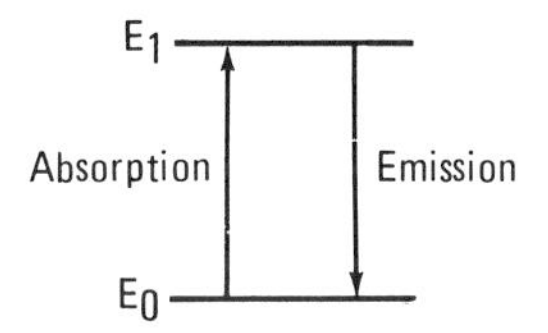

FIGURE 1.11 Energy level diagram depicting absorption of energy required for excitation from E_0 to E_1 and the emission of an equal amount of energy to descend from E_1 to E_0.

the molecule when the excited state returns to the ground state (*emission*). Thus we can see then that molecules have *quantized energy states* as a consequence of electrons occupying *quantized energy levels.* The family of all permitted electronic transitions constitutes the *electronic spectrum* of a molecule.

In many-electron atoms and molecules, the total energy is a complicated function of nuclear charges and interelectronic repulsions. Understandably, when two electrons occupy the same orbital, they experience stronger mutual repulsion because of their negative charge than when they are in separate orbitals. If the energy difference $\varepsilon_3 - \varepsilon_2$ in Figure 1.10a is sufficiently small, Figure 1.10b (with the corresponding change in ε_3 and ε_2 energy levels) would represent the ground state because its overall energy would be lower than that of Figure 1.10a. This is because of the substantial drop in electron–electron repulsion energy that would result compared with double occupation of the ε_2 level in Figure 1.10a. Although the only way to make a proper choice for the ground state is to carry out an accurate calculation or a definitive experiment, most molecules are "closed shell" (i.e., they are *diamagnetic* an even number of electrons and all of them are paired in molecular orbitals). In other words, each molecular orbital is filled up sequentially to a maximum of two electrons, starting with the orbital of lowest energy. Thus the energy separations among the orbitals are large compared with the electron–electron repulsion or pairing energy. One way an "open-shell" (*paramagnetic*) molecule can arise (i.e., one with one or more half-filled molecular orbitals) is by having an odd number of electrons. The last electron then half occupies the last orbital. A second way an open-shell molecule can occur, as mentioned earlier, is for the molecule to have an even number of electrons and for the last two orbitals to have an energy separation that is small compared with the interelectronic repulsion energy. A third way is for two or more orbitals to be "degenerate," i.e., to have the same energy. In such a case we fill those orbitals with one electron at a time. If the total number of electrons available for the degenerate set of orbitals is less than twice the number of orbitals in the set, the molecule will be open-shell. When we carry out these orbital filling procedures, we are following Hund's rule which states that *orbitals (in atoms or molecules) are half filled as long as the orbital energy separation is small compared with the repulsion energy experienced by two electrons with opposite spin occupying the same orbital.* Although the equations describing properties of many-electron systems are obviously complicated, they do have one feature in common; namely, that by observing the properties of the entire system it is not possible to tell which electron occupies which orbital. The electrons are said to be *indistinguishable.*

1.6. Flexibility of Orbital Sets

An interesting feature of many-electron systems, which is related to the indistinguishability of electrons, is the fact that the orbitals that are occupied in a particular state of a system are not unique. That is to say, one can often make certain alterations in the individual occupied orbitals without changing the values of any of the overall electronic properties of the system. This remarkable concept is frequently very useful in understanding the bonding in molecules. The result of applying this concept, known as *orthogonal transformation,* to a set of orbitals ϕ_1, ϕ_2, ϕ_3, for example, is that another set ψ_1, ψ_2, ψ_3 can be constructed from them in which each individual orbital is also normalized, and orthogonal (i.e., "orthonormal"). Appendix I.C shows how this can be done. These two sets of orbitals are equivalent insofar as observing the electronic properties of the system is concerned. For example, even though the *individual* orbital densities of the ϕ_n are different from *individual* orbital densities of the ψ_n ($n = 1, 2, , 3 \ldots$), the total density is independent of which orbital set we choose. The *individual orbital energies* are also different for the two sets but *the total energy is not dependent upon this choice.* In other words, *the state of the whole molecule or atom can be described by either set of orbitals.* An analogy here would be three cubical containers having volumes of 1, 2, and 3 cm^3 and each containing two gaseous atoms. Even though each individual gas density is different, the total gas density is 6 atoms/6 cm^3. By mechanical means, we could transform these boxes into three rectangular containers of equal dimensions (e.g., each having a volume of 2 cm^3). The total gas density would still be 6 atoms/6 cm^3.

If a given molecular or atomic state can be described by a variety of orbital sets (as long as these sets are orthonormal), which set shall we choose? One set that is very useful is called the *canonical* set. *Canonical orbitals are the most fundamental type, and they represent the various standing waves of a single electron as if it were moving alone in the total potential field generated by nuclei and electrons.* Consider the energies of a ground and first excited state given by Equation 1.4:

$$\begin{aligned} E_0 &= 2\varepsilon_1 + 2\varepsilon_2 + 2\varepsilon_3 \\ E_1 &= 2\varepsilon_1 + 2\varepsilon_2 + \varepsilon_3 + \varepsilon_4, \end{aligned} \tag{1.4}$$

and their difference, which is described by Equation 1.5:

$$E_1 - E_0 = \varepsilon_4 - \varepsilon_3. \tag{1.5}$$

Since all the orbitals of E_0 are equally occupied, we can perform an orthogonal transformation to obtain an equivalent set ψ_1', ψ_2', ψ_3' for which the energy can be written as

$$E_0 = 2\varepsilon_1' + 2\varepsilon_2' + 2\varepsilon_3'. \quad (1.6)$$

The orbitals ψ_1 and ψ_2 in E_1 contain equal numbers of electrons (two) while ψ_3 and ψ_4 each contain one. Each of the doubly occupied orbitals can be replaced by an equivalent set ψ_1'', ψ_2'', and each of the singly occupied orbitals can be replaced by an equivalent set ψ_3'', ψ_4'', which is, of course, different from ψ_1'', ψ_2''. The state energy for E_1 is

$$E_1 = 2\varepsilon_1'' + 2\varepsilon_2'' + \varepsilon_3'' + \varepsilon_4''. \quad (1.7)$$

The orbital energies ε_j' and ε_j'' will not be the same and so no simplifications can be made in the energy calculation in Equation 1.8:

$$E_1 - E_0 = 2(\varepsilon_1'' - \varepsilon_1') + 2(\varepsilon_2'' - \varepsilon_2') + \varepsilon_3'' + \varepsilon_4'' - 2\varepsilon_3'. \quad (1.8)$$

We can see then that the canonical orbitals (rather than other orthonormal equivalent orbital sets) furnish a simpler picture of the different energy states of a system and, in particular, of what occurs when molecules emit and absorb energy as a result of electronic transitions.

There are some problems, such as the analysis of chemical similarities between related molecules, for which orbital sets orthonormal to the canonical set are more useful. One such set is the *localized* orbital set in which orbitals are spatially separated insofar as possible. A good example is the π electron *density distribution* for butadiene in Figure 1.12. Note that the total density ρ is the same for both the canonical and the localized view, even though the individual orbital distributions in the two canonical orbitals ϕ_1 and ϕ_2 in (a) of Figure 1.12 differ from those of the localized orbitals ψ_1 and ψ_2 in (b) of this figure. In the localized approach each π bond has associated

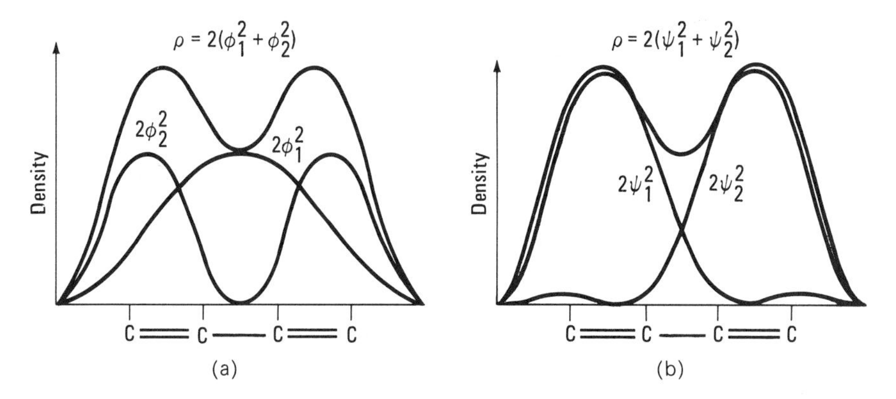

FIGURE 1.12 π Electron density distributions in butadiene provided by canonical (a) and localized (b) orbitals.

with it an orbital of identical shape. To describe the π orbitals in hexatriene ($CH_2{=}CH{-}CH{=}CH{-}CH{=}CH_2$) with localized orbitals, it turns out that all three π bonds have π orbitals that look similar to those in butadiene. Thus *localized orbitals are more enlightening when comparing ground states of similar molecules whereas canonical (more delocalized) orbitals are a more effective aid in comparing different energy states of the same molecule.* As will become clear in subsequent chapters, localized and delocalized orbital pictures can be developed for both π and σ bonding patterns.

Summary

In this chapter the ideas of traveling and standing waves have been used to introduce the concept of electron orbitals. As with standing waves, the energies of bound electrons are quantized. The meaning and some of the usefulness of the electron orbital normalization, orthogonality, and flexibility concepts was discussed and the meaning of canonical (delocalized) and localized orbitals was introduced.

Exercises

1. The crests of water waves in deep water do not curl over on themselves, but they do as they approach the shore. Suggest a reason for this.

2. Shake a length of garden hose or a long rope up and down to make a traveling wave. Why does the amplitude decrease as the wave moves away from your hand?

3. What happens to the frequency (pitch), energy, and wavelength of the wave a drum will produce when you stretch the surface?

4. What happens to the amplitude of a drum wave as the drum is struck with greater force?

5. Is the wave generated in a drum surface when it is struck a traveling or standing wave? Is the sound wave produced a traveling or a standing wave? (Give reasons for your answers.)

6. Two waves, $f(x,t) = \sin 2\pi[(x/\lambda) - (t/\tau)]$ and $f(x,t) = \sin 2\pi[(x/\lambda) + (t/\tau)]$, travel in opposite directions. Show by substituting numbers into the sum of these equations that their sum forms a plot of a standing wave.

7. The lowest two canonical orbitals in butadiene are the electron waves $\phi_1 = (2/5)^{1/2}\sin(\pi/5)x$ and $\phi_2 = (2/5)^{1/2}\sin(\pi/5)2x$, where the distance between two carbons is chosen as a unit of length, so that the distance L within which the waves are contained is 5 units long (see Figure 1.12). The corresponding two localized orbitals are $\psi_1 = 2^{-1/2}(\phi_1 + \phi_2)$ and $\psi_2 = 2^{-1/2}(\phi_1 - \phi_2)$. (a) By substituting numbers, make a plot simultaneously showing $\psi_1(x)$, $\psi_2(x)$, $\psi_1{}^2(x)$, $\psi_2{}^2(x)$, and $[2\psi_1{}^2(x) + 2\psi_2{}^2(x)]$. (b) Do the same as in (a) for $\phi_1(x)$, $\phi_2(x)$, $\phi_1{}^2(x)$, $\phi_2{}^2(x)$, and $[2\phi_1{}^2(x) + 2\phi_2{}^2(x)]$. (c) Show from the plots that the members of each set are orthogonal. (d) Saying that $\langle\phi_1|\phi_1\rangle = 1$ when ϕ_1 is normalized means that 1 is a

unitless number. From the fact that $\phi_1{}^2 = (2/L)(\sin^2 \pi/L)x$, what must be the dimensions of $\phi_1{}^2$? Draw on your plot of $\phi_1{}^2$ a rectangle that has exactly the same area as that under your curve. Find the numerical value of this area and justify why it is unitless.

8. Show that the numbers in Equation 13 satisfy Equations 10–12 in Appendix I.C.

9. Obtain Equation 15 in Appendix I.C. by the procedure suggested in Appendix I.C.

10. Scientists are now able to observe the shapes of electron clouds around molecules with a scanning tunneling electron microscope. Does this technique allow us to distinguish whether the electrons are localized or delocalized?

11. In hexatriene, the canonical orbitals are $\phi_n(x) = (2/7)^{1/2} \sin(\pi/7)nx$. In the ground state, the canonical orbitals ϕ_n and the localized orbitals ψ_n are related by the matrix shown below. Show that this transformation is orthogonal and demonstrate by making plots that ψ_1, ψ_2, ψ_3 are quite localized in the double bond regions. In your plots, let $L = 7$.

	ϕ_1	ϕ_2	ϕ_3
ψ_1	$1/2$	$(1/2)^{1/2}$	$1/2$
ψ_2	$(1/2)^{1/2}$	0	$-(1/2)^{1/2}$
ψ_3	$1/2$	$-(1/2)^{1/2}$	$1/2$

12. Verify Equations 10–12 in Appendix I.C.

Chapter 2

Atomic Orbitals

One of our major goals is to appreciate the shapes and energies of the orbitals that electrons can occupy in atoms and molecules because these are the properties that determine their chemistry. In contrast to the orbitals in a one-dimensional confined space or box that we examined in the previous chapter, atomic and molecular orbitals are three-dimensional standing waves. We first address ourselves to the various shapes of atomic *canonical* AOs (atomic orbitals) and their energies. Canonical AOs have *identical shapes* in all states and in all atoms. They differ only in their *sizes*. Secondly, we will introduce alternative sets of AOs called *hybrids*.

2.1. Shapes of Canonical AOs

The most important features of atomic orbitals are their three-dimensional lobes (where the amplitude function ψ is either positive or negative) and their nodes (where ψ is zero). Because of the *three-dimensional* character of lobes, the nodes that separate lobes of opposite sign must be two-dimensional surfaces. The nodal surfaces of *canonical* AOs are *spheres*, *cones*, or *planes*. We will come back to them shortly.

How big is an orbital? In other words, how large are their *spatial amplitudes*? Because all atomic (and molecular) orbitals extend to infinity, their boundary at infinity can be viewed as a two-dimensional surface, namely, a sphere with a radius $r = \infty$. We still can compare orbital sizes, however, because ψ decays rapidly with increasing r in an approximately exponential fashion, and the rate of decrease in ψ depends upon the orbital type. Thus $\psi(x, y, z) \cong Ke^{-r/\alpha}$ (when $r \gg \alpha$) where $r = (x^2 + y^2 + z^2)^{1/2}$ is the distance of the electron from the atomic (or molecular) center, and K and α are constants that are different for various orbitals. For a little more on spatial amplitudes, see Appendix II.A. For visualizing interatomic bonding, chemists prefer to think of orbitals as having sizes about as large as interatomic distances in molecules. Since orbital electron density falls off very rapidly with distance, disregarding only the outer 10% of an electron in any orbital brings the

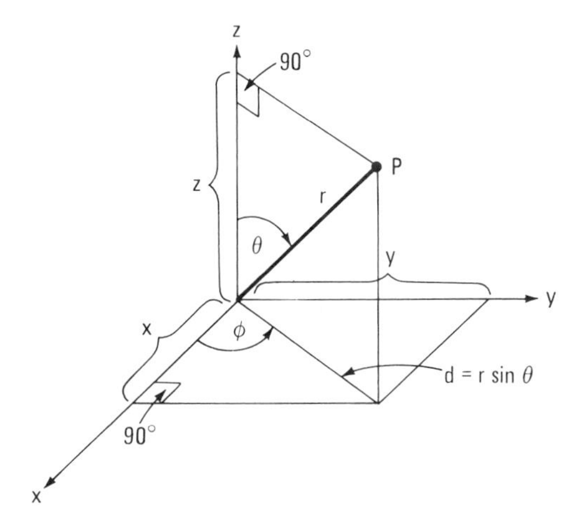

FIGURE 2.1 Polar coordinate system in which $z = r\cos\theta$, $y = r\sin\theta\sin\phi$, $x = r\sin\theta\cos\phi$, $\tan\phi = y/x$, $\cos\theta = z/r$, and $r = (x^2 + y^2 + z^2)^{1/2}$. The ranges of θ and ϕ are given by the relationships $0 \leq \theta \leq \pi$ and $0 \leq \phi < 2\pi$, respectively.

infinitely large orbital down in size so that the remaining 90% of its electron density is contained in a space which extends roughly to a neighboring atom in the molecule! The radii, which contain 90% of an orbital's electron density vary with the orbital (i.e., they are proportional to α). This allows us to make comparisons of the "chemical reach" of different orbitals.

To describe the nodal surfaces of atomic orbitals, we will use the polar coordinate system in Figure 2.1, which is very convenient to describe a point P on that surface. As mentioned earlier, AOs have radial (i.e., spherical), planar, and conal nodes. *Radial nodes* are described by equations of the type r = constant. These nodes are spheres with fixed radii, and we denote the number of such spherical nodes by N_r. Here, we use the convention that the boundary at infinity is counted as one spherical node. *Planar nodes* (also called *planar angular nodes*) are characterized by equations of the type ϕ = constant. These planes are perpendicular to the xy plane, they contain the z axis, and they form fixed angles ϕ with the xz plane. The number of such nodes is given by N_ϕ. *Conal nodes* (also called *conal angular nodes*) are described by equations of the type θ = constant. The symmetry axis of these cones is z and the angle between z and a line on the conal surface is a fixed angle θ. When $\theta < 90°$ the cone lies above the xy plane and when $\theta > 90°$ the cone lies below this plane. When $\theta = 90°$, the cone lies in the xy plane. This plane is a special case of a conal node. We will use the symbol N_θ to denote the number of conal nodes.

We now develop the concept that the nodal pattern as well as the shapes of the lobes are completely determined by the numbers N_r, N_θ, and N_ϕ. These numbers are related to the canonical AO quantum numbers n, l, and m by Equations 2.1:

$$n = N_r + N_\theta + N_\phi = \text{total number of nodes (including the one at infinity)}$$

$$l = N_\theta + N_\phi = \text{total number of conal angular and planar angular nodes} \quad (2.1)$$

$$m = N_\phi = \text{total number of planar angular nodes.}$$

The n, l, and m quantum numbers have the values $n = 1, 2, 3, 4, \ldots, l = 0, 1, 2, 3, \ldots, (n-1)$ for a given n, and $m = 0, 1, 2, 3, \ldots, l$ for a given l. If we know the values of n, l, and m, we can calculate the number of nodal surfaces by Equations 2.2.

$$\begin{aligned} N_\phi &= m \\ N_\theta &= l - m \\ N_r &= n - l. \end{aligned} \quad (2.2)$$

A common way of designating AOs is by the *nlm* system. For historical reasons, letters have become associated with l values (i.e., s for $l = 0$, p for $l = 1$, d for $l = 2$, f for $l = 3$, and then alphabetically starting with g for succeeding l values). Table 2.1 gives a breakdown of all orbitals from ($1s$) to ($4f$) using the *nlm* symbolism. Note that parenthesizing orbital and quantum number designations means that we are referring to its *spatial amplitude.* The meaning of m and m' and of the Cartesian labels will be discussed later.

What happens to an orbital shape as the number of nodes increases (i.e., as the energy rises)? The ground state of any one-electron atom is the $(1s) = (100)$ orbital. Aside from the spherical boundary at infinity, it has no other node. As mentioned earlier, the ($1s$) distribution is spherically symmetrical and in fact this is true for all s orbitals [i.e., ($1s$), ($2s$), ($3s$), etc.]. In other words all of them are functions of r alone and as we see from Figure 2.2 they all possess $(n - 1)$ *finite* radial (spherical) nodes (plus the node at infinity). The intersections of the amplitudes with the r axis correspond to the positions of the spherical nodes. Remember that the amplitude sign of any orbital lobe is arbitary and we could just as well have shown plots obtained by inverting each of the ones in Figure 2.2 with respect to the r axis.

The conal and planar angular nodes occur in certain patterns. *When there is more than one conal node, they are always arranged symmetrically above and below the xy plane.* Thus whenever there is a cone at $\theta =$ some angle β which is not $\pi/2$, then there is another cone which is its mirror image at $\theta = \pi - \beta$ on the other side of the xy plane. If $\theta = \pi/2$, the conal node in the xy plane is its own mirror image and in that case there is only one conal node.

TABLE 2.1 Atomic Orbitals and their Quantum Numbers

Orbital	n	l	m	N_ϕ	N_θ	N_r
$(1s0) = (1s)$	1	0	0	0	0	1
$(2s0) = (2s)$	2	0	0	0	0	2
$(2p0) = (2pz)$	2	1	0	0	1	1
$(2p1') = (2px)$	2	1	1	1	0	1
$(2p1'') = (2py)$	2	1	1	1	0	1
$(3s0) = (3s)$	3	0	0	0	0	3
$(3p0) = (3pz)$	3	1	0	0	1	2
$(3p1') = (3px)$	3	1	1	1	0	2
$(3p1'') = (3py)$	3	1	1	1	0	2
$(3d0) = (3d(3z^2 - r^2)) =$ "$(3dz^2)$"	3	2	0	0	2	1
$(3d1') = (3dxz)$	3	2	1	1	1	1
$(3d1'') = (3dyz)$	3	2	1	1	1	1
$(3d2') = (3d(x^2 - y^2)) =$ "$(3dx^2 - y^2)$"	3	2	2	2	0	1
$(3d2'') = (3dxy)$	3	2	2	2	0	1
$(4s0) = (4s)$	4	0	0	0	0	4
$(4p0) = (4pz)$	4	1	0	0	1	3
$(4p1') = (4px)$	4	1	1	1	0	3
$(4p1'') = (4py)$	4	1	1	1	0	3
$(4d0) =$ "$(4dz^2)$"	4	2	0	0	2	2
$(4d1') = (4dxz)$	4	2	1	1	1	2
$(4d1'') = (4dyz)$	4	2	1	1	1	2
$(4d2') =$ "$(4dx^2 - y^2)$"	4	2	2	2	0	2
$(4d2'') = (4dxy)$	4	2	2	2	0	2
$(4f0) = (4f(5z^3 - 3r^2z))$	4	3	0	0	3	1
$(4f1') = (4f(5z^2x - r^2x))$	4	3	1	1	2	1
$(4f1'') = (4f(5z^2y - r^2y))$	4	3	1	1	2	1
$(4f2') = (4f(x^2z - y^2z))$	4	3	2	2	1	1
$(4f2'') = (4f(xyz))$	4	3	2	2	1	1
$(4f3') = (4f(x^3 - 3y^2x))$	4	3	3	3	0	1
$(4f3'') = (4f(3x^2y - y^3))$	4	3	3	3	0	1

Whenever there is a cone with $\pi = \pi/2$, it automatically means that total number of such nodes (N_θ) is odd. As shown in Figure 2.3, the first instance of this occurs at the orbital (210) = (2pz). To learn how the angles θ are calculated from a wave function, see Appendix II.B.

Planar nodes where $\phi =$ constant are always arranged so that they divide the total angular range $0 \leq \phi < 2\pi$ into $2m$ equal parts, as shown in Figure 2.4. An additional property of these planar nodes is that for each value of N_ϕ (except zero), they have two independent arrangements in which the nodal planes of one orbital bisect the angles between the nodal planes of the other (Figure 2.5). The orbital that *does not* have a node in the xz plane is designated by m' and the orbital that *has* a node in this plane is labeled m''. The

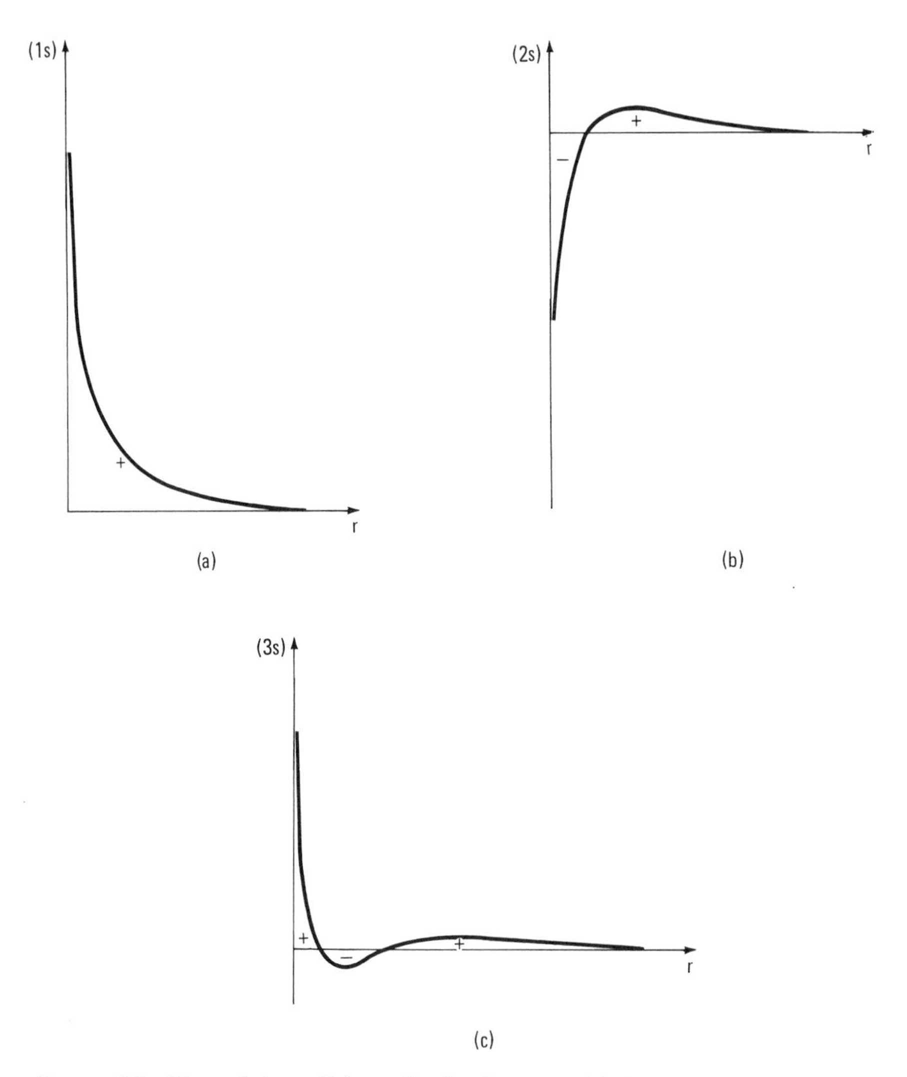

FIGURE 2.2 Plots of the radial amplitude of *s*-type orbitals as a function of *r*, the distance from the nucleus.

nodal planes for the arrangements in Figure 2.5 are in Table 2.2. The equations of the nodal planes can be expressed in terms of Cartesian coordinates. Such notation, which is included as part of the expressions in Figure 2.5, is also frequently used in the shorthand designations for the orbitals instead of m' or m''.

The orbital lobes (amplitudes) occur in the regions between the nodes. These lobes and nodes are shown for the atomic canonical orbitals ($2s$)–($5g$) in

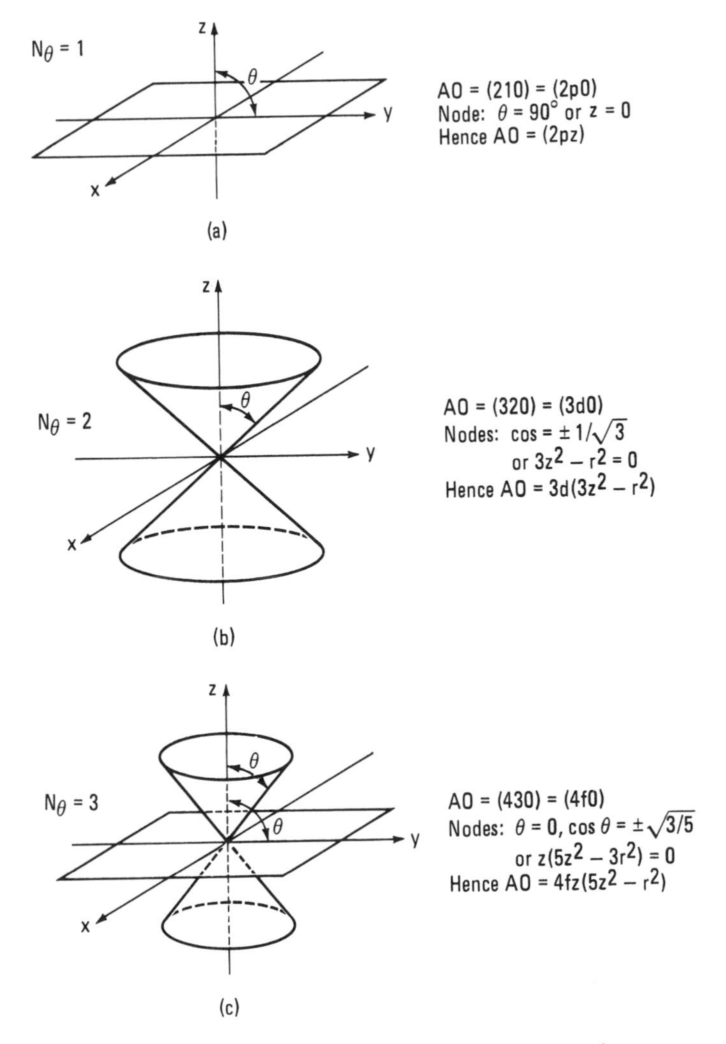

FIGURE 2.3 Conal angular nodes for (a) $(2pz) = (2p0)$, (b) $(3dz^2) = (3d0)$, and (c) $(4f0)$.

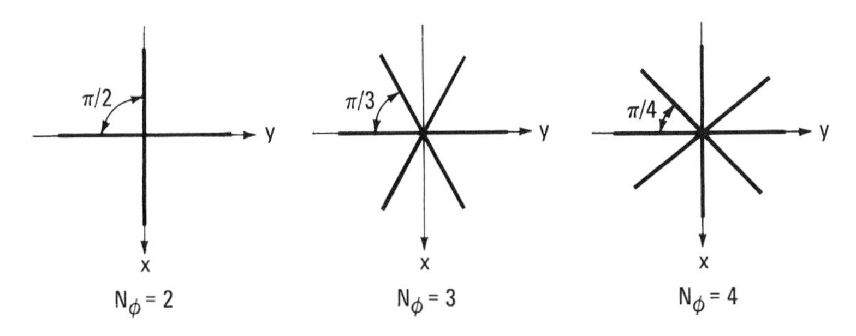

FIGURE 2.4 Drawings showing how planar angular nodes divide the angular range $0 \leq \phi < 2\pi$ into $2m$ equal parts.

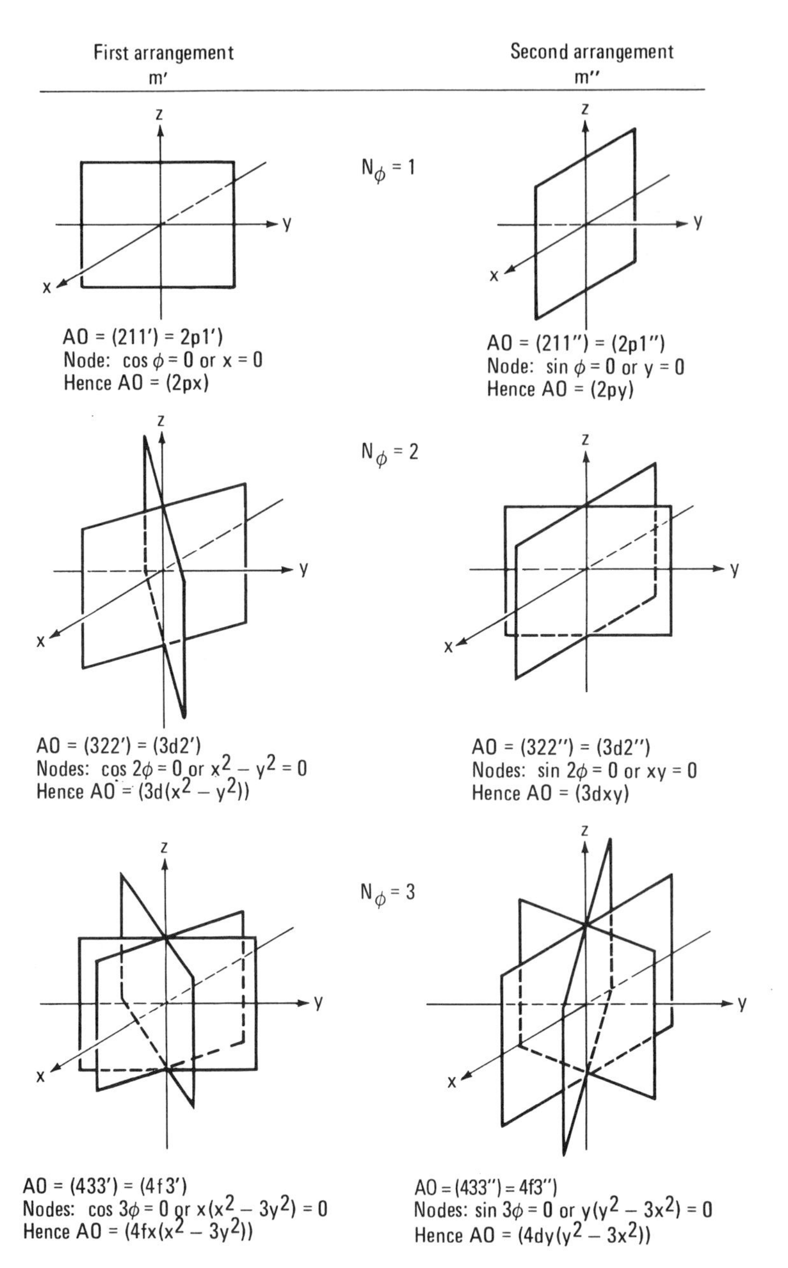

FIGURE 2.5 Drawing showing the two arrangements of vertical planar angular nodes for $N_\phi = 1, 2, 3$.

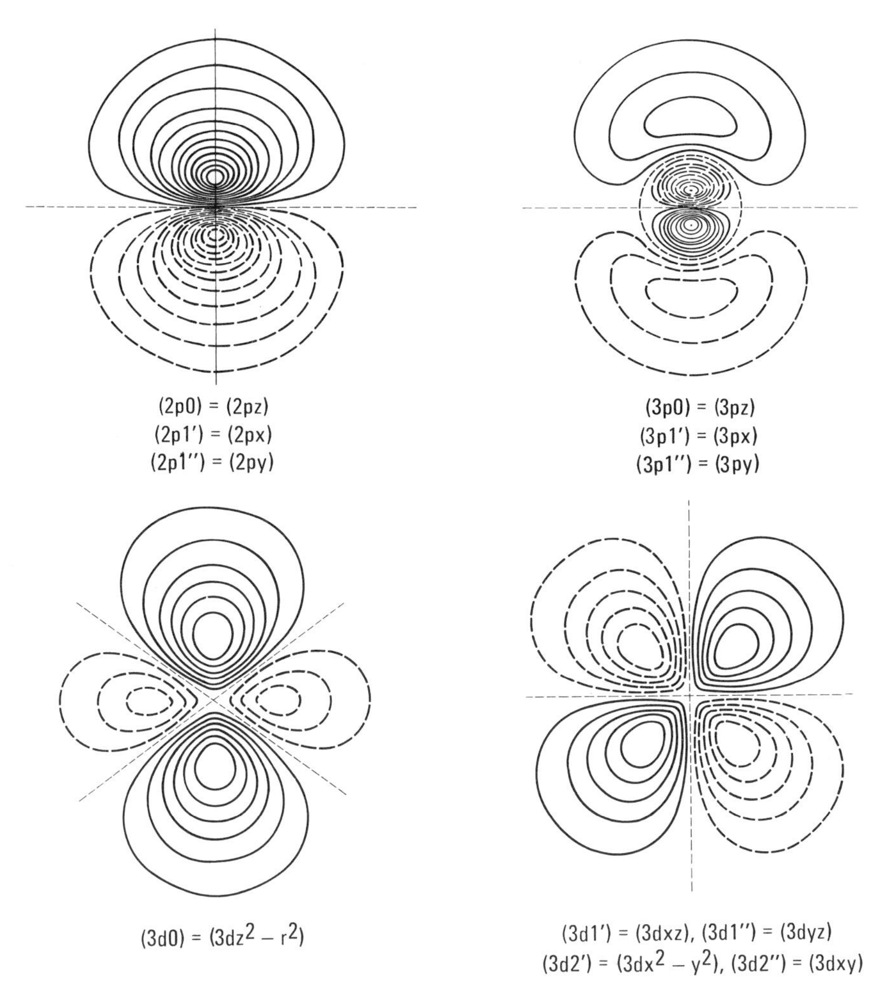

FIGURE 2.7 Two-dimensional contour plots of some hydrogen orbitals in which each contour has the same electron density. Regions where the wave function is positive are solid contours and where it is negative the contours are dashed lines. The nodes are represented by shorter dashed lines. Note the planar nodes in ($2px$), ($2py$), and in the set of four d orbitals; the conal nodes in ($2pz$), ($3pz$), ($3dz^2$), ($3dxz$), ($3dyz$); and the spherical node in ($3pz$), ($3px$), ($3py$). Because of the greater density in the smaller ($2p$) orbital (which is not drawn to scale), the contours are reduced in number by a factor of five compared to the larger ($3p$) and ($3d$) orbitals to avoid running too many contours into each other.

to stack a series of contour plots along y, which are parallel to the xz plane in order to obtain a three-dimensional picture of this orbital. This is a good time to familiarize yourself with Node Game, a software package you can access free of charge on a Mac or PC by following the easy instructions in section 2.10 of this chapter. By going to the Homework Drawing Board of Node Game, you can see colored three-dimensional pictures of a large selection of atomic orbitals in the "Generator Orbital" (**GO**) pic window.

2.2. Sizes of Canonical AOs

The sizes of orbitals depend on the total number of nodes and the magnitude of the nuclear charge. For the one-electron orbitals of hydrogen, the average orbital radius $\bar{R}$ is

$$\bar{R}_H(nlm) = n^2\left[\frac{3}{2} - \frac{l(l+l)}{2n^2}\right]a \tag{2.3}$$

where a is the Bohr radius. For more about the meaning of $\bar{R}_H(n, l, m)$ see Appendix II.C. The Bohr radius is given by $a = (h/2\pi)^2/me^2 = 5.29177 \times 10^{-9}$ cm and it represents the radius of the circular orbit of lowest energy for the electron in hydrogen as obtained by the classical Bohr model. Note that in such formulas m is the mass of the electron and not the quantum number m. It is also possible to define a radius R^* within which 90% of the electron is contained. For more about the meaning of R^* see Appendix II.D. In Table 2.4 are collected values of $\bar{R}$, R^*, and the ratio $R^*/\bar{R}$ for a range of hydrogen orbitals. Interestingly, the ratio decreases with n for a given l, and it rises with l for a given n value.

If we increase the nuclear charge Z while retaining only one electron, we proceed to He^+, then to Li^{2+}, etc. Stronger nuclear attraction is expected to shrink the hydrogenic orbitals. In fact, their $\bar{R}$ and R^* radii are related to those in hydrogen and *both these radii are inversely proportional to the nuclear charge*:

$$\begin{aligned}\bar{R}_Z(nlm) &= \frac{\bar{R}_H(nlm)}{Z}\\ R^*_Z(nlm) &= \frac{R^*_H(nlm)}{Z}.\end{aligned} \tag{2.4}$$

In atoms containing more than one electron, interelectronic repulsions occur. Thus if we consider two electrons having different n values, the electron with the higher n quantum number feels the attractive effect of Z to a lesser extent than the electron with the lower n value not only because it is further from the nucleus but also because the inner electron repels the outer one. In an approximate description, the latter effect merely changes the constant Z to a lower value (i.e., to an *effective* Z value) for the outer electron, but it does not change the orbital shapes or their nlm classifications. As a

TABLE 2.4 Various Measures for the Sizes of Orbitals of Hydrogen in Bohr Radii

Orbital[a]	$R = n^2$ [b]	$\bar{R}$ [c]	R^* [d]	$R^*/\bar{R}$
(1*s*)	1	1.5	2.66	1.77
(2*s*)	4	6	9.13	1.52
(2*p*)	4	5	7.99	1.60
(3*s*)	9	13.5	19.44	1.44
(3*p*)	9	12.5	18.39	1.47
(3*d*)	9	10.5	15.80	1.50
(4*s*)	16	24	33.62	1.40
(4*p*)	16	23	32.59	1.42
(4*d*)	16	21	30.32	1.44
(4*f*)	16	18	25.88	1.44
(5*s*)	25	37.5	51.69	1.38
(5*p*)	25	36.5	50.67	1.39
(5*d*)	25	34.5	48.50	1.41
(5*f*)	25	31.5	44.82	1.42
(5*g*)	25	27.5	38.52	1.40

[a] The *m* quantum number can have any value appropriate to the orbital in question.
[b] Appendix II.D.
[c] Equation 2.3.
[d] Equation 2.4.

result, outer electrons are said to be shielded from the nuclear charge. It is evident from Equation 2.4 that the orbital for such a shielded electron is larger than if the unshielded nuclear charge were acting only on the electron in that orbital.

2.3. Energies of Canonical AOs

The electrostatic potential energy $V(r)$ of an electron held to a nucleus of charge Z is given by

$$V(r) = -\frac{Ze^2}{r} \tag{2.5}$$

where r is the distance of the electron from the nucleus. This relationship describes the surface of a three-dimensional potential energy well, which is depicted in cross section in Figure 2.8. The potential energy $v(nlm)$ for an entire orbital $\psi = (nlm)$ can be shown to be expressed by Equation 2.6 (Appendix II.E).

$$v(nlm) = -Ze^2/R \tag{2.6}$$

where R turns out to be the Bohr radius in the case of the (1*s*) orbital.

Since $V(r)$ is negative, $v(nlm)$ is negative, which tells us that the electrostatic force is attractive. The kinetic energy $t(nlm)$ of the electron is always positive

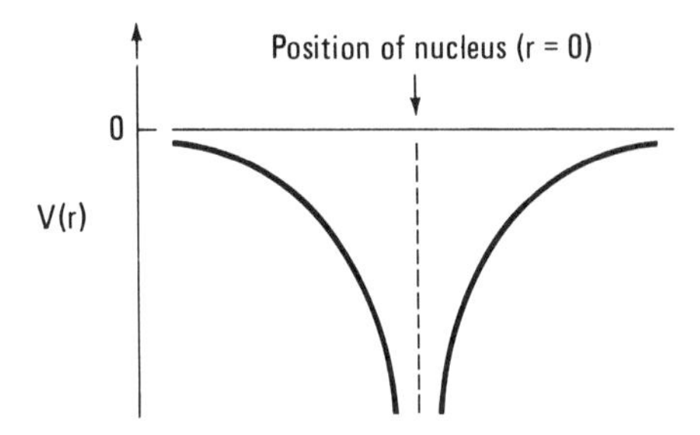

FIGURE 2.8 Potential energy well for an electron and a positively charged nucleus.

and the sum in Equation 2.7 gives us the total orbital energy $\varepsilon(nlm)$:

$$\varepsilon(nlm) = v(nlm) + t(nlm). \tag{2.7}$$

If $\varepsilon(nlm)$ is negative, the electron is bound (i.e., it is down in the potential well). If $\varepsilon(nlm) > 0$, the electron is essentially free to move out of the vicinity of the well. It can be shown (see Appendix II.F) that the total energy $\varepsilon(nlm)$ of the (1s) electron is

$$\varepsilon(nlm) = -e^2/2a \tag{2.8}$$

which corresponds to -13.6 electron volts (eV). This means that the ionization energy E_H or, in other words, the energy required to completely remove this electron from a gaseous hydrogen atom is positive (i.e., $E_H = e^2/2a$) and amounts to 13.6 electron volts (eV). The hydrogen one-electron orbital energies can then be expressed by Equation 2.9:

$$\varepsilon(nlm) = -\frac{e^2}{2a}\left(\frac{Z}{n}\right)^2 - 13.6\left(\frac{Z}{n}\right)^2 \text{ eV}. \tag{2.9}$$

This tells us that an increase in nuclear charge leads to tighter binding of the electron while an increase in the n value (i.e., more nodes) leads to looser electron binding. In addition to electron ionization, we note that electronic transitions among orbitals are also possible, and this phenomenon gives rise to the electronic spectrum of hydrogen.

Notice that the energies of hydrogenic orbitals depend only on the quantum number n, and not on l or m. The energies of orbitals of different l and m values for a given n value are therefore the same in one-electron atoms. Orbitals of the same energy are said to be *degenerate*.

As a consequence of Equation 2.9, the shielded orbital energy is smaller in absolute value (less negative) than an unshielded one. In other words, an electron in a shielded orbital is less tightly bound and therefore has a lower ionization energy. The magnitude of the shielding effect on an orbital depends upon the number of electrons occupying the orbitals that lie between the nucleus and the region of maximal density of the orbital in question. Another important feature of shielding is that it varies with l for a given

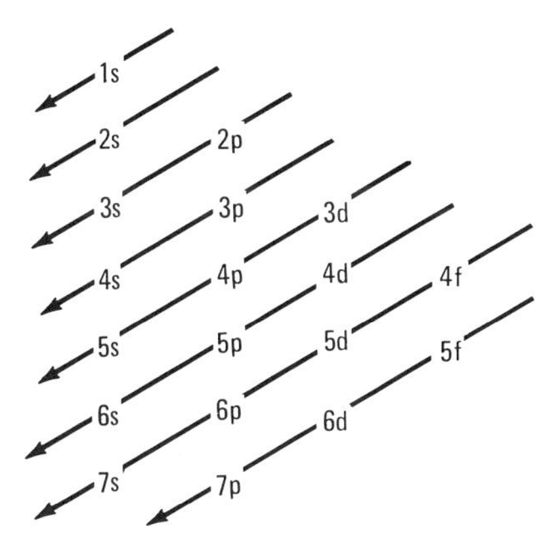

FIGURE 2.9 Mnemonic device for determining the filling sequence of atomic orbitals.

value of n. Consequently, there are substantial energy differences among the s, p, d, etc., orbitals within a given quantum shell. It is the shielding effect that gives rise to the orbital occupancy sequence represented by the memorization device (sometimes called the $n + l$ rule) shown in Figure 2.9. This ordering of the energy levels of the occupied orbitals turns out to be *nearly* the same in all atoms.

2.4. Hybrid Orbitals

In Chapter 1 we found that in many-electron molecules there are states we can express by alternative molecular orbital descriptions. We now expand this concept to atomic states. For example, in the Be $1s^2 2s^1 2pz^1$ excited state configuration we can replace the ($2s$) and ($2pz$) orbitals by two different orbitals of the form

$$\begin{aligned}(h_1) &= A(2s) + B(2pz) \\ (h_2) &= -B(2s) + A(2pz)\end{aligned} \tag{2.10}$$

where A is arbitrary and B is related to A by $A^2 + B^2 = 1$. We could also write $A = \cos\omega$ and $B = \sin\omega$ where ω is a parameter whose value lies between 0 and π. It can be shown for this excited state of Be, that

$$\begin{aligned}\rho(x, y, z) &= 2(1s)^2 + (2s)^2 + (2pz)^2 \\ &= 2(1s)^2 + (h_1)^2 + (h_2)^2\end{aligned} \tag{2.11}$$

where the coefficient 2 in the first term represents the fact that there are two electrons in the ($1s$) orbital. Remember that the symbol $1s^2$ as a configuration description does not have the same meaning as $(1s)^2$. The latter symbol denotes the square of the ($1s$) orbital wave amplitude function and represents a density. The orbitals (h_1) and (h_2) which are obtained by taking orthogonal linear combinations of different canonical atomic orbitals (Chapter 1) are called *hybrid orbitals.* Because hybrid atomic orbitals can provide valuable information about molecular orbitals, it is important to discuss some characteristics of hybrid atomic orbitals. It should be remembered, however, that hybrid orbital shapes depend very much on the orbital occupation (i.e., the state) whereas this is not true for canonical orbitals. This point will be discussed again later.

Sometimes it is convenient to rotate a set of canonical or hybrid equivalent orbitals in space to accommodate a molecular geometry whose bonding picture we wish to develop. This is done by taking linear combinations of the orbitals as is illustrated in detail in Appendix II.G for a set of three p orbitals.

2.5. Hybrid Orbitals through *s-p* Mixing

Mixing $(2s)$ and $(2p)$ AOs *in different ratios* is quite important to chemists. Let us mix the $(2s)$ and $(2px)$ orbitals to give a hybrid (h) in Equation 2.12:

$$(h) = A(2s) + B(2px). \tag{2.12}$$

The square of this hybrid wave function (h) can then be written as Equation 2.13 in which the shorthand notation $\langle(2s)|(2s)\rangle$ etc. stands for the overlap integral between two orbitals (Appendix I.B).

$$(h)^2 = A^2\langle(2s)|(2s)\rangle + B^2\langle(2px)|(2px)\rangle + 2AB\langle(2s)|(2px)\rangle \tag{2.13}$$

We would like the probability of finding an electron in our hybrid $(h)^2$ to be 1 (i.e., $(h)^2 = \langle(h)|(h)\rangle = 1$). The principle of orthonormality further tells us that $\langle(2s)|(2s)\rangle = 1$, $\langle(2px)|(2px)\rangle = 1$ and $\langle(2s)|(2px)\rangle = 0$. This means that Equation 2.13 can be simplified to Equation 2.14:

$$(h)^2 = A^2 + B^2 \tag{2.14}$$

If $A^2 = 0.8$ and $B^2 = 0.2$, we say that the hybrid has 80% $(2s)$ and 20% $(2p)$ character. Often A and B are expressed as $A = [a/(a+b)]^{1/2}$, $B = [b/(a+b)]^{1/2}$ where a and b are small integers or some other convenient numbers. The hybrid is then called a $(2s)^a(2p)^b$ hybrid and we can write Equation 2.12 as Equation 2.15:

$$h[(2s)^a(2p)^b] = [a/(a+b)]^{1/2}(2s) + [b/(a+b)]^{1/2}(2p). \tag{2.15}$$

A common convention is to abbreviate the orbital character of such a hybrid as $s^a p^b$. *Neither convention is to be confused with orbital occupations.* In Figure 2.10 are contour plots for fourteen $(2s)^a(2px)^b$ carbon hybrids. On the largest contour, h always has the value of either $+0.05a^{-3/2}$ or $-0.05a^{-3/2}$. The ratio of $a:b = A^2:B^2$ is given for each contour. Thus this ratio is $1:0$ for the pure $(2s)$ and $0:1$ for the pure $(2px)$. As more $(2px)$ is mixed into the $(2s)$ (i.e., $50:1$, $10:1$, $6:1$, etc.), the orbital becomes increasingly deformed, though it retains the x axis as its axis of symmetry. This deformation or "polarization" with

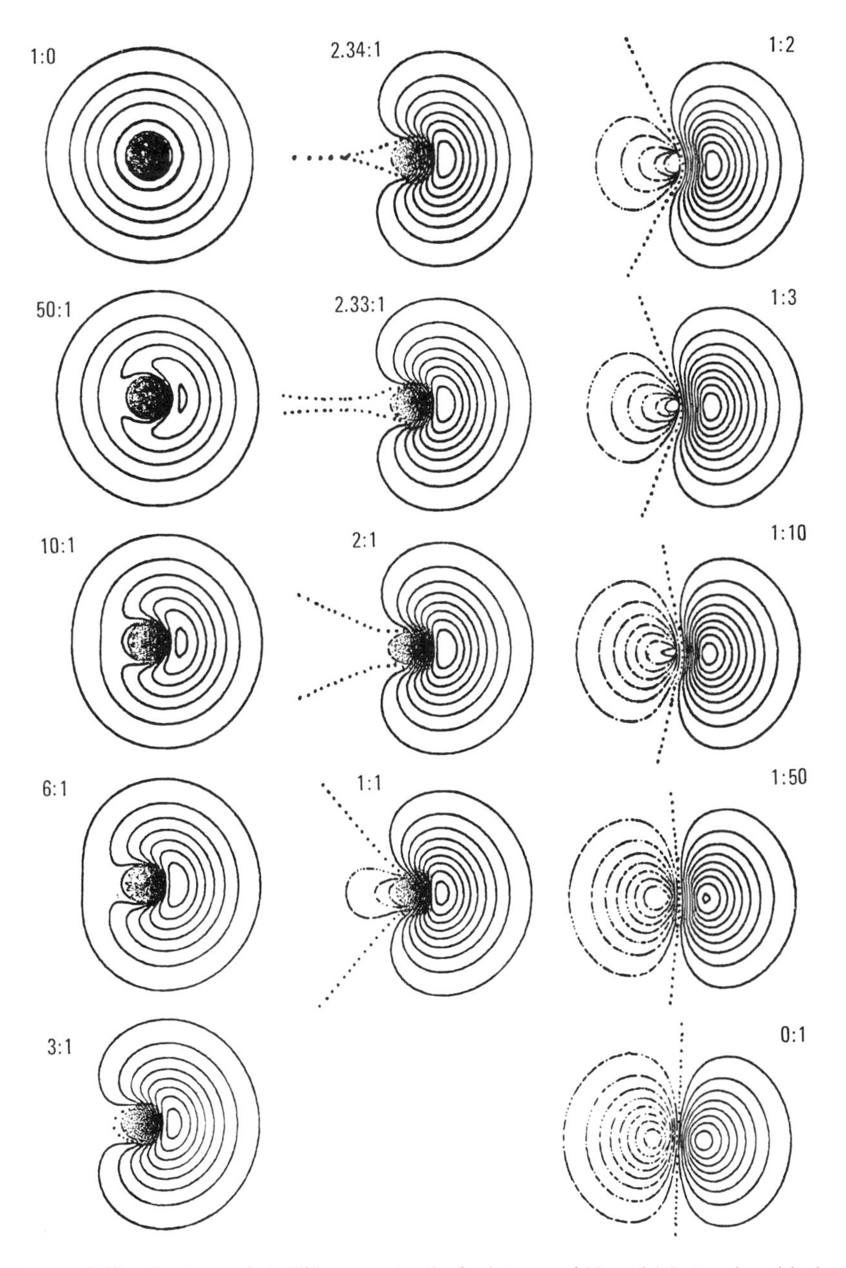

FIGURE 2.10 Contour plots ((h) = constant) of mixtures of (s) and (p) atomic orbitals showing the ratios $A^2 : B^2$. In these plots solid, dotted, and dashed lines denote regions where $(h) > 0$, $(h) = 0$, and $(h) < 0$, respectively. The absolute value of the smallest value of (h) is 0.05 Bohr$^{-3/2}$ and $\Delta(h)$ has the same value. (Courtesy of Dr. K. Ruedenberg.)

increasing p character is initially accompanied by elongation of the nodal sphere for $(a/b) = 0$. For $(a/b) = 2.34$ the node elongates to a single line in the $-x$ direction. This line for $(a/b) < 2.34$ opens into a cone-like surface, and as a/b decreases still further this surface flattens to a plane, which is the nodal plane of the pure $(2px)$ orbital.

For chemically bonded atoms that have more than one AO involved in bonding, it is often advantageous to construct hybrid AOs from a canonical AO set. For example, atoms in the second period frequently bind to four equivalent neighboring atoms. In the case of carbon (e.g., in CH_4) this set of AOs is composed of the AOs $(2s)$, $(2px)$, $(2py)$, and $(2pz)$. To help us understand chemical bonding, we distinguish between *core* AOs and *valence* AOs (VAOs). Ionization energies associated with VAOs are low enough to allow involvement of such AOs in chemical bonding whereas core AOs have higher ionization energies. The four canonical carbon VAOs can be linearly combined mathematically to form the "hybrid quadruple" or "(h) set" (h_1), (h_2), (h_3), (h_4). For more on these linear combinations, see Appendix II.H. In what directions do these four orthogonal hybrids "point" in space? As we show in Appendix II.I, this depends on how many of them we "mix" together and in what ratios they are mixed. We can, of course, have an infinite number of hybrid sets depending on the number of orbitals we linearly combine and on their relative contributions to the hybrid set. For chemical bonding, however, three important cases will now be considered, namely, the formation of *equivalent hybrid sets* of two, three, and four hybrid AOs by linearly combining an (s) with one, two, and three (p) orbitals, respectively. Equivalent orbitals are defined as being identical in shape but pointing in different directions. For example, a $(2px)$ orbital is equivalent to a $(2py)$ orbital. Their equivalency is demonstrated by the fact that rotation by 90° about the z axis converts one in to the other. Thus, orbital equivalency is characterized by rotation about a symmetry axis or reflection through a symmetry plane that carries one orbital into another. Equivalent hybrid AOs have identical s-p ratios and $a = b$ in Equation 2.15. If there are several equivalent hybrids, *the angle between any two of them must be the same.*

2.6. Equivalent s^1p^1 Hybrid Orbitals

Linearly combining an (s) with a (p) AO gives rise to two equivalent *digonal* hybrid AOs (d_1) and (d_2). They are called digonal because they lie 180° apart and point in opposite directions. The expressions for our hybrid quadruple now become

$$
\begin{aligned}
(h_1) &= 2^{-1/2}[(2s) + (2px)] = (d_1) \\
(h_2) &= 2^{-1/2}[(2s) - (2px)] = (d_2) \\
(h_3) &= (2py) \\
(h_4) &= (2pz).
\end{aligned}
\tag{2.15}
$$

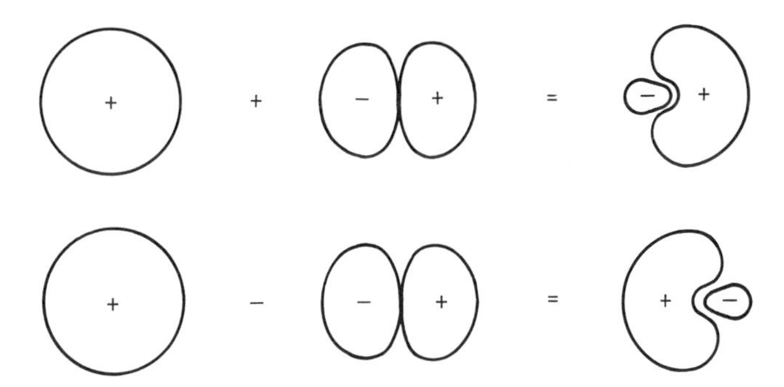

FIGURE 2.11 Outer contours of digonal orbitals obtained from linearly combining an (*s*) and a (*p*) orbital.

The $2^{-1/2}$ coefficient in the equations for (d_1) and (d_2) normalizes these hybrids (i.e., the sum of the squares of the orbital coefficients must equal 1). The fact that (d_1) and (d_2) indeed point in opposite directions can be pictorially visualized by linearly combining the orbital contours themselves as shown in Figure 2.11. In this figure it is seen that the positive lobe of the (2*p*) orbital is enhanced by the additive effect of the positive (2*s*) orbital while the negative lobe of the (*p*) orbital is partially cancelled by the positive (2*s*) orbital. To do this accurately, many density contours must be considered rather than just one. It is seen from Figure 2.11 *that the formation of hybrids results in a net localization of electron density in the large lobe.* In Appendix II.J is shown the transformation matrix for Equations 2.16. The energies of the two identical digonal hybrids are, of course, degenerate, and not surprisingly they lie at the average of the (2*p*) and (2*s*) energies (Figure 2.12).

We could, if we wish to do so, rotate our (*h*) quadruple set in space. The details of this process appear in Appendix II.K.

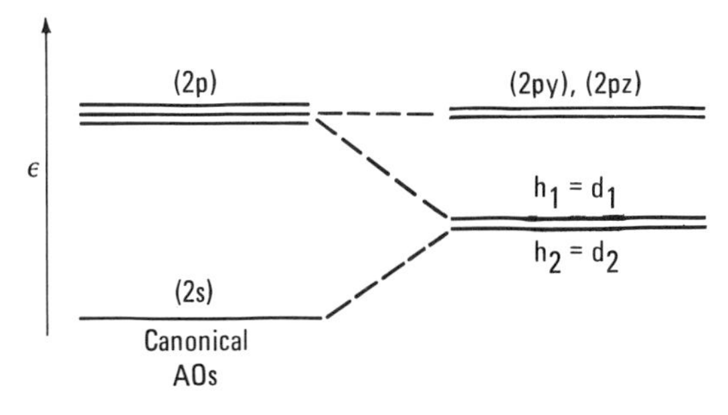

FIGURE 2.12 Diagram showing the relationship of the energies of the canonical and hybrid AOs when an (*s*) AO is mixed with a (*p*) AO in a 1 : 1 ratio.

2.7. Equivalent sp^2 Hybrid Orbitals

If we mix (2*s*) with *two* (*p*) orbitals, say, (2*px*) and (2*py*), three hybrid AOs are produced whose axes lie in the *xy* plane. That the axes of such a hybrid set lie in the *xy* plane is reasonable because the (2*px*) and (2*py*) contours when algebraically combined with the (2*s*) must give hybrid AOs that are symmetrical with respect to this plane which cuts through them. If the linear combinations are carried out so that the resulting three hybrids are identical (i.e., they each contain 1/3 (2*s*) character and 2/3 (2*p*) character) such (s^1p^2) orbitals point to the vertices of an equilateral triangle. This is why they are called *trigonal* hybrids. Notice that only by pointing the members of these hybrid orbitals to the vertices of an equilateral triangle can all three orbitals be identical (i.e., have identical geometrical environments). To demonstrate pictorially the linear combinations that give rise to hybrids more complicated than digonal ones is difficult, and this will not be done here. The orthogonal matrix which transforms three canonical AOs into the *more localized* trigonal hybrids is given in Appendix II.L. The energy of each of the trigonal hybrids relative to the canonical AOs is depicted in Figure 2.13. These energies (Equation 2.17) reflect the 1/3 *s* and 2/3 *p* character of these hybrids:

$$\varepsilon(tr_k) = 1/3\ \varepsilon(2s) + 2/3\ \varepsilon(2p) \tag{2.17}$$

2.8. Equivalent sp^3 Hybrid Orbitals

Mixing an (*s*) orbital equally with all three (*p*) orbitals gives rise to four identical hybrid AOs. It might be thought that four such hybrids could be pointed to the corners of either a square or a tetrahedron. The square can be ruled out, however, from the fact that the angle (Appendix II.I) between any two hybrids is not the same. The square can also be eliminated on simple geometrical considerations. To see this let us take the hybridization process in a stepwise manner. Recall that mixing an (*s*) with one (*p*) orbital, namely (*px*), gives us digonal hybrids pointed in opposite directions along *x*. Mixing

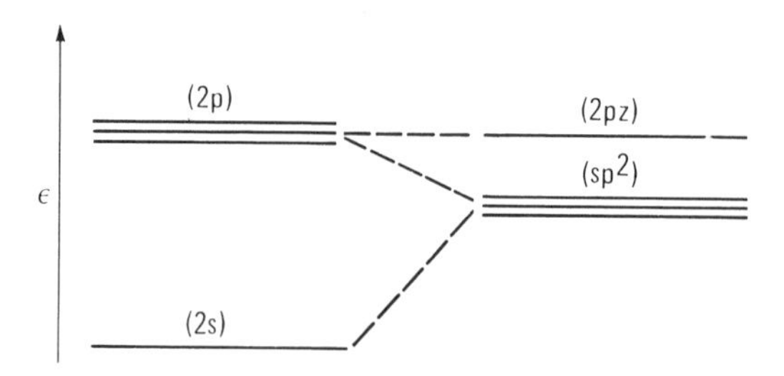

FIGURE 2.13 Energies of the (sp^2) trigonal hybrid set of orbitals.

in an additional (p) orbital, namely (py), with the digonal AOs leads to trigonal hybrids whose axes for geometrical reasons must lie in the xy plane. If we now mix into the (sp^2) set the (pz) orbital, whose axis lies *perpendicular* to the axes of the trigonal AOs, it is impossible to obtain four hybrids whose axes are all in the xy plane since the coefficient of (pz) would be necessarily zero. The only other choice that preserves the *identical* character of all four hybrids is to require the hybrids to point to the vertices of a tetrahedron. The orthogonal transformation of the canonical orbitals that gives such tetrahedral hybrid (th) AOs is given in Appendix II.M.

2.9. Intermediate Hybrid Orbitals

It is often necessary for bonding purposes to consider hybrids that are *intermediate* in character between the *equivalent* hybrids just discussed. Intermediate hybrids can be visualized as continuous transitions from one orbital set to another, for example, (s) plus added (p) character until (sp) is reached, (sp) plus (p) character to (sp^2), and (sp^2) plus (p) to (sp^3). Such transitions can be accomplished without violating orthogonality conditions as is shown in Appendix II.N.

In a similar manner, transitions between other regular hybrid sets can occur. In Figure 2.14 is shown in pictorial form an example involving an (s) and three (p) orbitals. Paths (a) and (h) of this figure show two orientations of the digonal set and in (e) and (f) are depicted two orientations of the tetrahedral set. (To see how the tetrahedral hybrids are reoriented in Figure 2.14e and f, see Appendix II.O.) The transition from digonal (a) to trigonal (c) involves bending the two digonal orbitals (d_+) and (d_-) in (a) backwards in the xy plane to become (h_2) and (h_3) in the intermediate (b). Meanwhile, (p_1) gains (s) character [at the expense of (d_-) and (d_+)] to become (h_1) in (b). When the angle α in (b) reaches 120°, the trigonal hybrids (tr_k) in (c) are born. If we now imagine that the (tr_k) in (c) are bent upwards in the $+z$ direction to give us (h_1), (h_2), (h_3) in (d), (p_2) in (c) reverses its direction and gains (s) character to become (h_4) in (d). When $\alpha + 90°$ in (d) opens up to 109°28′ we have the (th_k) in (e). One can also imagine a transition from the digonal set in (h) to the tetrahedral set in (f) via the intermediate in (g). Here (d_-) and (d_+) (which are 180° apart) are bent down [at a faster rate than the opening of the angle between (p_1) and (p_2)]. The orbitals (p_1) and (p_2) gain (s) character [at the expense of the (d_+) and (d_-) hybrids] until the angles α and β in (g) become equal (109°28′) in (f).

2.10. Accessing Node Game

The address on the www is: http://www.public.iastate.edu/~jverkade/nodegame.html.

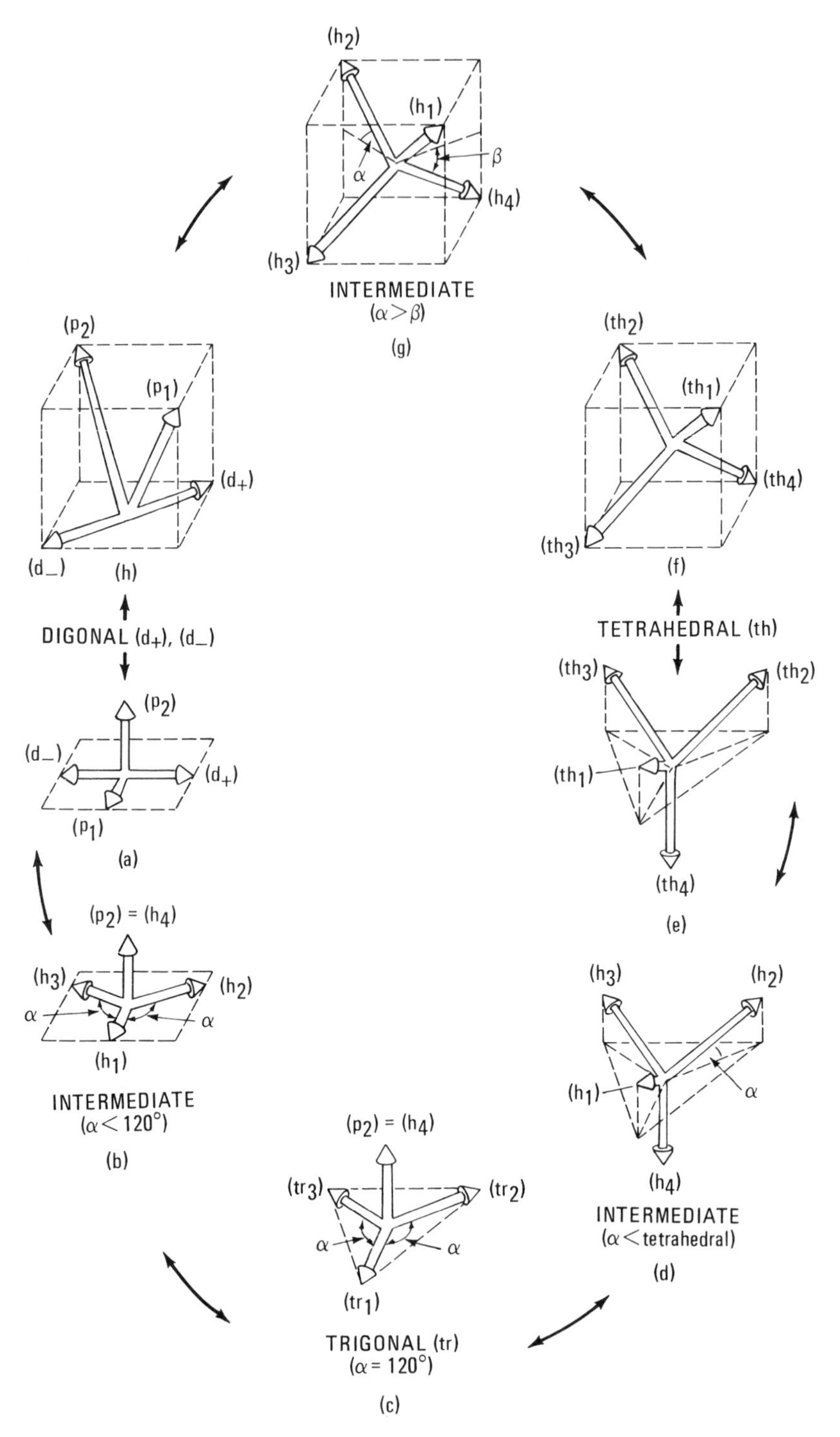

FIGURE 2.14 Pictorial representation of progressions from one hybrid orbital set to another.

The instructions you will need for using Node Game are included in the program. In subsequent chapters you will find further opportunities for using this interactive instructional aid.

Summary

We have learned how to obtain the nodal and lobal configurations (i.e., the shapes) of canonical AOs by converting spherical to conal and planar nodes. We introduced three different measures of orbital sizes (i.e., R, $\bar{R}$, and R^*) and briefly examined factors that affect the energies of canonical AOs. We found that the device in Figure 2.9 gives us a simple way to remember the sequence in which AOs are occupied by electrons for the vast majority of atoms and ions. We then saw how hybrid AO sets can be formed from canonical AOs. When such hybrid AOs are formed by equally mixing, say, a pair of ($2p$) AOs, two equivalent ($2p$) AOs are produced which may be oriented in space along any set of mutually orthogonal directions. By similarly mixing an (s) AO with one or more (p) AOs, hybrid sets of the digonal, trigonal, and tetrahedral types are formed which, in addition to possessing orientational flexibility, also localize electron density to a greater extent than their canonical parent AOs. Finally, we addressed the possibility of unequal mixing of (s) and (p) AOs and discovered that equivalent hybrid sets are related by series of intermediate hybrids.

Exercises

1. Sketch all the $6h$ atomic orbitals and label the θ, ϕ, and r nodes.
2. How many spherical nodes (not counting the node at infinity) are present in the ($5d0$), ($5f1''$), and ($6p1'$) orbitals?
3. Make a plot of $V(r)$ in eV as a function of r for H and Li^{2+}. Label the r axis with the appropriate units of length.
4. Sketch the ($1s$) orbital of hydrogen in HF and the ($2p$) AOs of the fluorine with one of the ($2p$) AOs pointing toward the hydrogen. Indicate which ($2p$) orbital(s) is orthogonal to the hydrogen ($1s$).
5. Verify the second line of Equation 2.11.
6. Print the atomic orbitals in the GO Menu of "Node Game" Homework Drawing Board, and sketch in any conal and planar nodes they may have.
7. Calculate the potential energy and the total energy for the $1s^0 2s^1$ and the $1s^0 2pz^1$ states of the hydrogen atom.
8. Verify Equation 25 in Appendix II.I.
9. Justify Figure 2.12 from Equation 26–28 in Appendix II.I. In a similar manner justify Figure 2.13.

10. Show that Equations 2.15, 30 in Appendix II.K, and 32 in Appendix II.L are orthogonal for any pair of functions in each.

11. Verify Equation 34 in Appendix II.M by substituting Equation 35 and making use of the values of $\mathbf{g}_k$ in Figure 6 in the same appendix.

12. Show from trigonometric considerations that the $\mathbf{f}_k$ ($k = 1 - 3$ in Appendix II.L) for the (sp^2) hybrids stem from the geometric relationship of these orbitals.

13. What are α and β in Equation 23 in Appendix II.I for digonal hybrids?

CHAPTER 3

Diatomic Molecules

Because homonuclear diatomic molecules are the simplest of all molecules, we can use them to illustrate some important concepts that are common to all molecules, namely, molecule formation and molecular bonding. In this chapter we introduce the generator orbital (GO), which is used here and in subsequent chapters as a device to generate in a pictorial way delocalized and localized molecular orbital (MO) pictures for a wide variety of molecules. The GO device is also employed to generate the normal vibrational modes of these molecules.

3.1. Molecule Formation and Motions

If the centers of atoms A and B in a diatomic molecule are separated by a finite distance R, they experience an internuclear repulsion, an attraction between the nucleus of one and the electrons of the other, and an interelectronic repulsion. In addition to these electrical forces, there is a "kinetic force" due to the R-dependence of the kinetic energy. All these forces combine to create an effective potential $U(R)$. A repulsive potential [for which $U(R) > 0$ everywhere] is depicted in Figure 3.1a. In this case, two atoms held at R_1 and then released will move apart. At the same time the potential energy $U(R_1)$ the system has at R_1 will be converted to kinetic energy until at $R(\infty)$ all of the potential energy is converted to kinetic energy and the atoms continue to separate with respect to one another with a kinetic energy $(mv^2/2) = U(R_1)$. For all intents and purposes $U(\infty)$ corresponds to $R \geq$ to about 10^{-4} cm. If, on the other hand, two atoms separated by $R(\infty)$ approach one another with a kinetic energy corresponding to $U(R_1)$, they will decrease their relative velocity until at R_1 they are halted. However, they immediately fly apart again since the potential curve describing their interaction is overall repulsive. The curve in Figure 3.1b describes a potential function, which for a range of internuclear distances is attractive ($U(R) < 0$). That is, a pair of atoms held at an internuclear distance of R_0 is unable to move apart since the atoms have no kinetic energy to expend in climbing the wall

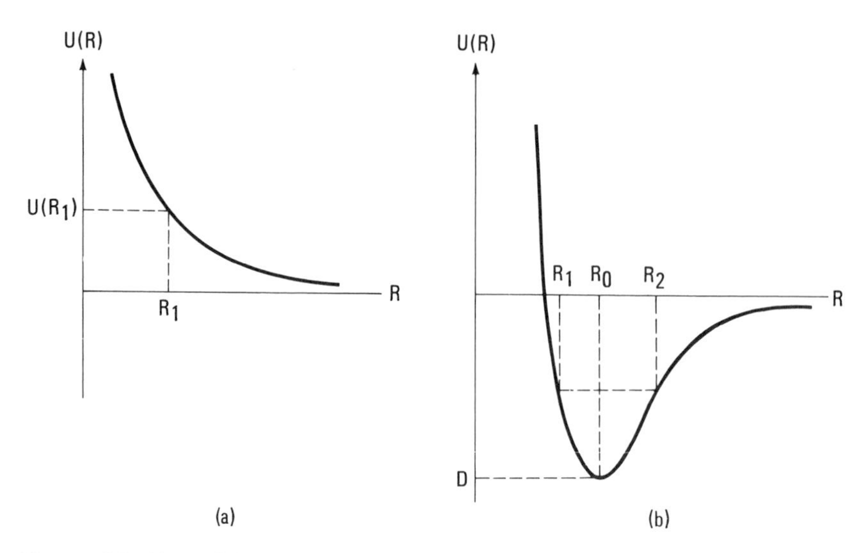

FIGURE 3.1 Potential energy $[U(R)]$ as a function of distance (R) for a repulsive (a) and an attractive (b) interaction between two atoms. Since the direction of approach of the atoms is arbitrary, the two-dimensional potential diagrams should be spun around the $U(R)$ axis to visualize the potential energy surface.

in either direction. In such a situation we say that R_0 is the *equilibrium distance*, or the *equilibrium bond length*, of the molecule. For a chemical bond of average strength, R_0 is about 0.8 to 0.9 times the sum of the atomic radii as defined by $R(nlm) = an^2/Z^*_{nlm}$ where Z^*_{nlm} is the effective nuclear charge as a result of shielding effects. Weak bonds are generally longer and strong bonds shorter.

Suppose now that we impart motion to the two atoms at R_0 such that they move toward one another in Figure 3.1b with a kinetic energy of $mv_0{}^2/2$. This kinetic energy is transformed into potential energy until the atoms stop at R_1 where $U(R_1) - U(R_0) = mv_0{}^2/2$. They then accelerate up to R_0 and then decelerate up to R_2 whereupon they go back to R_0, and the whole cycle repeats itself over and over again. This oscillation constitutes a molecular vibration. Of course, if the kinetic energy initially imparted exceeds the energy difference $U(\infty) - U(R_0) = -U(R_0) = D$, then the atoms will fly apart to infinity. The energy D is called the *dissociation energy*. The stronger the bond, the larger its dissociation energy. Bonds that are weak, average, and strong have dissociation energies of about 2, 5, and 8 eV, respectively. Weak bonds tend to be easily stretched and compressed compared to strong bonds, and their oscillations also tend to be larger than those of strong bonds. These characteristics are reflected in Figure 3.2 in which the potential well is seen to be wider than in Figure 3.1b. Molecular vibrations tend to be quite small (ca. 0.1 Å, 1 Å $= 10^{-8}$ cm) relative to bond lengths, which are on the order of one

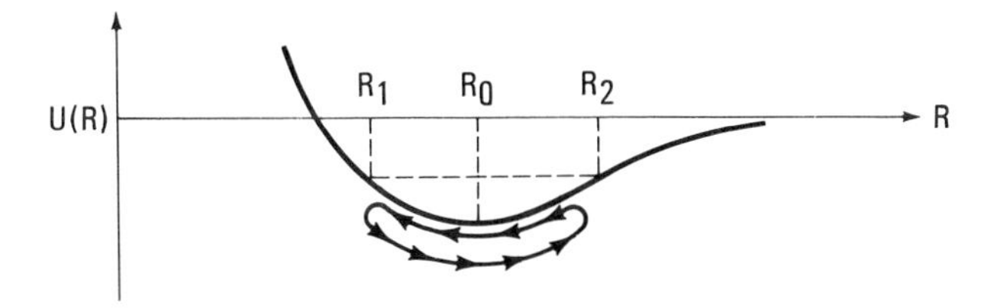

FIGURE 3.2 A potential well for a relatively weakly bonded diatomic molecule. The amplitude of the vibration is seen to be comparatively large.

or two angstroms. It should be realized that molecular vibrations are independent of any translational or rotational motions of the molecules as a whole.

What is the origin of the potential well for stable molecules? In the classical particle picture, electrons whirl around the two fixed nuclei just as they do around an atom at rest. For the purposes of this discussion, let us make the simplifying assumption that the two nuclei are at rest separated by R_0. The electronic motion then determines $U(R_0)$, which is equal to the total (i.e., kinetic plus potential) energy of all electronic motions plus the internuclear repulsion. This is summarized in Equation 3.1:

$$U(R_0) = E_{el}(R_0) + \frac{e^2 Z_A Z_B}{R_0} \tag{3.1}$$

wherein eZ_A and eZ_B are the nuclear charges. This relationship is also true for $R \neq R_0$ if the nuclei are held at R, except that R_0 in Equation 3.1 is replaced by R. The total electronic energy $E_{el}(R_0)$ must be the dominant contribution to $D = -U(R_0)$ at R_0, since the repulsion term $e^2 Z_A Z_B/R_0$ is always positive, whereas $E_{el}(R_0)$, of course, must be negative for stable molecules.

3.2. Generator and Molecular Orbitals

The electronic structure and energy of a molecule is determined by the occupied MOs. What do MOs look like? Like AOs, they possess lobes and nodes. The shapes of the nodes of AOs are a consequence of the *spherical* symmetry of the potential energy of an electron around *one* nucleus. Because molecules have more than one nucleus, the potential energy of an electron will not have spherical symmetry and so the patterns of its MO nodes and lobes will be less than spherically symmetrical. For additional discussion of precisely what is meant by symmetry and its connection to generator and molecular orbitals, see Appendix III.A.

To visualize the nodal patterns of MOs, we employ a device that consists of *placing an imaginary AO at the center of a molecule. For a diatomic molecule this point is midway between the nuclei. We then examine its nodal pattern.*

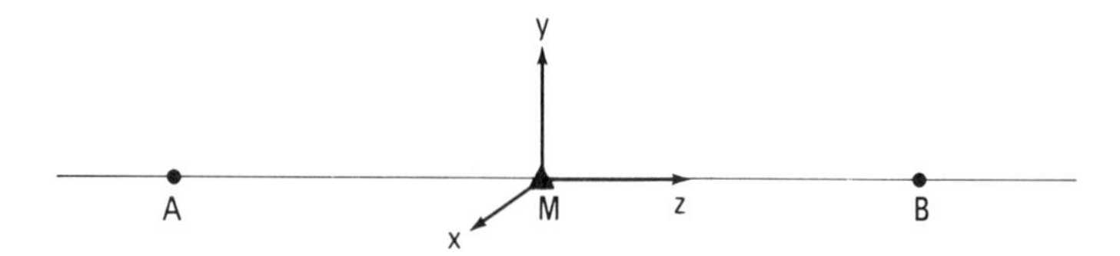

FIGURE 3.3 Axis system for GOs at M, the center of symmetry of a homonuclear diatomic potential provided by nuclei A and B.

By "imaginary" we mean that this AO exists only in our minds. (The mathematical meaning of imaginary is not relevant here.) We will refer to these imaginary AOs as *generator orbitals* or GOs. Before generating diatomic MO nodal patterns from them, let us examine the behavior of some GOs in the diatomic potential. Imagine a ($2px$), a ($2py$), and a ($2pz$) AO at the center M in Figure 3.3 of a homonuclear diatomic molecule. From now on, we will generally assume that the coordinates of the GOs are oriented so that the z axis lies along the internuclear axis. We will also frequently use a triangle to denote the GO center when that location is not obvious. It is clear that both the ($2px$) and ($2py$) GOs feel the same environment from the two nuclei but that the ($2pz$) GO must feel this environment differently since its lobes are directed toward the nuclei rather than being oriented in a plane perpendicular to the internuclear axis. Thus the energy degeneracy of these three orbitals in the atom is only partially retained in diatomic molecules. Similarly the ($3d0$), ($3d1'$), ($3d1''$), ($3d2'$), and ($3d2''$) GOs lie at three different energies in the diatomic potential, whereas they are fivefold degenerate in the atom: the degenerate pairs are now ($3d1'$), ($3d1''$) and ($3d2'$), ($3d2''$). The distinction between different values of m is therefore an important one energetically and we employ these symmetry labels for both GOs and MOs. In diatomics, orbitals with $m = 0, 1, 2$ are said to be of the σ, π, and δ types, respectively. We will also use these Greek symbols to designate the symmetries of orbitals in polyatomics.

Let us apply the GO concept to the generation of the MOs in N_2. Each of the A and B atoms of N_2 contains a set of ($1s$), ($2s$), ($2px$), ($2py$), and ($2pz$) AOs. A fundamental rule is that *we must construct as many MOs as there are AOs available for use.* Recall that a similar conservation rule applies to hybrids in that the number of hybrid AOs must equal the number of canonical AOs we linearly combine. Since we have a total of ten AOs in N_2, we can form ten MOs. Using the coordinate system in Figure 3.4, we first place the simplest possible GO, namely, a ($1s$) AO, at the molecular center. The advantages of directing the z axes along the internuclear bond axis in the manner shown will become clear later. If we now place a ($1s$) AO on each nitrogen, we find that the ($1s$) GO has the same symmetry *in the diatomic potential* as a pair of *identical* ($1s$) orbitals, one on each nitrogen. To understand that this is so, image a point q anywhere in the ($1s$) GO. As shown in Figure 3.5, in which the orbitals are shown in cross-sectional views, an equivalent point q' for that

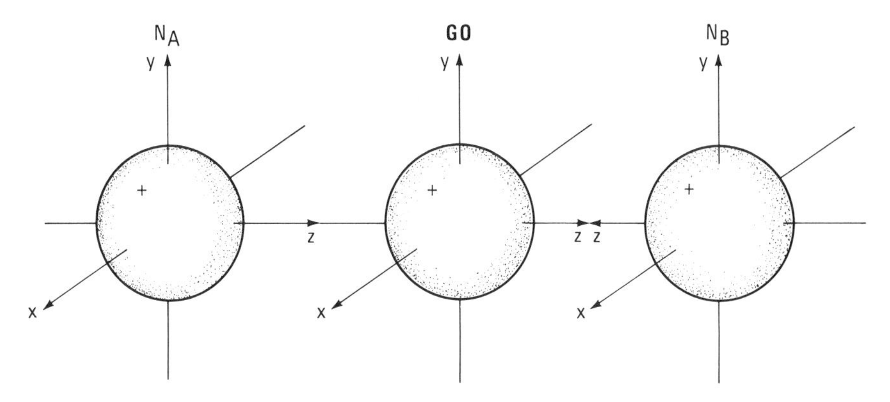

FIGURE 3.4 A sketch of a (1*s*) GO and two (1*s*) nitrogen AOs in N_2.

point *q* can be found by rotating *q* around the *z* axis by any angle, by reflecting *q* through the *xy* plane, by reflecting *q* through any plane containing the molecular axis, or by inverting it through the molecular center. At these equivalent pairs of points, the (1*s*) GO has the same value. By inspection it can be seen that the same is true for the combination of the nitrogen (1*s*) orbitals in which each nitrogen AO has the same size and sign. In Figure 3.6 we see that the (1*s*) GO also has the same symmetry in the diatomic potential as a pair of identical (2*s*) orbitals placed on the nitrogens with their signs as shown. When AOs possess spherical nodes, we will by convention always place the same sign in the outer lobe as we have in the GO. We also see that the symmetry of the (1*s*) GO is the same as that of the pair of nitrogen (2*pz*) AOs, as in shown in Figure 3.7a, whether we point the negative (2*pz*) lobes both inward or both outward. Since the (1*s*) GO is taken to have a positive amplitude, we adopt the convention that the (2*pz*) lobe closest to the GO takes the same amplitude sign as the GO. Of course we could have reversed all the orbital signs in Figure 3.4. In fact, we could even have used

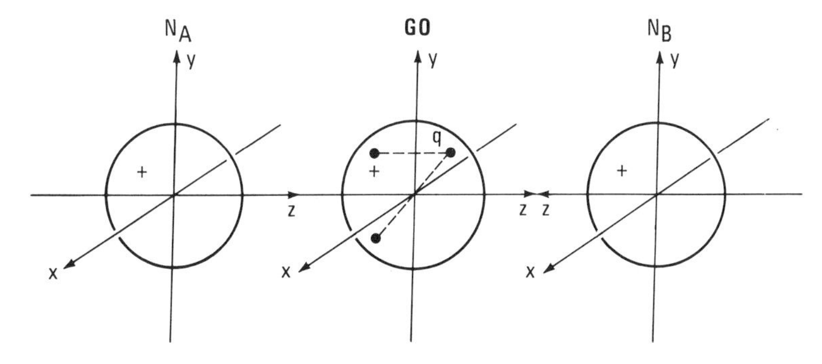

FIGURE 3.5 A depiction of the identity of the symmetries of a (1*s*) GO and a pair of (1*s*) AOs of the diatomic molecule N_2 in the field of the two nuclei.

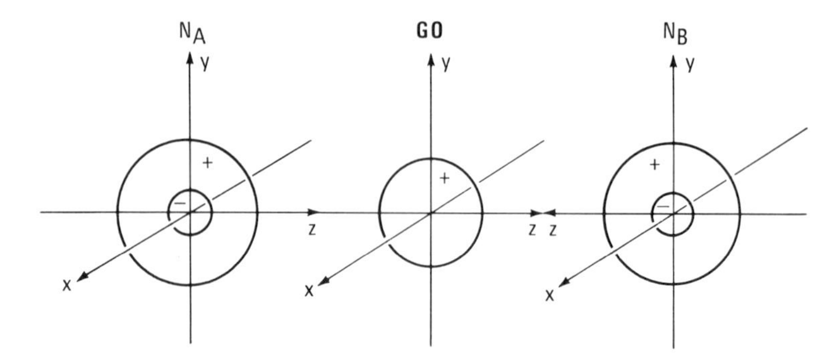

FIGURE 3.6 A depiction of the identity of the symmetries of a (1*s*) GO and a pair of (2*s*) AOs of the diatomic molecule N_2 in the field of the two nuclei.

opposite amplitude signs for the central GO and the pair of nitrogen orbitals in this figure without destroying the symmetry equality of the GO and the nitrogen (1*s*) orbitals. We adopt the convention, however, that the (1*s*) GO and the associated (1*s*) AOs on the atoms always have positive sign amplitudes.

Up to now we have seen nitrogen AO arrangements that are *permitted* by the (1*s*) GO. In other words, these permitted AO arrangements have the same symmetry in the diatomic potential as the GO. In Figure 3.7b, we see a nitrogen AO arrangement that is *not permitted* by a (1*s*) GO, since no matter how we arrange the signs in the (2*py*) AOs, their symmetry in the diatomic potential is not the same as that of the (1*s*) GO. This can be seen by imagining a point q anywhere in this combination of the (2*py*) AOs and inverting it through the molecular center. The sign of the wave function at q' is seen to be opposite sign to that at q. While it is true that rotating the (2*py*) AOs in Figure 3.7b by 180° about the z axis does not change the wave function, note that now reflection of a point q through the xz plane reverses the sign of the

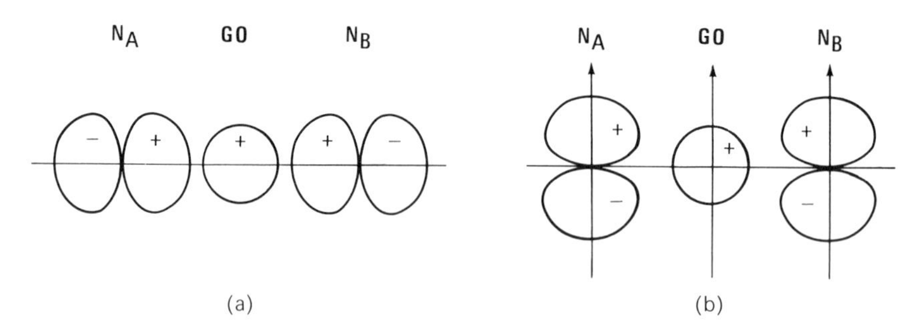

FIGURE 3.7 A depiction of the identity of the symmetries of a (1*s*) GO and a pair of (2*pz*) AOs of a diatomic molecule N_2 in the field of the two nuclei (a). In (b) it is seen that the symmetries of the (1*s*) GO and a pair of (2*py*) AOs are not identical in a diatomic potential.

wave function. Another way of saying the same thing is that this combination of the $(2py)$ AOs possesses a node in the xz plane, and the $(1s)$ GO does not have such a node. Therefore, the symmetries of the nitrogen $(2py)$ AOs and the $(1s)$ GO cannot be compatible.

From the three *permitted* nitrogen AO arrangements for N_2 we have generated thus far, we can form three MOs that can be written as linear combinations of atomic orbitals (LCAOs):

$$\begin{aligned}(\sigma_{1s,1s}) &= N_1[(1sA) + (1sB)]\\ (\sigma_{2s,2s}) &= N_2[(2sA) + (2sB)]\\ (\sigma_{2pz,2pz}) &= N_3[(2pzA) + (2pzB)].\end{aligned} \tag{3.2}$$

Although it is not apparent from Figures 3.4–3.7, the nitrogen AOs are sufficiently close to interact with each other, and the algebraic addition of their contours implied in Equation 3.2 is depicted in Figure 3.8. LCAOs of this type are called *delocalized* MOs because the electron density contours are free to encompass more than one nucleus (here, two nuclei). Equally valid for the description of the bonding in diatomics are *localized* MOs, which, as we will see later, can be obtained by hybridization of the delocalized MOs. From Figure 3.8 it is seen that the delocalized MOs described by Equation 3.2 are of the σ type because their constituent canonical AOs have $m = 0$. It may be noted here that the advantages of directing the z axes as shown in the coordinate system we have adopted is that the $(\sigma_{2pz,2pz})$ MO can be written as a sum (Equation 3.2) as can the $(\sigma_{1s,1s})$ and $(\sigma_{2s,2s})$ MOs.

Having considered the symmetry effect of the (1s) GO on all the available nitrogen AOs, we find that we still must generate seven more MOs. To do this, we progress to the next nodally more complex GO which is of course the (2s) GO. However, the (2s) has the same symmetry in the diatomic potential as a $(1s)$ GO since it differs from the $(1s)$ only by the presence of a spherical node. GOs with spherical nodes are not required for generating MOs in molecules of the type Z_n (e.g., rings, cages and linear and bent Z_3 species) or of the type EZ_n (e.g., a central atom E bonded to nZ atoms) because the presence of such a node does not change the symmetry properties of an orbital in the potential of these types of geometries. As we will see later, this is also true when we generate the valence orbitals for a central atom E in EZ_n geometries. For example, a $(4s)$ VAO on E is generated by a $(1s)$ GO at the

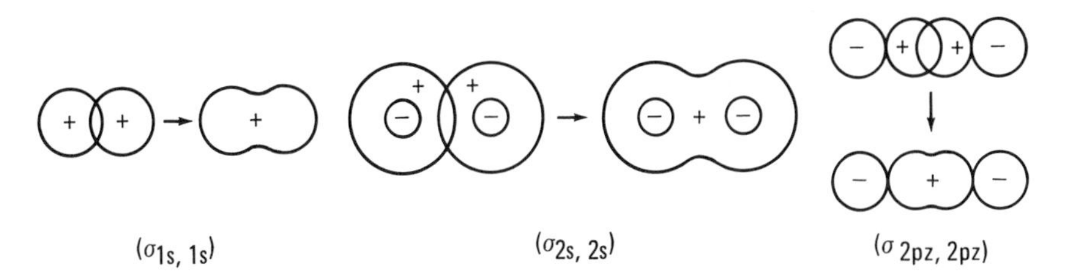

FIGURE 3.8 AO interactions in N_2 to form σ molecular orbitals.

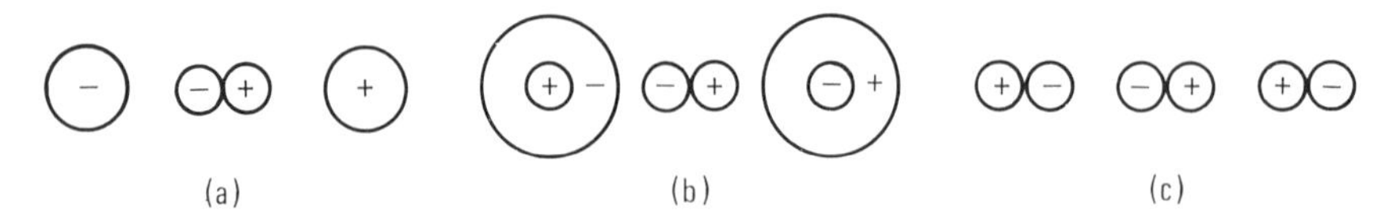

FIGURE 3.9 AO arrangements in N_2 permitted by the (2*pz*) GO.

same location, because the presence of three spherical nodes in the (4*s*) VAO does not change the spherical symmetry of its outer, or for that matter inner, signed wave function regions. Similarly, we can use a (2*px*) GO to generate a (3*px*) VAO on E. When we examine bonding patterns in polymers and solid state materials, however (Chapter 12), GOs with spherical nodes will be useful. Thus at this point in our treatment of the delocalized bonding in N_2, we can skip the (2*s*) GO and move on to the (2*p*) GO set.

Arbitrarily beginning with a (2*pz*) GO, we see from Figure 3.9 that only three AO arrangements are permitted by the (2*pz*) GO. Note that in visualizing these AO arrangements, we use the convention discussed earlier that *the amplitude sign of the AO lobe nearest to a GO lobe is the same as the amplitude sign of that GO lobe.* If the AO has a node (as in Figure 3.9b and c), then as we cross that node from the lobe nearest to the GO to the one further away, we of course change the sign of the amplitude.

The three AO arrangements in Figure 3.9 become MOs as shown in Figures 3.10a–c. A noteworthy feature of these three MOs is that, in contrast to the previous three, they contain a central planar node perpendicular to the internuclear axis because algebraic differences of amplitude contours are taken. This node is, of course, a consequence of the planar node in the (2*pz*) GO. As we will come to realize more fully later, an MO without an internuclear node is a *bonding* MO (BMO) and an MO possessing such a node is an *antibonding* MO (ABMO). The latter type is designated by an asterisk, as shown in Figure 3.10.

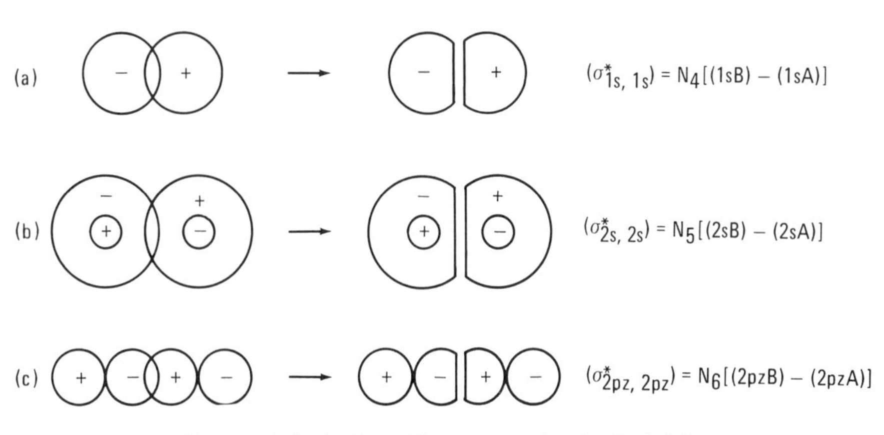

FIGURE 3.10 MOs of N_2 generated by the (2*pz*) GO.

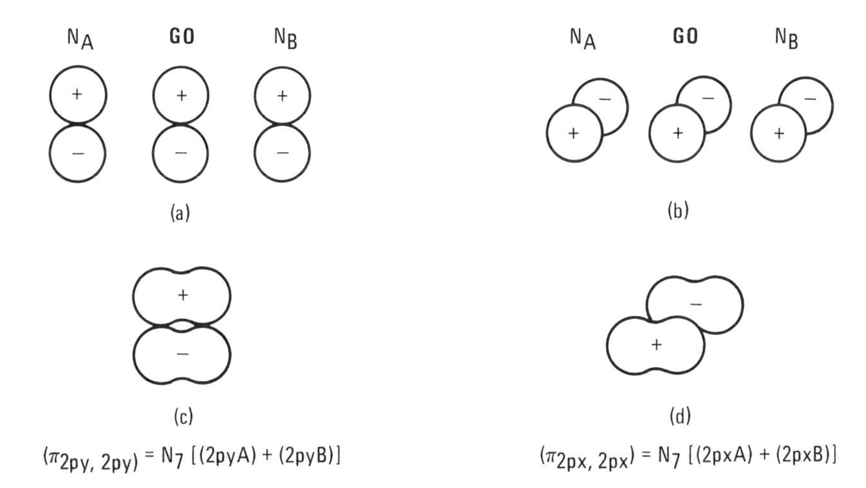

FIGURE 3.11 AO arrangements (a), (b) and MOs (c), (d) generated by the (2*py*) and (2*px*) GOs, respectively, in N_2.

Moving on to the (2*px*) and (2*py*) GOs, we see from Figures 3.11a and b that the symmetry of these GOs allows them to generate AO arrangements only among the (2*py*) and (2*px*) nitrogen AOs. These AO arrangements become the BMOs shown in Figures 3.11c and d. Since these MOs are generated by a pair of (2*p*) GOs that are energetically degenerate in the diatomic potential, the MOs are also degenerate and thus have identical normalization coefficients in their expressions. It should be realized that these are BMOs since they contain no internuclear node *perpendicular* to the internuclear axis.

The remaining two MOs can be generated by a pair of (3*d*) GOs, as shown in Figure 3.12. Having now generated all ten MOs for N_2, we can ask what effect the remaining (3*d*) GOs and also the higher GOs have on the nitrogen AOs. It is always true that *once all the permitted AO arrangements are generated by GOs, all other GOs either generate no permitted AO arrangements or they reproduce permitted arrangements already obtained with the simpler GOs.*

Expressions for the normalization constants N_1 to N_8 in our diatomic nitrogen MOs can be obtained by realizing that these constants allow the integral over all space occupied by the electron density to equal 1 (Appendix III.B).

The qualitative LCAO treatment we have been developing permits us to estimate the positions of the MO energy levels relative to each other in most instances. There are mathematical models that yield more accurate expressions for MOs. These expressions often contain substantial contributions from more than two AOs in each MO, however. Although the LCAO approach is only an approximation, it is a useful one in most cases. Earlier, we saw that the orbital energies for all atoms [except hydrogen for which $\varepsilon(2s) = \varepsilon(2p)$] are in the order $\varepsilon(1s) < \varepsilon(2s) < \varepsilon(2p)$. It is reasonable then that the

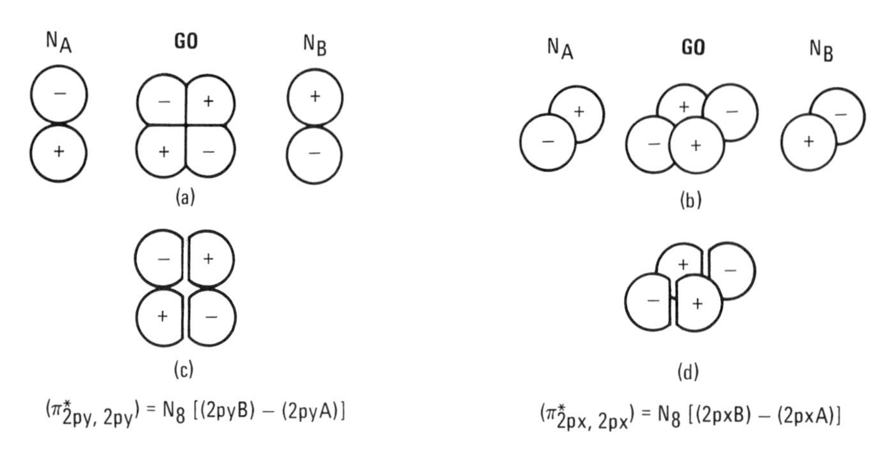

FIGURE 3.12 AO arrangements [(a), (b)] and MOs [(c), (d)] generated by the $(3dyz)$ and $(3dxz)$ GOs, respectively, in N_2.

energies of the MOs formed from these AOs lie in the order $\varepsilon(\sigma_{1s,1s}) < \varepsilon(\sigma_{2s,2s}) < \varepsilon(\sigma_{2pz,2pz})$. What about the relative positions of the BMO and ABMO formed from the same AOs, for example, $\varepsilon(\sigma_{1s,1s})$ versus $\varepsilon(\sigma^*_{1s,1s})$? Although both have about the same electron density very close to the nuclei, the ABMO has a nodal plane at its center perpendicular to the internuclear axis. As we have seen before with AOs, the orbital energy rises as the number of nodes increases, and as expected $\varepsilon(\sigma_{1s,1s}) < \varepsilon(\sigma^*_{1s,1s})$. Moving now to the MOs involving the nitrogen $(2p)$ AOs, we conclude from the above discussion that

$$\varepsilon(\pi_{2px,2px}) = \varepsilon(\pi_{2py,2py}) < \varepsilon(\pi^*_{2px,2px}) = \varepsilon(\pi^*_{2py,2py}). \tag{3.3}$$

Furthermore, we would expect that because the overlap in a π or π^* MO is smaller than in a σ or σ^* MO constructed from p orbitals in the same set (Figure 3.13), the energy difference would be larger between $\varepsilon(\sigma_{2pz,2pz})$ and $\varepsilon(\sigma^*_{2pz,2pz})$ than between $\varepsilon(\pi_{2py,2py})$ and $\varepsilon(\pi^*_{2py,2py})$. Thus better overlap in a BMO leads both to stronger bonding (but only up to a point, as we shall see later) and to stronger antibonding. We would expect then that the order of energies is as shown in Figure 3.14 wherein abbreviations for the MOs have been employed. There is evidence, however, that for most second-row diatomics in their ground states (Li_2 through N_2) the (σ_p) orbital lies *above* the (π_p), as shown in Figure 3.15a. When nuclear charges are low, the $(2s)$ and $(2p)$ AOs lie closer in energy. (Recall that in hydrogen these energies are actually the same.) Thus it is conceivable that considerable mixing of the $(2s)$ and $(2pz)$ orbitals can occur so that the resulting σ-type MOs are no longer of pure $(2p)$ or of pure $(2s)$ character. As we have seen earlier, such mixing of $(2s)$ with $(2p)$ on the atoms is expected to produce (sp) hybrids pointing in opposite directions on the z axis of each atom of nitrogen. To a first approxi-

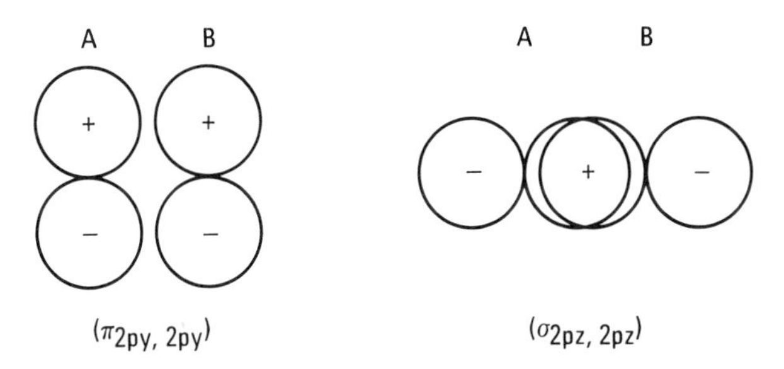

FIGURE 3.13 The smaller orbital overlap of a π BMO compared to a σ BMO constructed from (p) AOs having the same principal quantum number.

mation, each hybrid on an atom could be thought of as an (s^1p^1) hybrid, although these hybrids are probably not equivalent for homonuclear diatomics since the chemical environment *between* the atoms is not the same as at the *outside ends* of the molecule. The consequence of such mixing is that (σ_{2s}) and (σ_{2s}^*) are stabilized (decrease in energy) and (σ_p) and (σ_p^*) are destabilized, as illustrated in Figure 3.15b. Thus the bracketed MOs in Figure 3.15 split apart because each bracketed pair is similar in energy and they

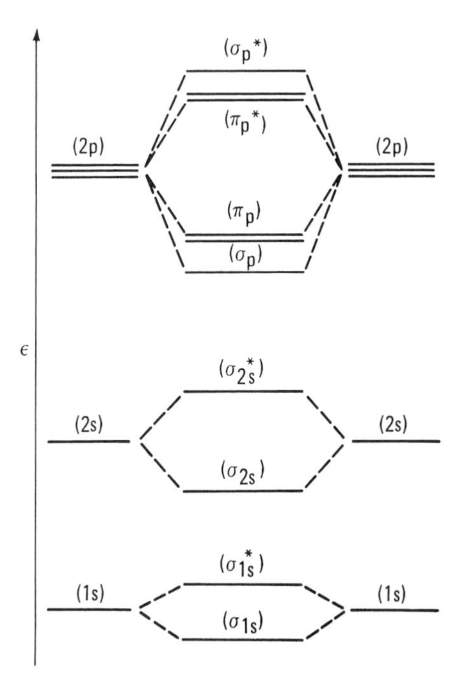

FIGURE 3.14 Energy level diagram for a diatomic molecule, showing the interaction of the atomic orbitals on nuclei A and B to form bonding and antibonding molecular orbitals.

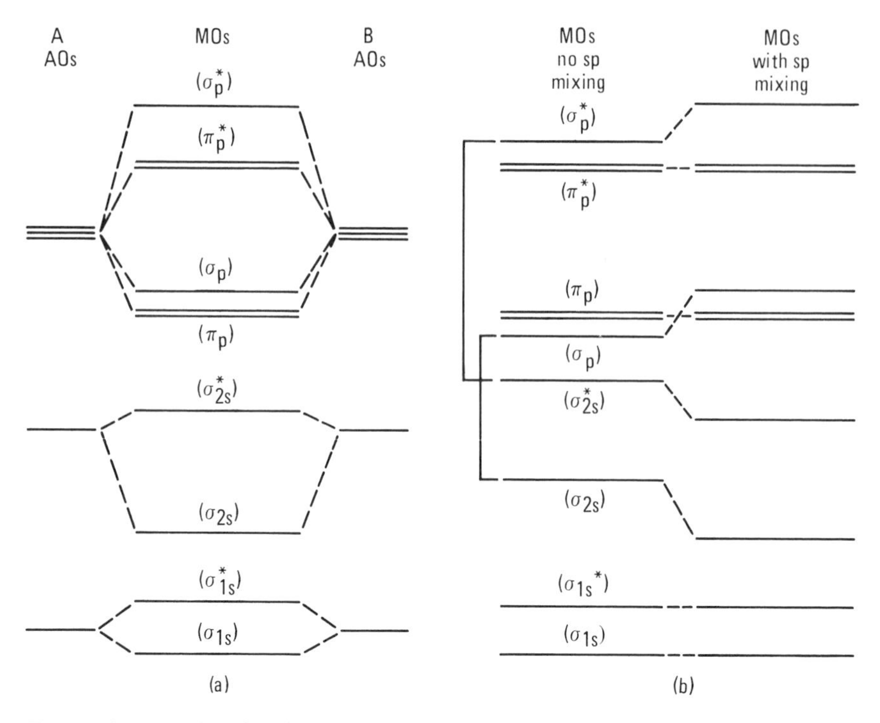

FIGURE 3.15 MOs of a diatomic molecule when there is s–p mixing (a). In (b), the MOs that split apart under s–p mixing are identified by connecting brackets.

have the same symmetry properties in the diatomic potential. This is evident from the fact that each such pair is generated by the same GO. Going across the periodic table from left to right, this splitting decreases because the separation of the valence (2s) and (2p) AOs increases. As a consequence of this process, (σ_p) apparently drops in energy below (π_p) for O_2 and F_2 (Figure 3.14).

It should be apparent that we can use GOs to generate MOs from the inward- and outward-pointing pairs of hybrids on N_A and N_B since they behave like (2pz) AOs as far as their symmetry is concerned. This process is the subject of a problem at the end of the chapter.

Because the GO approach to bonding will be consistently employed throughout the remaining chapters, a summary of the rules developed so far governing the use of GOs is now given:

1. Recognize that the number of valence AOs (VAOs) equals the number of MOs to be formed. Although we included the core (1s) AOs in second-row diatomics, their overlap is relatively inconsequential because of their small size compared to the (2s) and (2p) AOs. *From now on we will generally ignore MO formation from core AOs.*

2. Determine pictorially the permitted VAO combinations by placing first the (1*s*), then the (2*p*), (3*d*), etc., GOs at the center of the molecule and then matching the signs of their wave functions in nearest lobes of VAO sets composed of the *same* VAOs from one VAO set at a time. Thus only AOs of the same symmetry combine to form MOs [e.g., (2*px*A) combines with (2*px*B) but not with (2*py*B) or (2*s*B)]. Also, only AOs having similar energies in the free atoms combine to form MOs [e.g., (2*s*A) forms an MO with (2*s*B) but not with (1*s*B)]. When the permitted number of AO combinations (Rule 1) has been obtained, the generation process is complete.
3. Draw the LCAOs from the permitted VAO combinations and formulate the algebraic sums. Identify the MOs according to their σ or π nature by the absence or presence, respectively, of a planar node containing the internuclear axis, and determine whether or not the MOs are bonding or antibonding by the absence or presence, respectively, of a planar node perpendicular to the internuclear axis between the atoms.

Before moving ahead, a point needs to be made about filling electrons into MO energy level diagrams. As in the case of atomic orbital energy levels, electrons are paired with opposite spins into MO levels as long as the energy input necessary to place the second electron of the pair into the next orbital (the excitation energy) is larger than the energy necessary to pair the electrons (the electron pairing energy) in the same orbital. If the excitation energy is less than the electron pairing energy, which is certainly the case in a degenerate set of orbitals, each successive electron added remains unpaired in a separate orbital until each orbital in the entire degenerate set is half-occupied and all the spins are lined up. Electrons added after that go into each orbital of the degenerate set with spins opposite to that of the electron already there (Hund's rule).

Exercise 1 in the Practice Exercises section of Node Game takes you through the three steps given above in a pictorial way for the H_2 molecule. By repeating this exercise two or three times in a row, you will find the material that follows easier to learn and apply.

3.3. Generator Orbitals and Molecular Motions

Above zero degrees Kelvin, all molecules undergo translational and rotational motions in space, as well as vibrate internally. The translational and rotational motions can be thought of as having Cartesian components. Thus all molecules have three translational degrees of freedom and can therefore move along the x, y, and z directions (or some linear combination of these directions). Similarly, all molecules, except linear ones, also possess three rotational degrees of freedom. Linear molecules have only two rotational degrees of freedom because rotations around the linear axis cannot be monitored. The remaining degrees of freedom belonging to molecules are vibrational modes. Another approach to molecular motions is to think of a molecule as a set of n *independent* atoms (i.e., no bonds). Each atom then has three degrees of freedom corresponding to translation of its center of mass. Now think of fastening the atoms together with bonds. The $3n$ translational degrees of freedom formerly available to the assembly of unconnected n atoms become reclassified, since only the molecule as a whole can now translate, which accounts for three out of the total $3n$ degrees of freedom. With bonds connecting all the atoms together, only the molecule as a whole can rotate, thus accounting for another three degrees of freedom if it is nonlinear. The rest of the degrees of freedom (i.e., $3n - 5$ for linear and $3n - 6$ for all other molecules) must be of the vibrational type. As we will see, vibrations result either from alternately stretching and compressing bonds or bending them back and forth.

GOs provide a convenient pictorial way to visualize what the vibrational modes of molecules look like. As for translational and rotational degrees of freedom, a vibrational mode also has components along the Cartesian coordinates. Thus we can think of the components of such a motion as a vector directed in either the positive or negative direction along x, y, or z. Let us consider such possible atomic motions to be directed in the positive direction along the axes as depicted for N_2 in Figure 3.16. We will refer to such motions of atoms as atomic vectors (AVs). We realize, of course, that any atomic motion must reverse itself when its kinetic energy is expended in climbing the walls of the molecule's potential well by stretching and compressing the bond or by bending it. *All of the degrees of freedom (i.e., the translational plus rotational plus vibrational modes) are linear combinations of the atomic vectors (LCAVs) generated by GOs.*

Let us use N_2 as an example for generating the $3n - 5 = 1$ normal vibrational modes in any homonuclear diatomic molecule. We begin by noting that the *total* number of degrees of freedom is always equal to the total number of AVs, which in the present case is six. This parallels Rule 1 for generating MOs, which says that the total number of VAOs equals the number of MOs. Next we parallel Rule 2 for MOs and place a (1s) GO in the center of the molecule with sets of AVs of the same symmetry type, as shown in Figure 3.17a–c. *Vectors along the Cartesian axes behave as* (p) *orbitals.* That is, (p) orbitals have vectorial directionality in that they possess a positive and negative end and so they can be represented by an arrow. It is

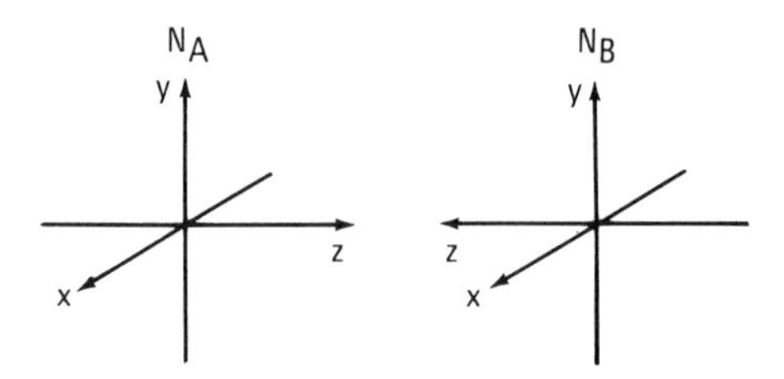

FIGURE 3.16 The axis systems chosen for the atomic vectors of the atoms in N_2.

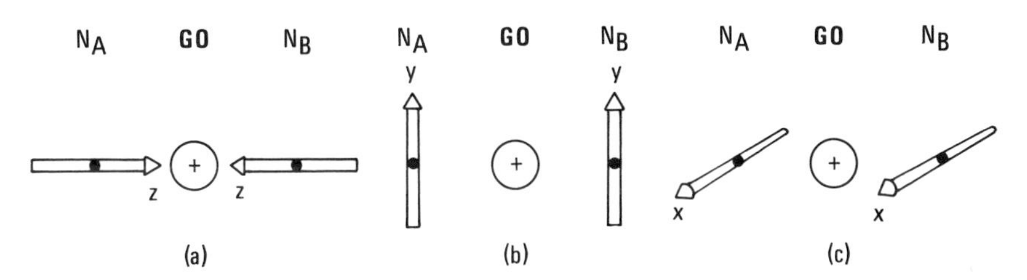

FIGURE 3.17 Atomic vector (AV) combinations in N_2 permitted (a) and not permitted [(b), (c)] by a (1*s*) GO.

therefore easily seen that Figure 3.17a represents a permitted AV combination whereas no combination of vectorial directions represented in either Figure 3.17b or c match the symmetry of the (1*s*) GO. Moving on to the (2*pz*) GO, we see in Figure 3.18 that a permitted combination is generated with the *z* atomic vectors. As with the placement of wave function signs in (*p*) orbitals, the orientation of the vectors is arbitrary. You should show that for symmetry reasons, no permitted combination can be generated by the (2*pz*) GO with the *y* and *x* AVs. Having generated two permitted AV combinations, we look to higher GOs to generate the remaining four shown in Figure 3.19. It may be noted in Figures 3.19c and d that the (3*d*) GOs are abbreviated with arrow drawings in which the double-headed arrows denote the positive lobes and the doubly tailed arrows denote the negative lobes of the corresponding (3*d*) GOs. The motion depicted in Figure 3.17a (which is the sum of the GO and the permitted AV combination) is clearly the vibrational mode since both atoms are simultaneously moving toward the molecular center. Of course, these atoms will move apart again, and we could generate that motion by employing a minus sign for the (1*s*) GO wave function. Notice the parallism with Rule 3 here. The motions in Figures 3.18 and in Figure 3.19a and b are translational modes of the molecule in the *z*, *y*, and *x* directions, respectively. It is also easy to see that the motions represented by Figure 3.19c and d are rotational ones around the *x* and *y* axes, respectively. Of course, each of these motions can be reversed by taking the linear combination of the GO and the corresponding AV combination as a difference. In visualizing all of these motions, notice that the GO is only a learning device and not part of the molecule. Note that no combination of AVs can represent a rotational motion around the *z* axis for a linear molecule and indeed such a rotational motion does not exist.

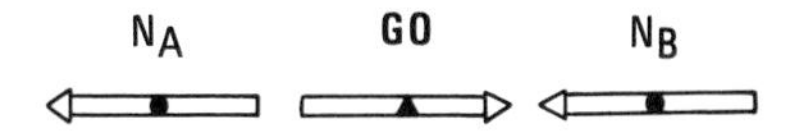

FIGURE 3.18 AV combination in N_2 generated by a (2*pz*) GO.

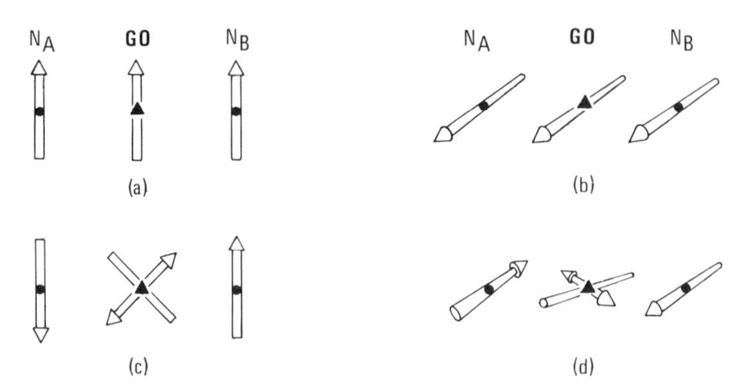

FIGURE 3.19 AV combinations in N_2 generated by the (a) ($2py$), (b) ($2px$), (c) ($3dyz$), and (d) ($3dxz$) GO.

Use the Node Game Homework Drawing Board to carry out for yourself the pictorial steps we took to generate the AV combinations in Figures 3.17, 3.18, and 3.19 for a diatomic molecule. Then delete the GOs in your pictures and label each drawing appropriately as vibration, rotation, or translation.

It may appear that the GO approach to the bonding and molecular motion of diatomics is trivial. It should be realized, however, that the LCAOs and LCAVs for species such as tetrahedral CH_4 or octahedral CoF_6^{3-} cannot be generated without GOs unless one happens to be well acquainted with group theory, a subject that actually turns out to be much easier to learn after one has gained an appreciation for the power of GOs. The pictures in this book and those generated by Node Game are two-dimensional representations of three-dimensional objects. If you would like three-dimensional models to help you visualize the GO approach to molecular bonding and motion, see Appendix III.C for a set of simple plans that use inexpensive materials.

3.4. Bond Strengths

A stretching vibration can be thought of as an oscillation of two masses separated by a spring with a force constant k, which characterizes the stiffness of the spring. The frequency ν of vibration is given by Equation 3.4:

$$\nu = 3906 \times 10^{10}[k(m_A{}^{-1} + m_B{}^{-1})]^{1/2} \text{ sec}^{-1} \tag{3.4}$$

in which m_A and m_B are the atomic weights of the atoms in atomic mass units and k, the force constant, is in units of mdyne Å. Values for ν range from 10^7 to 10^8 sec^{-1}, which corresponds to infrared radiation. In general, strong bonds tend to be stiff (i.e., relatively large k values) and the atoms tend to vibrate rapidly with small amplitudes. The converse is true for weak bonds.

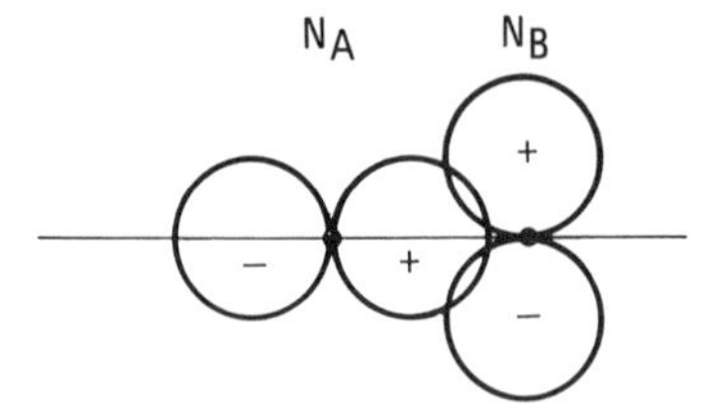

FIGURE 3.20 The net zero overlap of two (2*p*) orbitals of different symmetry in a diatomic potential.

Two critically important factors in determining the strength of a bond are the similarities of the energies of the neighboring-atom orbitals, which can interact because of their identical symmetries in the potential field, and the degree of overlap of these orbitals which is limited from reaching its maximum value by internuclear repulsive forces. (For more on this, see Appendix III.B.) For example, core (1*s*) electrons on atoms N_A and N_B of N_2 are at the same energy and are of the same symmetry in the diatomic potential, but they do not overlap nearly as strongly as the (2*s*) valence orbitals owing to the larger distance between the outer contours of the (1*s*) orbitals. The resultant σ and σ^* MOs are also split apart more in the case of the (2*s*A)–(2*s*B) interactions than in that of the (1*s*A)–(1*s*B) interactions. It is now easy to see why we do not consider AOs of different symmetry when forming LCAOs in molecules. For example, as shown in Figure 3.20, a (2*pzA*) and a (2*pyB*) have zero overlap.

Let us have a look at the simplest molecular species there is, namely, H_2^+. That is, it has the minimum number of atoms to be a molecule and it is bound by only one electron. It is interesting to see that a plot of the potentials for the $(\sigma_{1s,1s})$ and $(\sigma^*_{1s,1s})$ orbitals of the H_2^+ ion (Figure 3.21) reveals that

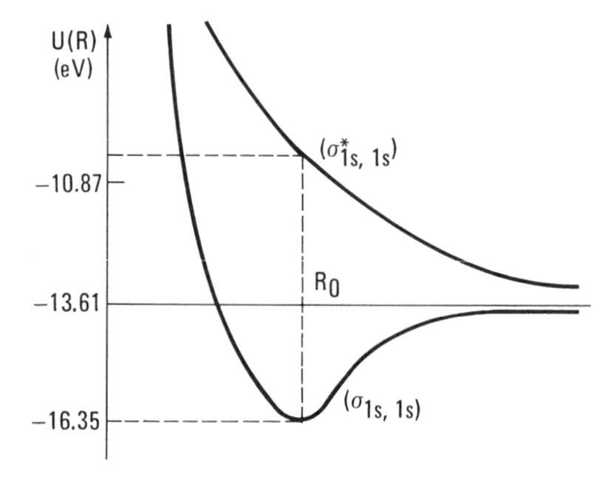

FIGURE 3.21 Plots of the potential energy of the BMO and ABMO of H_2^+ versus internuclear distance. For any given internuclear distance, the ABMO is more destabilized in energy than the BMO is stabilized.

TABLE 3.1 Bond Strengths in Homonuclear Diatomic Molecules

Molecule	Bond Order	Dissociation Energy (eV)	Bond Length (Å)	Force Constant (mdyne/Å)
H_2^+	0.5	2.74	1.06	1.4
H_2	1	4.48	0.74	5.1
He_2^+	0.5	3.1	1.08	3.1
He_2	0	no chemical bond		
Li_2	1	1.03	2.67	0.25
Be_2	0	no chemical bond		
B_2	1	3.0	1.63	3.5
C_2^+	1.5	5.5	1.34	
C_2	2	6.5	1.24	9.3
N_2^+	2.5	8.72	1.12	19.5
N_2	3	9.76	1.09	22.4
N_2^-	2.5			13.2
O_2^+	2.5	6.48	1.12	16
O_2	2	5.12	1.21	11.4
O_2^-	1.5		1.28	5.6
F_2^+	1.5	3.19	1.28	
F_2	1	1.56	1.42	4.5
Ne_2	0	no chemical bond		

the ABMO has a *repulsive* potential. The energy $U(R) = 13.61$ eV in this figure represents H_2^+ dissociated into H and H^+. In other words, promotion of the electron in the $(\sigma_{1s,1s})$ BMO to the partner ABMO results in destabilization of the H_2^+ molecule and the component H and H^+ species fly apart. It is for this reason that we call the $(\sigma^*_{1s,1s})$ an *antibonding* MO. In fact, ABMOs in general are somewhat *more* antibonding than the corresponding BMOs are bonding (cf. Figure 3.21). Thus each occupied BMO strengthens the bond and each occupied ABMO slightly more than offsets its equally occupied partner BMO. These considerations now allow us to define the concept of *bond order.* A single bond is afforded by double occupancy of a BMO (as in the ground state of H_2). Consequently, H_2^+ is held together by a half-bond. In general, we can define bond order by Equation 3.5.

$$\text{Bond Order} = \tfrac{1}{2}[(\text{Total No. e}^-\text{'s in BMOs}) - (\text{Total No. e}^-\text{'s in ABMOs})] \tag{3.5}$$

In Table 3.1 are collected the bond orders, dissociation energies, bond lengths, and force constants for some diatomic species. *In general, dissociation energies and force constants increase and bond lengths decrease as bond orders rise.*

3.5. Heteronuclear Diatomic Molecules

Unlike the MOs in homonuclear diatomics, the MOs in heteronuclear diatomics must necessarily be obtained by interacting nonidentical VAOs. This means the interacting atomic VAOs are of similar symmetry, but the *princi-*

TABLE 3.2 Valence Atomic Orbital Potential Energies[a] in eV

Atom	1s	2s	2p	3s	3p	4s	4p
H	13.6						
He	24.5						
Li		5.5					
Be		9.30					
B		14.0	8.3				
C		19.5	10.7				
N		25.5	13.1				
O		32.4	15.9				
F		46.4	18.7				
Ne		48.5	21.6				
Na				5.2			
Mg				7.7			
Al				11.3	6.0		
Si				15.0	7.8		
P				18.7	10		
S				20.7	12		
Cl				25.3	13.7		
Ar				29.3	15.9		
K						4.3	
Ca						6.1	
Zn						9.4	
Ga						12.6	6.0
Ge						15.6	7.6
As						17.6	9.1
Se						20.8	11.0
Br						24.1	12.5
Kr						27.5	14.3

[a] It should be recalled that these are negative energies since they represent attractive potentials between an electron and a shielded nuclear charge.

pal quantum numbers may be different; in general, they are VAOs of different energies. It can also mean that VAOs of *different l quantum numbers* interact if they are sufficiently close in energy. Sometimes the choice of VAO pairs is not obvious, and it is further complicated by the relative degrees of VAO overlap, which can only be determined by complex calculations. When there is strong competition between two VAOs on an atom, the MOs gain partial character of *both* contributing VAOs on that atom. For simplicity, we will assume that the VAOs closest in energy will interact as long as they have the proper symmetry. In Table 3.2 are some data that will be helpful to us in this regard.

Let us consider LiH as an example. The VAOs on Li include the (2*s*) and the (2*p*) set. Should we allow the (H1*s*) to interact with (Li2*s*) or with a (Li2*p*) pointed toward the (H1*s*) [i.e., (Li2*pz*)]? The energy of the Li (2*p*) AOs is not given in Table 3.2 because this table applies to the ground states of atoms and in the case of Li the (2*p*) level is unoccupied in the ground state. If we excite an electron from the (2*s*) to a (2*p*) orbital, the (2*p*) energy is about 8 eV. On

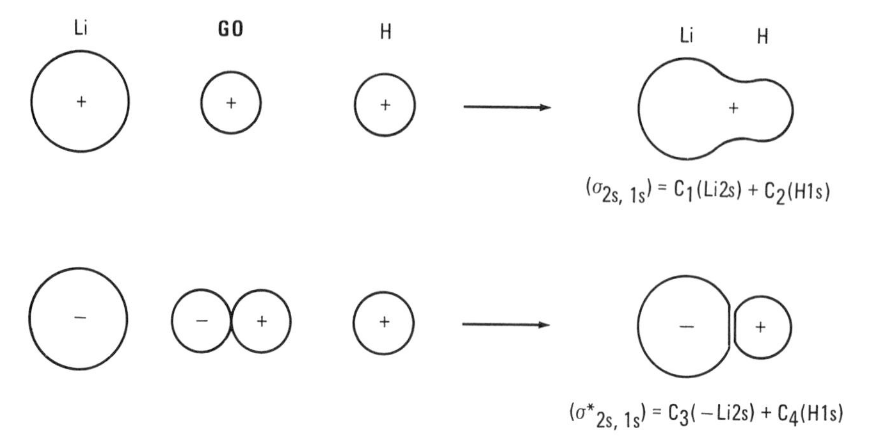

FIGURE 3.22 The generation of the MOs in LiH from (H1*s*) and (Li2*s*) using GOs.

the basis of relative energies, therefore, we would wish to interact the (H1*s*) orbital with a (Li2*pz*) orbital, although the difference in energy between the (2*s*) and (2*p*) levels in Li is relatively small compared to the gap between these levels in, say, the atoms boron through neon. Thus we might also expect some interaction between (H1*s*) and (Li2*s*). As we saw earlier, we must also consider the overlap criterion. That is, how does the overlap between (H1*s*) and (Li2*pz*) compare to that between (H1*s*) and (Li2*s*)? From calculations, it is known that the overlap is actually better between (H1*s*) and (Li2*s*). The resultant bonding in LiH according to this calculation involves a 64% contribution by (Li2*s*) and only a 36% contribution by (Li2*pz*). This mixture represents the compromise between energy and the overlap criteria. How, then, do we make a decision regarding orbitals to use in future examples? For reasons which will become clear later, this is not a severe problem.

Since the (Li2*s*) orbital is the main contributor to the bonding in LiH, let us, for the sake of simplicity, assume that it is the only orbital on lithium which interacts with (H1*s*). Note that in a *heteronuclear* diatomic molecule, a point reflected across the *xy* plane or inverted through the midpoint of the molecule does not feel the same potential, regardless of which pair of orbitals we choose [including (Li2*s*) and (H1*s*)] because the nuclear charges are different. Thus we must be aware that we are dealing now with a molecule having less symmetry than a homonuclear diatomic. In Figure 3.22 we see how the two MOs arise from the generation of the VAO combinations permitted by the GOs. The corresponding MO energy level diagram is drawn in Figure 3.23. There are several important points to note from these figures:

1. Determination of the magnitude of the MO splittings is beyond the scope of this book. Therefore we will always be very qualitative about such splittings in our MO energy level sketches.

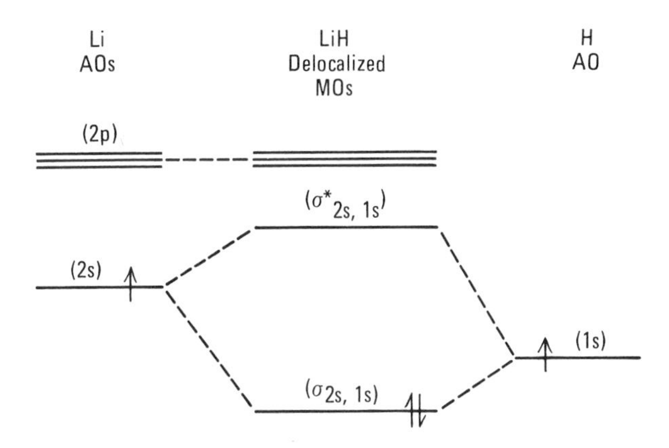

FIGURE 3.23 Delocalized MO energy level diagram for LiH.

2. Notice that separate constants (C_1 through C_4) are used to describe contributions from the VAOs in the MO expressions. These constants are positive numbers that contain a normalization constant as well as a fraction representing the contribution of that particular VAO to the MO. As a consequence, the BMO has more hydrogen (1*s*) character than lithium (2*s*), whereas the opposite is true for the ABMO because of the orthogonality requirement. In other words, the BMO electron density is more concentrated near the H nucleus than the Li nucleus. A bonding electron excited to the ABMO conversely would spend most of its time near the Li nucleus.
3. The electron pair bond in LiH is polar in that the electron density is shifted toward the hydrogen because its electronegativity exceeds that of lithium. Recall that the electronegativity of an atom reflects its tendency to withdraw electron density from neighboring atoms in a molecule. In homonuclear diatomics, the electronegativities are the same, so the bonds are nonpolar.
4. For our qualitative MO diagrams we will generally ignore the fact that ABMOs are somewhat more antibonding than the corresponding BMOs are bonding.

3.6. Localized MOs for Diatomics

In previous discussions we learned that the canonical delocalized orbitals in conjugated molecules such as butadiene could be transformed into localized orbitals having maximum amplitudes between carbon atoms. We also saw that the canonical AOs (*s*), (*p*), (*d*), etc., could be recast into more localized hybrid sets (*sp*), (sp^2), etc. We will now show qualitatively that the delocalized diatomic canonical MOs can also be localized. The mathematical treatment will not be given here, but the results can be visualized from Figure 3.25. For a molecule such as Li_2, which contains one pair of valence electrons ($n = 1$ in Figure 3.24), localization is meaningless and in fact not possible. Thus the electron density is still concentrated between the two nuclei. For the hypothetical Be_2 molecule $n = 2$, and we recall that the first two canonical MOs [$(\sigma_{2s,2s})$ and $(\sigma^*_{2s,2s})$] would have to be occupied. The two linear combinations of these two MOs obtained by superposing them are easily visualized to lead to σ-type orbitals, which concentrate electron density along the internuclear axis but outside of the hypothetical molecule rather than between the nuclei. These are designated σ_l^{lp} and σ_r^{lp} where the subscripts stand for one orbital pointing to the left and the other to the right. Such electron pairs produce no bonding since there is no significant electron density between the nuclei. These electron pairs are associated largely with only one nucleus and are therefore called *lone pairs* (lp). The resulting lack of bonding between the two Be nuclei in the localized bonding view leads us to the same conclusion reached earlier with the delocalized MO picture, namely that Be_2 is not a stable molecule.

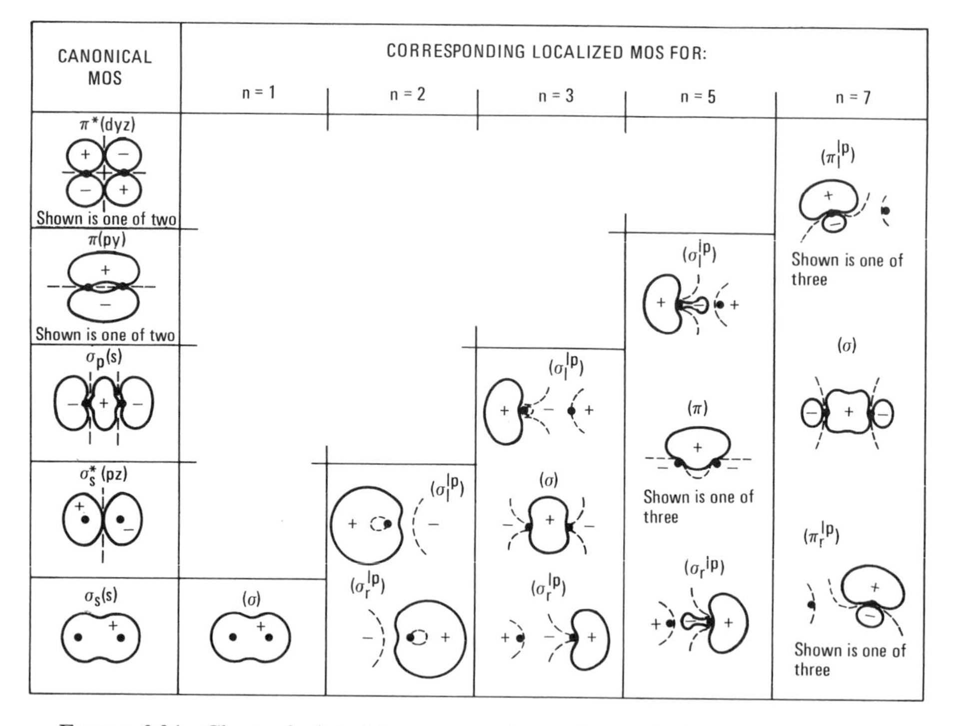

FIGURE 3.24 Chart of pictorial representations of canonical and corresponding localized MOs for homonuclear diatomics containing *n* pairs of valence electrons (adapted from W. England, L.S. Salmon and K. Ruedenberg, *Topics in Current Chemistry*, *23*, 31, (1971).

Before moving to $n = 3$, a word of clarification on lone pairs is in order. Lone pair orbitals are predominantly (s) in character because neutralization of the charge on one nucleus is more efficiently done by a spherical (s) orbital than by an orbital with directionality such as a (p). Bonding electron pairs tend to reside in orbitals having more (p) character because such AOs are more directional and bonding electrons are better able to neutralize the repulsive positive charges of two nuclei by residing in an MO made up of AOs that concentrate the electron density between the two nuclei. This becomes increasingly true for atoms toward the right of the periodic table for which the difference in energy between the valence (s) and (p) AOs increases (Table 3.2).

The molecule B_2 is an example of a system in which $n = 3$. However, from Figure 3.24, B_2 is seen to have two unpaired electrons, one in each (π_p) level. Difficulties arise with the localized approach when orbitals are singly occupied and we will therefore concern ourselves with only the delocalized approach in such cases. Excitation of ground state B_2 does give rise to an excited state in which the unpaired electrons are paired in the (σ_p) level, however. We can then form three localized MOs and the permitted linear combinations of delocalized MOs which produce them are shown in Figure 3.25.

It is interesting to note that in cases where there are lone pairs as well as bond pairs, the lone pairs generally lie higher in energy than the bond pairs and are in fact more easily ionized. Indeed, the lone pair MOs are seen in Figures 3.24 and 3.25 to contain a node *between* the nuclei giving them some *antibonding* character. What are the energies of the localized MOs for our excited B_2 molecule? Recall that the sum of the energies of the occupied orbitals must remain constant whether the orbitals are canonical or hybrids

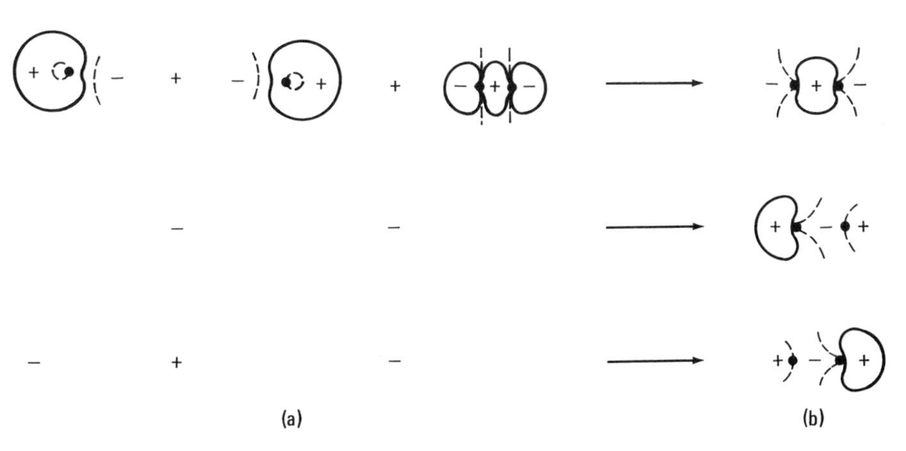

FIGURE 3.25 Combinations of delocalized MOs (a), which give the localized MOs in (b).

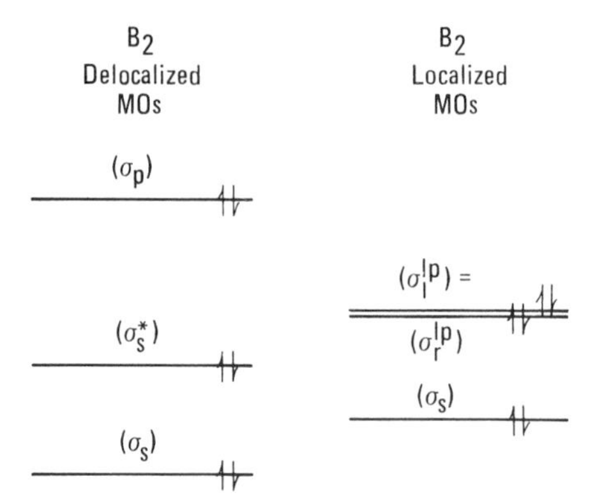

FIGURE 3.26 Delocalized and localized MO energy level diagrams for the lowest excited state of B_2.

(Chapter 2.5). Because of this, the delocalized and localized MO energy level diagrams for the excited state of B_2 might look like those in Figure 3.26.

The ground state of C_2 ($n = 4$) has no unpaired electrons, and the visualization of its localized MOs is left as an exercise. Since the ground state of O_2 ($n = 6$) contains unpaired electrons, we will move on to the ten- and fourteen-electron cases of N_2 and F_2, respectively. In the fifth and sixth columns of Figure 3.24 are the orthogonal localized MOs obtained by linearly combining the first five and all seven canonical MOs, respectively. Notice that N_2 is bonded by three equivalent "banana" bonds which are localized forms of the delocalized (σ_p) and two (π_p) MOs. We also observe that whenever both members of a pair of bonding and antibonding MOs are occupied in the delocalized bonding view (e.g., N_2 and F_2), we obtain a pair of *slightly antibonding* lone pair orbitals in the localized approach. The slightly antibonding nature of lone pair orbitals on adjacent atoms arises from the fact that delocalized ABMOs are more antibonding than the partner delocalized BMOs are bonding. The lone pair orbitals in N_2 and F_2 are shaped roughly like atomic (sp) and (sp^3) hybrids, respectively, and later we will return to this point.

From the localized view of diatomics we can see that lone pairs and bond pairs appear to repel one another. For example in the excited state of B_2 each boron is surrounded by two electron pairs (one lone pair and one shared bond pair) and these pairs are concentrated in lobes on each atom which are 180° apart. In F_2 there are three lone pairs and a shared bond pair around each atom and they point *roughly* to the apices of a tetrahedron. This tetrahedral geometry is approximate because lone pairs have more (s) character. Therefore the angle they make with each other is $> 109°28'$ and their angle with the bond pair is $< 109°28'$. The idea of electron pair "repulsions" will be useful to us later in predicting molecular geometries.

The localized view of diatomics (as well as of more complicated molecules, as we shall see) can also be generated from the delocalized view by a GO device similar to the one developed for the delocalized view. Since for our purposes only fully occupied delocalized MOs can be localized, *we begin in Step 1 of this process by identifying the GOs that give rise to the occupied delocalized MOs.* To simplify the process, *we do this after all BMO–ABMO pairs have been localized as lone pairs on the atoms.* Taking F_2 as an example, we see that after localizing the three lone pairs on each atom, only one occupied MO is left and it is a BMO generated by an (*s*) GO. As we have already seen with H_2 (Figure 3.24), localization of one occupied delocalized MO does not alter its appearance. Moving to N_2 as an example we note that an (*s*)-generated σ and two (*p*)-generated π BMOs remain after localizing the lowest-energy pair of MOs as lone pairs. The three BMOs are generated by (*s*), (*px*), and (*py*) GOs. Step 2 is to hybridize the GOs that generate occupied delocalized MOs and use the GO hybrid set as a template at the molecular center to generate localized MOs. It is seen from Figure 3.27a that the symmetry of such a GO set would "call in" the lobes of suitably oriented (sp^2) VAOs on the nitrogens. Since we do not know the precise admixture of (*s*) and (*p*) character in the nitrogen hybrids, it is not difficult to see that by adding some (*pz*) character to the nitrogen (sp^2) VAOs (Figure 3.27b), the banana bonds depicted in Figure 3.24 could arise through overlap of the main lobes of the nitrogen hybrid VAOs shown in Figure 3.27b. In fact, we indeed expect such an enrichment in (*pz*) character in these orbitals because, as we have seen earlier, the lone pairs on each nitrogen tend to have more (*s*) than (*p*) character and therefore the hybrids in which they are housed must have less than 50% (*pz*) character.

Using the Node Game Homework Drawing Board, see if you can duplicate the pictures shown in Figure 3.27 for the localization of the three delocalized bonds in N_2. Then remove the hybrid GOs and move the two hybrid sets in Figure 3.27 closer together so that the arrowheads almost touch, thus revealing the "banana" bonds more clearly.

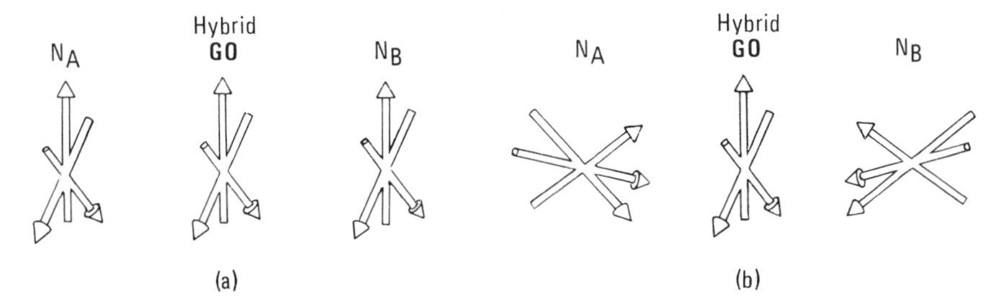

FIGURE 3.27 The use of GOs in N_2 to generate localized "banana" bonds composed of (*spxpy*) VAOs on each atom (a) and (*spxpy*) hybrids to which (2*pz*N) character has been added (b). The banana-like appearance of each bond is more apparent in (b) because of the angular overlap of the nitrogen hybrids.

3.7. Stable and Unstable Diatomic Molecules

A substantial number of diatomic molecules are sufficiently stable at room temperature and pressure to be isolated and stored indefinitely. Familiar examples include the halogens, O_2, N_2, CO, and H_2. All of these gases are important in industrial reactions that lead to a variety of important products ranging from pharmaceuticals to automobiles. One diatomic, namely NO, even plays important roles in our body's chemistry. This molecule has recently been found to be a neurotransmitter, a blood vessel dilator, and a blood pressure regulator. While we normally think of antiviral drugs as being complicated organic molecules, NO is found to be a potent inhibitor of pox viruses and *herpes simplex* type 1. Stable cationic diatomics are also known as in KrF^+ and $Br_2{}^+$ salts.

Unstable ions such as $N_2{}^+$, $Cl_2{}^-$, $Br_2{}^-$ and $I_2{}^-$ have been observed in mass spectrometric experiments and species such as $F_2{}^-$, ClO, BrO, SiO, GeO, PO, S_2, and Te_2 have been isolated and spectroscopically studied at low temperature in frozen inert gas matrices. Like O_2, S_2 is paramagnetic. Unstable diatomics such as CN, CS, SiC, SiO, SO, NS, NP, OH and CH have been detected in interstellar space.

Summary

After considering some of the factors leading to the formation of a diatomic molecule, the generator orbital device for visualizing delocalized MOs in diatomics was introduced and three rules were developed for its application. It was seen that the order of delocalized MO energies in diatomics is determined by the degree of (s)–(p) mixing. The GO device was then applied to molecular motions in a diatomic molecule and rules parallel to those for GO generation of delocalized MOs were presented. The criteria of VAO energy and overlap as they relate to MO formation and bond strength were briefly examined and a definition of bond order was given. An approach to delocalized MOs in heteronuclear diatomics was then elaborated and some observations were made on characteristics of heteronuclear bonding. We concluded our discussion by developing two steps for using GOs to generate localized MOs from delocalized ones.

In doing the exercises in this and the following chapters, use the Node Game Homework Drawing Board wherever you can. Node Game is an interactive tool that helps teach you how to sketch the allowed linear combinations of atomic orbitals (LCAO's) that make up the delocalized molecular orbitals (MO's) for inorganic and organic molecules without having to use group theory.

Exercises

1. Generate drawings showing how the degeneracy of the ($3d$) GO set is broken when these AOs are at the center of a diatomic potential. Would the results be different or the same for the ($4d$) AOs? Do the degeneracies of either set of orbitals break differently in a heteronuclear diatomic potential?

2. Show that two of the (d) GOs do not generate permitted AO arrangements for N_2. Which (d) GO(s) generates an MO already generated by a lower GO?

3. Using the GO approach, generate pictures of the permitted AO arrangements, sketch an MO energy diagram showing the appropriate orbital occupation, and give the overall bond order for Rb_2, He_2^+, BO, BeH, and diamagnetic O_2.

4. Of the species in Problem 3, which have nonpolar and which have polar bonds? For the latter, show the direction in which the electron density is polarized.

5. Account for the fact that removal or addition of an electron to N_2 weakens the bond.

6. Account for the fact that removal of an electron from O_2 strengthens the bond.

7. Account for the fact that the bonds in N_2^-, O_2, F_2^+, and F_2 are slightly weaker than the bonds of equal bond order in N_2^+, C_2, C_2^+, and B_2, respectively.

9. Considering the drawings of the localized MOs in Figure 3.24, draw the localized MOs for ground state C_2. Using the GO device, pictorially generate the localized view of C_2 from the delocalized view and compare the results with the first part of this problem. Draw the delocalized MO energy level diagram for C_2 and right beside this diagram draw the corresponding diagram for the localized view you obtained. Describe the relative s and p character of all the localized orbitals in C_2.

10. Account for the observation that the Br–Br bond distance in $Br_2(Sb_3F_{16})$ is 2.18Å, whereas in Br_2, it is 2.27Å.

11. Account for the decreasing bond dissociation energy trend: N_2 (941 kJ/mol), NP (613 kJ/mol), PP (485 kJ/mol).

12. Generate a delocalized MO view for the Cl_2^- anion, which has been observed in low-temperature reactions of sodium metal and chlorine.

13. Evidence has recently been presented supporting the presence of an N_2^- ion in the complex cation shown below. Generate a delocalized MO view for the N_2^- anion.

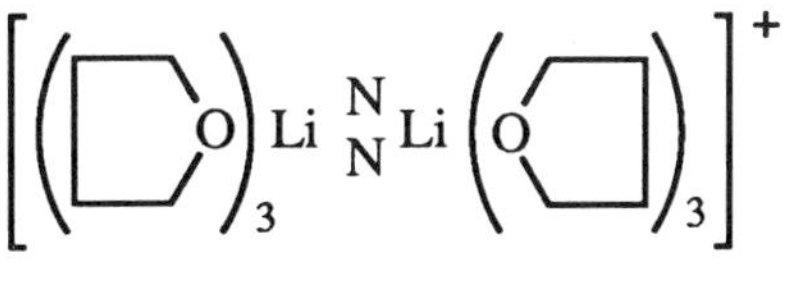

$$\left[\left(\mathrm{O}\right)_3\mathrm{Li}\ \overset{\mathrm{N}}{\mathrm{N}}\ \mathrm{Li}\left(\mathrm{O}\right)_3\right]^+$$

Chapter 4

Linear Triatomic Molecules

Triatomic molecules are most frequently bent; less common are the linear and (the quite rare) triangular arrangements. Stable molecular geometries with precise bond angles and bond lengths can be predicted by determining quantum mechanically the molecular energy as a function of atomic positions and then finding the minima in this function. Sophisticated calculations are required for this purpose, however. For large molecules and compounds of heavier elements the computations are still too large even for modern computers to achieve substantial accuracy. Since one of our main purposes is to develop pictorial delocalized and localized views of bonding in polyatomic molecules, we must first know their geometries. The concept most often used for this purpose is the "valence state electron pair repulsion" (VSEPR) approach, and we shall in future chapters use this idea extensively. *To justify the use of this tool, however, we will assume that we know the molecular geometry* for the examples in this chapter and defer the VSEPR rules to the next chapter.

4.1. Linear FHF^-

This stable anion is an example of *hydrogen bonding* in which the hydrogen is exactly midway between the two neighbor atoms. The hydrogen bonding in this ion is the strongest known (27 kcal/mole, whereas most hydrogen bonds are in the 1–10 kcal/mole range). We shall see that the exceptional stability of FHF^- can be understood in terms of chemical bonding.

For the delocalized view of this ion we begin by noting that we have a central atom between two peripheral atoms. Thus we have *two different* sets of atoms and we must examine *separately* the behavior of their VAO sets in the presence of GOs placed at the molecular center (i.e., at the hydrogen location) in the axis system shown in Figure 4.1. A convenient way to track our progress in developing delocalized views of molecules is to set up a "generator table" as shown in Table 4.1. The VAOs available in FHF^- are those listed under the column headed VAO Equivalence Sets. Each of the

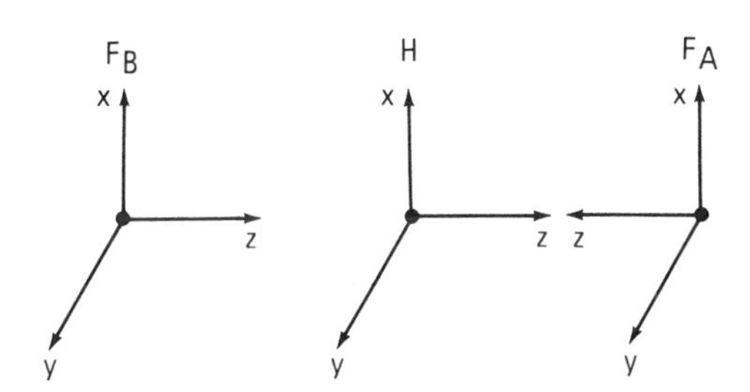

FIGURE 4.1 Axis system chosen for FHF$^-$.

TABLE 4.1 Generator Table for FHF$^-$

	GOs					
VAO Equivalence Sets	s	pz	px	py	dxz	dyz
H = (H1s)	n					
Fs = ($F_A 2s$), ($F_B 2s$)	n	n				
F$p\sigma$ = ($F_A 2pz$), ($F_B 2pz$)	n	n				
F$p\pi$ = ($F_A 2px$), ($F_A 2py$), ($F_B 2px$), ($F_B 2py$)			n	n	n	n

four sets are *equivalence* sets because for any point in the linear triatomic potential an equivalent point is generated by reflection through the xy plane, by inversion through the center, and by rotation around the z axis. In other words, VAOs within equivalence sets are those which are interchanged by a *symmetry operation.* In fact, the symmetry of the linear triatomic potential is seen to be the same as that of the diatomic potential. Each VAO equivalence set in Table 4.1 is denoted by a *set label* (i.e., H, Fs, F$p\sigma$, F$p\pi$) and an explicit listing of those VAOs which are equivalent in the potential of the molecule. The three equivalence sets arising from VAOs on each of the fluorines duplicate those we found for the F_2 molecule, and the symbols σ and π have the same meaning as for diatomic molecules. The GOs in Table 4.1 are denoted in abbreviated form and we will use this notation from now on. In molecules that are more complex than diatomics, it is convenient to view MOs in terms of the way in which they are generated. Because we generate them using the symmetry of GOs, we will refer to the MOs as *symmetry orbitals* (SOs). We will now see how this helps us generate the bonding, antibonding, and nonbonding MOs of FHF$^-$. Taking first the H = (H1s) equivalence set, we find that an s GO pictorially generates the symmetry orbitals (SOs) in Equations 4.1–4.3 in the H, Fs, and F$p\sigma$ equivalent sets:

$$\tilde{\sigma}(s|H) = (\mathrm{H}1s) \tag{4.1}$$

$$\tilde{\sigma}(s|\mathrm{F}s) = N_1[(\mathrm{F_A}2s) + (\mathrm{F_B}2s)] \tag{4.2}$$

$$\tilde{\sigma}(s|\mathrm{F}p\sigma) = N_2[(\mathrm{F_A}2pz) + (\mathrm{F_B}2pz)]. \tag{4.3}$$

In these equations we use the symbol $\tilde{\sigma}$ (GO|*Equiv. Set*) to denote a *symmetry orbital* $\tilde{\sigma}$ *which is generated by the specific GO in the indicated equivalence set.* As we will see, SOs can contain one or more atomic orbitals. We will also see that SOs, either alone or when combined with other SOs, make up the MOs of the molecule. Notice that in the case of $\tilde{\sigma}(s|\mathrm{H})$ in Equation 4.1, the SO *is* the (1*s*) AO on hydrogen. *It is a general rule that the VAOs on central atoms in molecules are SOs*, because their nuclei coincide with the centers of the GOs, which generate these VAOs. The remaining SOs appearing in Equations 4.4–4.9 are pictorially generated by the indicated GOs.

$$\tilde{\sigma}(pz|\mathrm{F}s) = N_3[(\mathrm{F_A}2s) - (\mathrm{F_B}2s)] \tag{4.4}$$

$$\tilde{\sigma}(pz|\mathrm{F}p\sigma) = N_4[(\mathrm{F_A}2pz) - (\mathrm{F_B}2pz)] \tag{4.5}$$

$$\tilde{\pi}(px|\mathrm{F}p\pi) = N_5[(\mathrm{F_A}2px) + (\mathrm{F_B}2px)] \tag{4.6}$$

$$\tilde{\pi}(dxz|\mathrm{F}p\pi) = N_6[(\mathrm{F_A}2px) - (\mathrm{F_B}2px)] \tag{4.7}$$

$$\tilde{\pi}(py|\mathrm{F}p\pi) = N_5[(\mathrm{F_A}2py) + (\mathrm{F_B}2py)] \tag{4.8}$$

$$\tilde{\pi}(dyz|\mathrm{F}p\pi) = N_6[(\mathrm{F_A}2py) - (\mathrm{F_B}2py)]. \tag{4.9}$$

The symbol "n" in Table 4.1 means that the SOs we have generated are all *nonbonding* SOs. Recall that in the case of diatomics, SOs are either BMOs or ABMOs because the atoms are sufficiently close to each other to favor VAO overlap. Thus we can write Table 4.2 as a generator table for F_2 wherein "b" means bonding and "a" means antibonding. Because none of the fluorine SOs in Table 4.1 overlap significantly with each other, any electronic charge in these very delocalized MOs would be confined to the fluorine atoms and so these MOs are nonbonding as indicated by the symbol "n". The SO on hydrogen is also nonbonding since it is composed of an orbital on only one atom. Since we are neglecting overlap between non-neighboring atoms, the normalization constants N are all equal to $2^{-1/2}$.

If all the SOs in Table 4.1 are nonbonding, how can we account for the unusual stability of FHF^-? In Table 4.1 we have a set of two SOs generated by *pz*. Linearly combining these will not split them enough to gain a strongly bonding and a strongly antibonding MO since these SOs involve non-neighboring atoms. However, we also have a set of three SOs generated by *s*. Linearly combining these will be profitable in terms of bonding

TABLE 4.2 Generator Table for F_2

	GOs					
VAO Equivalence Sets	*s*	*px*	*py*	*pz*	*dxz*	*dyz*
$\mathrm{F}s = (\mathrm{F_A}2s), (\mathrm{F_B}2s)$	b			a		
$\mathrm{F}p\sigma = (\mathrm{F_A}2pz), (\mathrm{F_B}2pz)$	b			a		
$\mathrm{F}p\pi = (\mathrm{F_A}2px), (\mathrm{F_A}2py), (\mathrm{F_B}2px), (\mathrm{F_B}2py)$		b	b		a	a

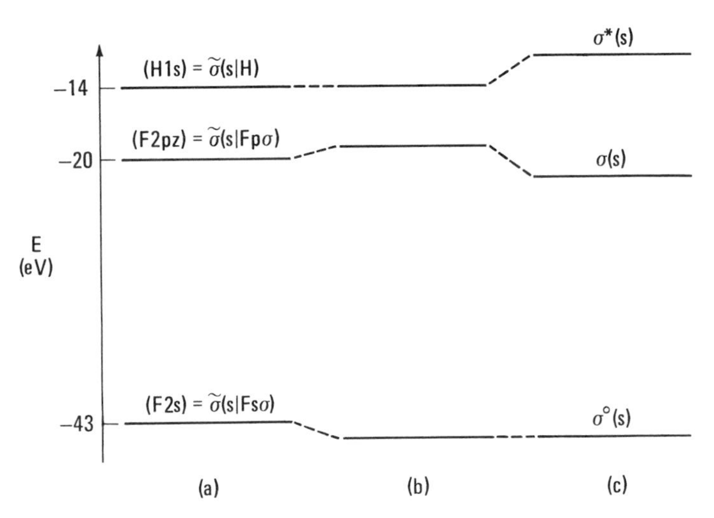

FIGURE 4.2 Energy level diagram showing the energies of the FHF^- SOs before interaction (a), after interaction of the two lowest-energy *s*-generated SOs (b), and after interaction of the two highest ones (c).

because neighboring atoms are involved in the resulting MOs. The MOs can be visualized from Figure 4.2. In Figure 4.2a are shown the relative energies of the (H1*s*), (F2*pz*), and (F2*s*) VAOs. Since there is negligible overlap between VAOs *in the SOs*, it is clear that these VAO energies are approximately equal to the SO energies. The relatively large separation of the two fluorine SOs of lowest energy causes their interaction to be small (Figure 4.2b) but the interaction between the two SOs closest in energy will be relatively large (Figure 4.2c), giving rise to a BMO [$\sigma(s)$] and an ABMO [$\sigma^*(s)$]. The lowest-energy MO $\sigma^0(s)$ remains largely nonbonding (i.e., NBMO), and we denote this with the superscript shown. Expressions for these three MOs can be written as seen in Equations 4.10–4.12

$$\sigma^*(s) = a_1^*\tilde{\sigma}(s|\mathrm{F}s) + a_2^*\tilde{\sigma}(s|\mathrm{F}p\sigma) - a_3^*\tilde{\sigma}(s|\mathrm{H}) \tag{4.10}$$

$$\sigma(s) = a_1\tilde{\sigma}(s|\mathrm{F}s) + a_2\tilde{\sigma}(s|\mathrm{F}p\sigma) + a_3\tilde{\sigma}(s|\mathrm{H}) \tag{4.11}$$

$$\sigma^0(s) = a_1^0\tilde{\sigma}(s|\mathrm{F}s) + a_1^0\tilde{\sigma}(s|\mathrm{F}p\sigma) + a_3^0\tilde{\sigma}(s|\mathrm{H}) \tag{4.12}$$

wherein a_2^*, a_3^*, a_2, a_3, and $a_1{}^0$ are relatively large positive constants causing the terms associated with them to be dominant contributors to their respective MOs. Confirmation of $\sigma^*(s)$ as an ABMO comes from the presence of two nodes perpendicular to the molecular axis, one lying between each pair of atoms. These nodes arise from domination of the last two terms in Equation 4.10. The $\sigma(s)$ MO has no internuclear nodes. Although $\sigma^0(s)$ also has no internuclear nodes, it may be noted that if we include the *internal* nodes of the VAOs there are zero, two, and four nodes in the $\sigma^0(s)$, $\sigma(s)$, and $\sigma^*(s)$

MOs, respectively, which is in accord with the rise in energy of these MOs in the same order. It should also be appreciated that $\sigma^0(s)$ is largely nonbonding mainly because a_3^0 is small.

Since the $a_1\tilde{\sigma}(s|Fs)$ term is the main contributor to $\sigma^0(s)$ [owing to the low energy of (F2s)] we can for all intents and purposes write $\sigma^0(s)$ as Equation 4.2. At this point we also note that pz generates an SO in the VAO equivalence set Fs, namely, Equation 4.4. As we saw in the case of diatomics having more than four electrons, the fully electron-occupied BMO–ABMO pair formed from *core* (s) VAOs can be recast into lone pairs on each atom. The same can be done for the two LCAOs in Equations 4.2 and 4.4 which, as we shall see shortly, are also occupied MOs. While there is also a splitting to be expected between the two pz-generated SOs in Equations 4.4 and 4.5, the interaction is expected to be small owing to the relatively large energy gap between the (F2s) and (F2p) VAOs. Thus $\tilde{\sigma}(pz|Fp\sigma)$, like $\tilde{\sigma}(pz|Fs)$, is a nonbonding MO and we give the former the label $\sigma_p^0(pz)$.

We are now ready to draw the orbital energy level diagram in Figure 4.3. First we note that there are sixteen valence electrons; one for H, seven each for the F atoms, and one for the overall negative charge (the latter electron having been arbitrarily placed in the F VAO set in Figure 4.3). These sixteen electrons occupy the lowest eight MOs. The lowest two MOs are atom-localized lone pairs. The next MO is a BMO [$\sigma(s)$], *which is delocalized over three atomic centers.* Since it is the only BMO and it is doubly occupied,

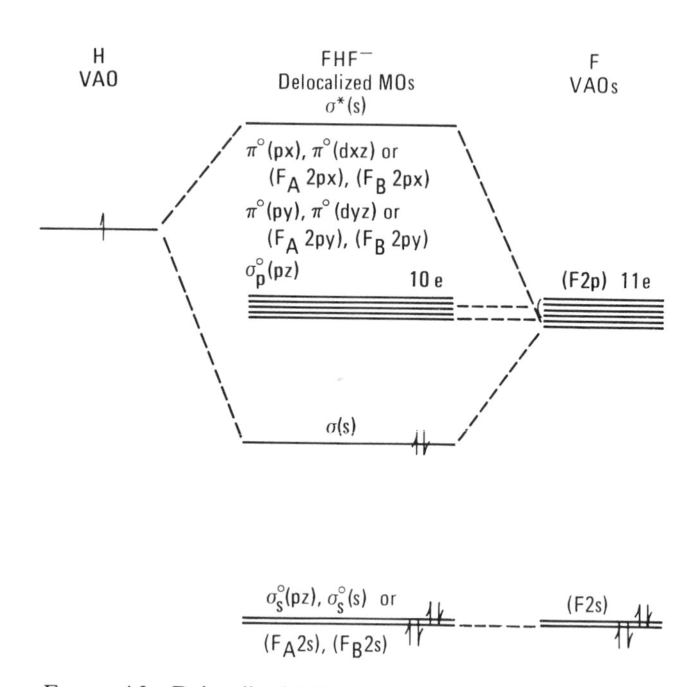

FIGURE 4.3 Delocalized MO energy level diagram for FHF$^-$.

we have two electrons for both atom links, or an average bond order of 1/2. Next in the diagram is a set of five *degenerate* fully occupied NBMOs which have no contributions from the hydrogen. Each of these NBMOs is still a 3-center 2-electron MO; four of them are of the π type and one is of the σ type. Although each of these NBMOs is made up of only fluorine contributions, it is a mistake to think of them as containing an *unpaired* electron on each fluorine. Moreover, it has been shown experimentally that FHF^- has no unpaired electrons. These NBMOs are simply orbitals with a relatively large degree of thinning out of the electron density in the region of the H atom lying between the main MO orbital lobes.

The concept of paired electrons in these NBMOs is parallel to using the Pauli principle in doubly occupied atomic orbitals such as *p*, which have no electron density at the nucleus because the nucleus lies between two lobes and the lobes are separated by a node.

Use the Node Game Homework Drawing Board to draw the SOs for FHF^- in Table 4.1. Also draw the LCAOs that correspond to the $\sigma(s)$ and $\sigma^*(s)$. Follow the steps in the appropriate column of Table 4.7 at the end of the chapter.

The localized view of FHF^- is obtained by the same procedure as that used for diatomics. After first localizing the (F2*s*) electrons as a lp on each fluorine (see Figure 4.3), we see that there are six *occupied* MOs left; one bonding and the rest nonbonding. In N_2, we saw that localization of all BMOs at once (i.e., one σ and two π-type occupied MOs) was convenient. Faced now with MOs of two types, namely, bonding and nonbonding, we will find it convenient to examine MOs in the same equivalence set in pairs. Let us begin with $\pi^0(px)$ and $\pi^0(dxz)$ generated from (F_A2px) and (F_B2px). Hybridization of GOs, which give rise to these occupied delocalized π^0 MOs, leads to two *pxdxz* GOs, as shown schematically in Figure 4.4a. Placing the

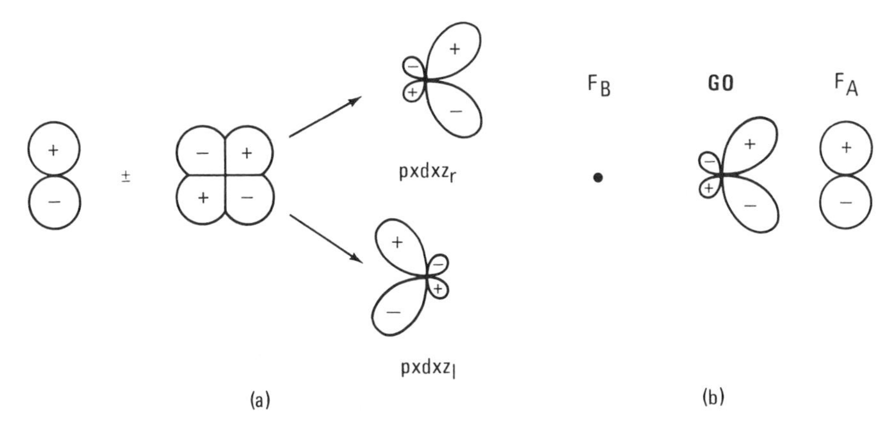

FIGURE 4.4 The formation of $pxdx_r$ and $pxdxz_l$ hybrid GOs (a) and the use of the $pxdxz_r$ hybrid GO in calling in (F_A2px) (b).

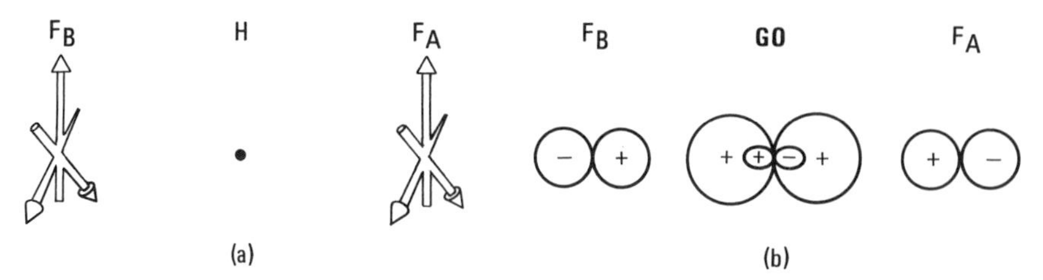

FIGURE 4.5 Hybrids of the (sp^2) type for fluorine lone pairs in FHF^-. In (b) is shown the calling in of the fluorine $(2pz)$ VAOs by the spz_r and spz_1 GOs.

$pxdxz_r$ GO at the center of the FHF^- ion is seen to call in the $(F_A 2px)$ AO in Figure 4.4b. Similarly, $pxdxz_1$ calls in $(F_B 2px)$. Analogously, hybridization of the delocalized $\pi^0(py)$, $\pi^0(dyz)$ pair leads to calling in of $(F_A 2py)$ and $(F_B 2py)$ via a pair of *pydyz* hybrid GOs. Of course, we can always localize still further the pair of occupied $(2p)$ orbitals called in on each fluorine by hybridizing them with the $(2s)$ lone pair on each of these atoms. By doing this we have an sp^2 hybrid set on each fluorine, as depicted in Figure 4.5a. Finally we consider the delocalized MO arising from the $Fp\sigma$ VAO set, namely $\sigma(s)$ and $\sigma_p^0(pz)$. The hybrid GO templates constructed from the s and pz GOs which generate $\sigma(s)$ and $\sigma_p^0(pz)$ yield the hybrid GOs spz_r and spz_1, which are seen in Figure 4.5b to call in the $(2pz)$ VAOs on each fluorine. Because there is no (pz) AO in the valence shell of hydrogen, however, we are unable to generate an $(spz)_r$ and an $(spz)_1$ VAO on the hydrogen. This means only the $(H1s)$ is a contributing VAO from hydrogen in the two localized orbitals. If we did have a (pz) VAO on a central atom (say, X), we could have formed an occupied localized two-center MO between $(Xspz)_r$ and $(F_A 2pz)$ and another between $(Xspz)_1$ and $(F_B 2pz)$. Examples of this type will be seen later. But how do we deal with the present case? It can be shown (Appendix IV) that *when the central atom does not cooperate fully in responding to the hybrid GO template by supplying a matching set of hybrid VAOs, then pure two-center localized MOs cannot be formed.* In other words, although the MO formed is *mostly* localized as a two-center BMO between the central atom and one of the peripheral atoms, there is a *small antibonding* contribution between the central atom and the *other* peripheral atom. *This can be visualized* (Figure 4.6) *by recognizing that in such cases the sign of the VAO on the lesser contributing atom is chosen so as to produce a node between it and the VAO contributed by the central atom.* This observation should not be viewed as a limitation or flaw in the GO approach to visualizing MOs. More complex bonding views also account for partially localized bonds in a similar way. By this procedure, we maintain the orthogonality of the orbitals. Note that there are no bonding–antibonding contributions in the MO in Figure 4.4b, for example, because that MO *does not* include a contribution from the central atom.

Since the two partially localized MOs, which are denoted by $\hat{\sigma}_A$ and $\hat{\sigma}_B$ in Figure 4.6b result from equal mixing of a delocalized BMO and an NBMO

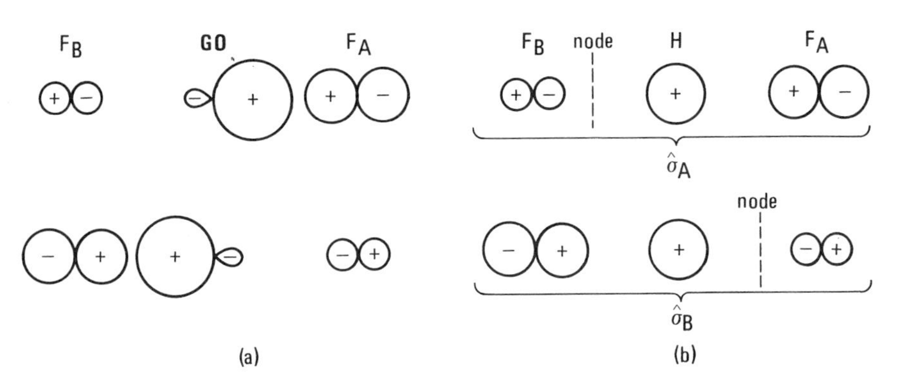

FIGURE 4.6 Orbital contributions from the fluorines in FHF^- generated by a pair of GOs (a) and the orbital contribution from hydrogen in this ion generated by these GO hybrids (b).

(see also Appendix IV), each localized MO must have a bond order of less than 1. Thus each partially localized MO ($\hat{\sigma}_A$ and $\hat{\sigma}_B$) is concluded to have a bond order of 0.5 because the delocalized view clearly showed that the three-center BMO has an average bond order of 0.5 per link. The energy level diagram for the localized bond picture is given in Figure 4.7. Notice that because of the large electronegativity difference between hydrogen and fluorine, the electron density in the BMO tends to concentrate on the fluorines, giving rise to considerable *polarity* (*ionicity*) in the HF links.

The eight localized MOs can be represented by the familiar electron dash formulation in Figure 4.8a, provided we bear in mind that the two bond dashes are really only half-bonds. Usually, of course, we associate a dash with a full electron pair bond. That would drive us to the impossible conclusion that hydrogen fully shares four electrons in the Lewis sense. Alternately, we can represent the bonding in FHF^- by the *average* of the conventional electron dash structures in Figures 4.8b and c. We see then that our conclusions regarding the localized view of FHF^- can be abbreviated by an electron dash structure in which six of the eight localized MOs are fluorine lp's while two pairs of electrons are delocalized in 3-center bonds. In fact, as a general rule, whenever the number of *occupied delocalized MOs* equals the total number of VAOs in the VAO equivalence sets giving these MOs, then the corresponding *localized MOs are identical with the VAOs themselves*. This is true irrespective of the number of VAOs in the equivalence sets. This rule permits considerable simplification of our future discussions. In FHF^- we could thus start with the electron dash structure and *preassign* our lone pairs as fluorine canonical VAOs in the delocalized view and as hybrid VAOs in the localized view. In FHF^- the lone pairs are on non-neighboring atoms so that mutual overlap is negligible. Therefore, the orthogonalizing admixtures from the "other" atom are of little consequence and the lone pairs

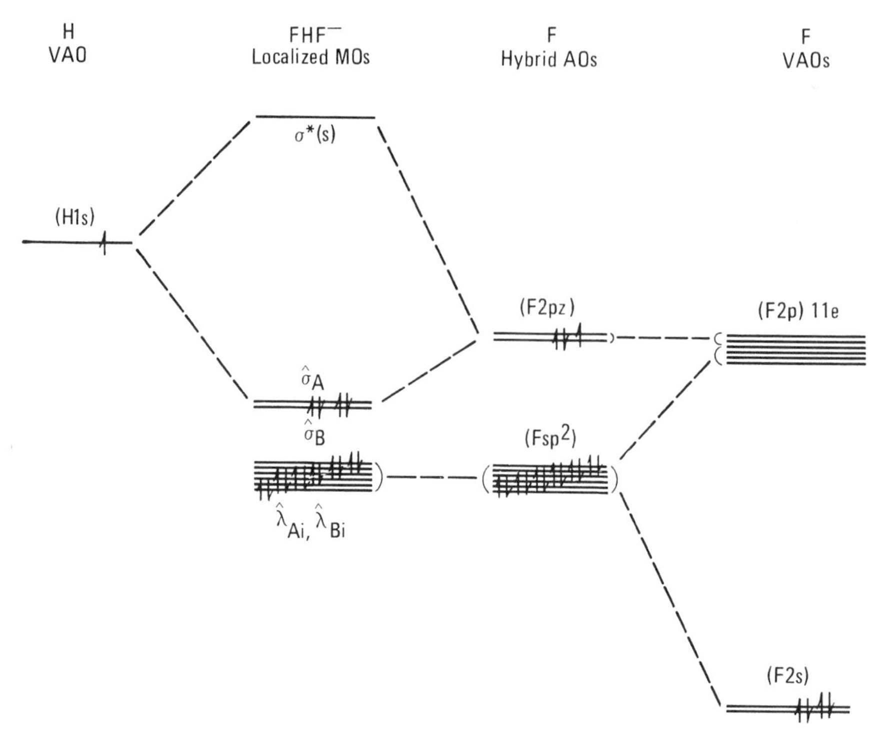

FIGURE 4.7 Localized MO energy level diagram for FHF^-.

are for all intents and purposes purely nonbonding. Recall that in F_2 the lone pairs are on neighboring atoms, where overlap is more important, and correspondingly there are slight antibonding effects between lone pairs from neighboring atoms known as *"nonbonded" repulsions.* As a result of the preassignment of lone pairs in FHF^-, only the (F_B2pz), (H1*s*), and (F_A2pz) VAOs would be left to consider in the bonding treatments.

Repeat the bond localization process for FHF^- using the Node Game Homework Drawing Board to draw pictures resembling those in Figures 4.4, 4.5, and 4.6. To represent a smaller fluorine *p* VAO contribution in Figure 4.6 using the arrow in the menu, you can use "white out" on part of the arrow tail in your printed out drawings. Follow the steps in the appropriate column of Table 4.7 at the end of the chapter.

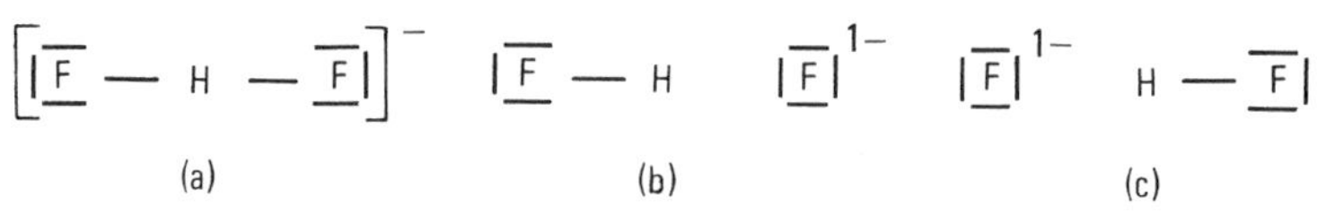

[|F — H — F|]⁻ |F — H |F|¹⁻ |F|¹⁻ H — F|

(a) (b) (c)

FIGURE 4.8 Electron dash structures for FHF^-

4.2. FXeF

Although procedures for drawing electron dash structures have not yet been discussed, let us adopt the one shown for FXeF in Figure 4.9. This electron dash structure indicates that we can localize three lone pairs on each atom. Using the same axis system as for FHF^- (Figure 4.1), each fluorine in the delocalized view would use its (2*s*), (2*px*), and (2*py*) VAOs for lone pairs and the xenon would similarly utilize its (5*s*), (5*px*), and (5*py*) VAOs. In the localized view, each of these lone-pair VAO sets could be hybridized to (sp^2) VAOs. From our treatment of FHF^-, it is clear that we then have only the (F2*pz*) VAOs to consider for bonding if three lone pairs are allocated to each fluorine atom. In the valence shell of xenon we have a (5*pz*) orbital as well as a set of (5*d*) VAOs. Since the extent of (*d*) orbital participation in σ bonding is, in general, controversial, we will examine the bonding in FXeF both excluding and including an (Xe5*d*) VAO. While (*d*) orbitals are present in levels where *n*, the principal quantum number, exceeds 2, they lie sufficiently high in energy above the (*p*) VAOs of the same principal quantum number that their inclusion as VAOs in nonmetal compounds is quite questionable. Because of the shielding effect, however, (*d*) orbitals do become important in bonding interactions in the transition metals.

For the delocalized view of FXeF after preassigning lone pairs and *excluding (Xe5d) participation* we can write Generator Table 4.3. The three SOs generated by the appropriate GOs are given in Equations 4.13–4.15:

$$\tilde{\sigma}(s|\mathrm{F}p\sigma) = N_1[(\mathrm{F_A}2pz) + (\mathrm{F_B}2pz)] \tag{4.13}$$

$$\tilde{\sigma}(pz|\mathrm{F}p\sigma) = N_2[(\mathrm{F_A}2pz) - (\mathrm{F_B}2pz)] \tag{4.14}$$

$$\tilde{\sigma}(pz|\mathrm{Xe}) = (\mathrm{Xe}5pz). \tag{4.15}$$

The first SO, $\tilde{\sigma}(s|\mathrm{F}p\sigma)$, must be an NBMO since it has no contribution from the Xe atom. That is, it is the only SO generated by the *s* GO. Since the

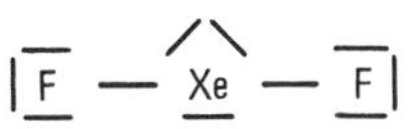

$$|\overline{\underline{\mathrm{F}}} - \overset{\wedge}{\underline{\mathrm{Xe}}} - \overline{\underline{\mathrm{F}}}|$$

FIGURE 4.9 Electron dash structure for FXeF.

TABLE 4.3 Generator Table for FXeF

	GOs	
VAO Equivalence Sets	*s*	*pz*
$\mathrm{Xe}p\sigma = (\mathrm{Xe}5pz)$		n
$\mathrm{F}p\sigma = (\mathrm{F_A}2pz), (\mathrm{F_B}2pz)$	n	n

second and third SOs are generated by the same GO, they will interact to form a BMO and an ABMO. The delocalized MOs are thus

$$\sigma^0(s) = \tilde{\sigma}(s|\mathrm{F}p\sigma) \tag{4.16}$$

$$\sigma(pz) = a_1\tilde{\sigma}(pz|\mathrm{Xe}) + a_2\tilde{\sigma}(pz|\mathrm{F}p\sigma) \tag{4.17}$$

$$\sigma^*(pz) = a_1^*\tilde{\sigma}(pz|\mathrm{Xe}) - a_2^*\tilde{\sigma}(pz|\mathrm{F}p\sigma). \tag{4.18}$$

Since the lone pairs were preassigned, we have four electrons left and so only the 3-center bonding and nonbonding MOs are occupied. The MO energy level diagram in Figure 4.10 indicates that again, as in FHF^-, one electron pair binds the molecule in a (polar) three-center bond and the other electron pair resides in an NBMO. The essential difference in the bonding between FHF^- and FXeF is that an (s) VAO is used on the central atom in FHF^- while a (pz) VAO is used in FXeF.

Use the Node Game Homework Drawing Board to draw the FXeF SOs in Table 4.3. Then remove the GOs and label the appropriate MO drawings BMO, NBMO or ABMO. Follow the steps in the appropriate column of Table 4.7 at the end of the chapter.

For the localized view of FXeF [without Xe ($5d$) participation] we begin by localizing the fluorine lone pairs in (sp^2) fluorine VAOs and the xenon lone pairs in xenon (sp^2) VAOs. The remaining occupied delocalized MOs are $\sigma(pz)$ and $\sigma^0(s)$ from which we obtain a pair of spz_r, spz_l GOs with which to generate localized MOs in the same way as we did for FHF^-. In the

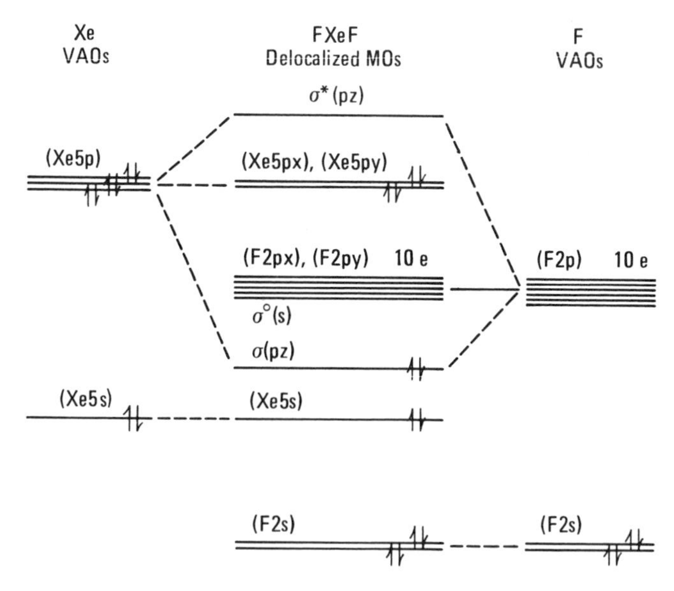

FIGURE 4.10 Delocalized MO energy level diagram for FXeF excluding xenon ($5d$) VAO participation.

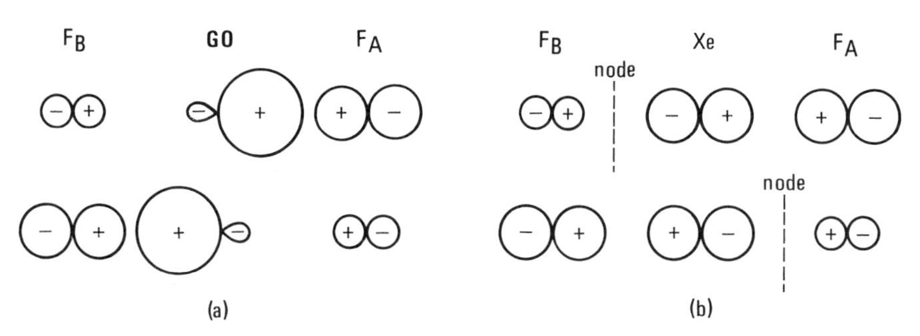

FIGURE 4.11 Orbital contributions from the fluorines in FXeF generated by a pair of *spz* GOs (a) and the orbital contribution from xenon in this molecule generated by these GO hybrids (b).

present case, however, only the (Xe5*pz*) VAO responds to the pair of hybrid GOs [since (Xe5*s*) contains a preassigned lp] and again the central atom VAO is involved in two partially localized MOs, as depicted in Figure 4.11b. Notice in the SOs in Figure 4.11a and the MOs in Figure 4.11b that the (2*pz*) VAO of the fluorine atom having the smaller contribution has the *opposite* sign compared to its counterpart 2*pz* in Figure 4.6 for the case of FHF^-. As we saw from our localized view of FHF^-, this is to preserve orthogonality of the orbitals. In other words, to have some antibonding character in the MOs of Figure 4.11b, which now involve a (*pz*) VAO on the central atom, the signs of the VAOs must be as shown (see also Appendix IV). In FHF^-, the central atom provided an (*s*) VAO which requires the VAO signs to be as depicted in Figure 4.6.

In the localized MO energy diagram shown in Figure 4.12 we must recognize that the energy of the localized BMOs $\hat{\sigma}_A$ and $\hat{\sigma}_B$ is the average of the localized $\sigma(pz)$ and $\sigma^0(s)$ MOs of Figure 4.10. This arises from slightly antibonding nature of these localized BMOs which produces a bond order of 0.5 in each MO.

Use the Homework Drawing Board in Node Game to draw the steps in localizing the delocalized MOs in FXeF. Follow the steps in the appropriate column of Table 4.7.

As with FHF^-, it is not possible to draw an electron dash structure for FXeF where each dash corresponds to a lone pair or a two-center two-electron bond. The diagram in Figure 4.9 is deficient in that it implies the availability of five VAOs on xenon when we included only four. Furthermore, we saw that each of the two localized MOs $\hat{\sigma}_A$ and $\hat{\sigma}_B$ must use VAOs on *all three atoms* because *both* MOs must use one and the same VAO on the xenon and still be mutually orthogonal. As in FHF^-, therefore, the bonding in FXeF must be symbolized by the average of the two ionic electron dash structures shown in Figure 4.13a and b.

Should we allow (Xe5*d*) participation in the VAO set of xenon? Sophisticated calculations on expanded-octet (hypervalent) molecules consistently

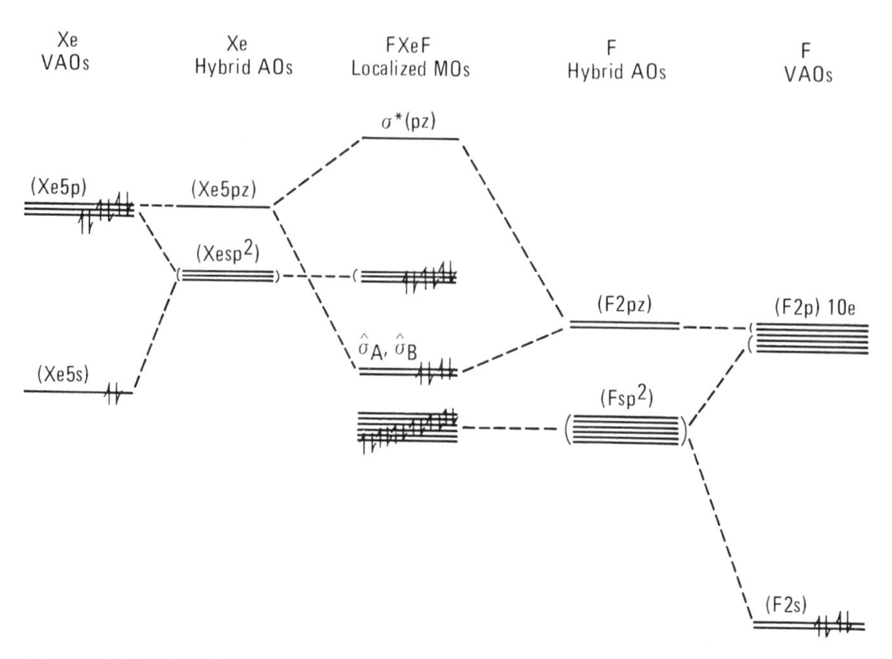

FIGURE 4.12 Localized MO energy level diagram for FXeF without xenon (5*d*) VAO participation.

indicate that introducing valence *d* orbitals is not really warranted because their participation is low owing to their relatively high energies. Therefore to use them in describing the localized bonding in such molecules is unrealistic. However, we will do so anyway to make a point that will emerge a little later. It is easily visualized that all five of the (Xe5*d*) VAOs are generated by the full set of five *d* GOs. However, four of the (Xe5*d*) VAOs generated by the corresponding GOs are of the wrong symmetry to interact with any of our two fluorine SOs (Table 4.3) and so these SOs, namely, (Xe5dx^2-y^2), (Xe5*dxz*), (Xe5*dyz*) and (Xe5*dxy*), will be nonbonding unoccupied VAOs. As shown in Figure 4.14, the only *d* GO (and hence the Xe VAO it generates) which has an appropriate symmetry to interact with a fluorine SO (Table 4.3) is (Xe5dz^2). It is clear that dz^2 generates exactly the same SO generated by the *s* GO in the F$p\sigma$ equivalence set. Which GO shall we use then? Or shall we include both? Including both is unreasonable because it introduces redundancy. It makes sense, though, to use the GO that generates the most SOs.

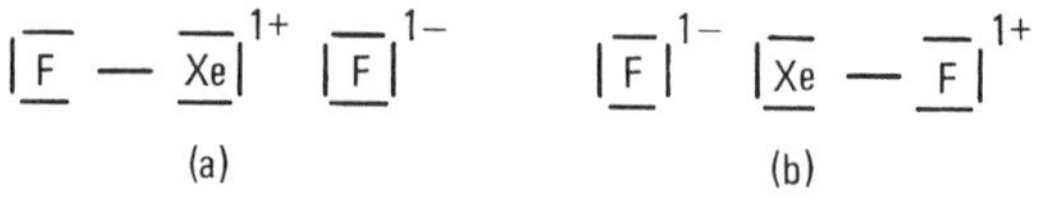

$|\overline{\underline{F}} - \overline{\underline{Xe}}|^{1+}\ |\overline{\underline{F}}|^{1-}$ $|\overline{\underline{F}}|^{1-}\ |\overline{\underline{Xe}} - \overline{\underline{F}}|^{1+}$

(a)   (b)

FIGURE 4.13 Ionic electron dash structures for FXeF.

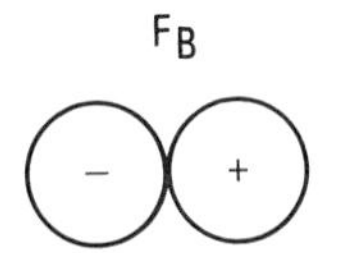

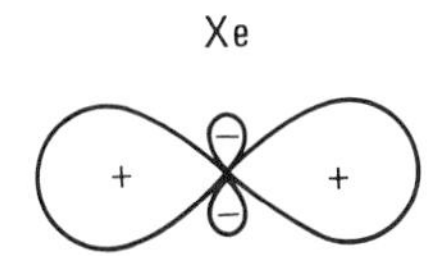

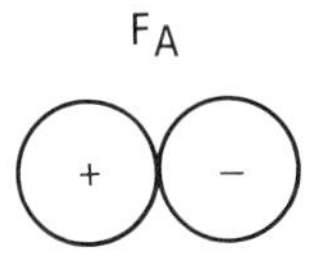

FIGURE 4.14 The generation by a $3dz^2$ GO of an SO in FXeF composed of a $(2pz)$ VAO on each fluorine. The $3dz^2$ GO also generates the $(\text{Xe}5dz^2)$ VAO, which is also an SO. Even though the latter AO contains two spherical nodes (not shown), its symmetry is the same as a $3dz^2$ GO.

TABLE 4.4 Generator Table for FXeF [Including $(\text{Xe}5dz^2)$]

	GOs	
VAO Equivalence Sets	dz^2	pz
$\text{Xe}p\sigma = (\text{Xe}5pz)$		n
$\text{Xe}d\sigma = (\text{Xe}5dz^2)$	n	
$\text{F}p\sigma = (\text{F}_A 2pz), (\text{F}_B 2pz)$	n	n

Based on this reasoning the FXeF generator table assumes the form given in Table 4.4. Taking linear combinations of the SOs generated by the same GO leads to two BMO–ABMO pairs: one pair generated from dz^2 and the other from pz. The delocalized MO energy level diagram is shown in Figure 4.15.

The localized view of FXeF is obtained by generating two localized BMOs from the delocalized ones via their pz and dz^2 GOs. Taking linear combinations of these GOs leads to a pair of digonal hybrids $pzdz_l^2$ and $pzdz_r^2$, depicted in Figure 4.16. The directionality of these hybrids (given by the largest lobe in each case) generates two localized MOs, one between Xe and F_A and the other between Xe and F_B. In other words these GOs generate two digonal (Xe*pd*) hybrids, each of which combines with the (F2*pz*) VAO toward which it points forming a localized two-center two-electron MO concentrated in each Xe–F link. A schematic drawing of these localized bonds together with the (sp^2)-hybridized lone pairs is shown in Figure 4.17. The localized MO energy level diagram shown in Figure 4.18 tells us that the localized MOs $\hat{\sigma}_A$, $\hat{\sigma}_B$ are two-center two-electron fully bonding MOs (i.e., the bond order is 1.0) which lie at the average of the energies of the $\sigma(pz)$ and $\sigma(dz^2)$ delocalized MOs in Figure 4.15.

Why then include xenon *d* VAOs in XeF_2 if it is not realistic to do so? The point to appreciate here is that many freshman textbooks employ sp^3d^x localized hybridization schemes to describe all expanded-octet molecules, implying a bond order of 1.0 in all the bonds. This is misleading because such participation is fractional at best.

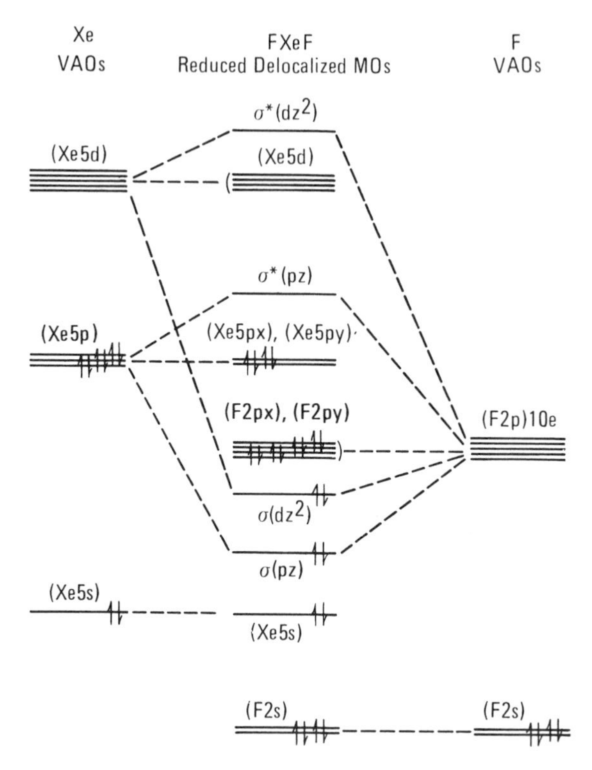

FIGURE 4.15 Delocalized MO energy level diagram for FXeF including ($Xe5dz^2$) participation.

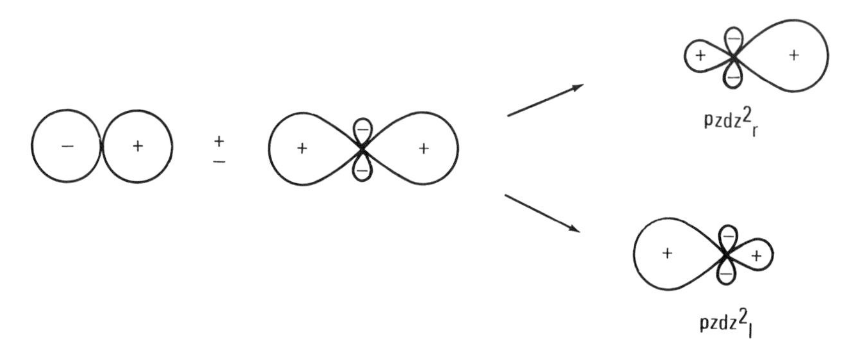

FIGURE 4.16 Formation of two $pzdz^2$ GO hybrids from pz and dz^2 orbitals.

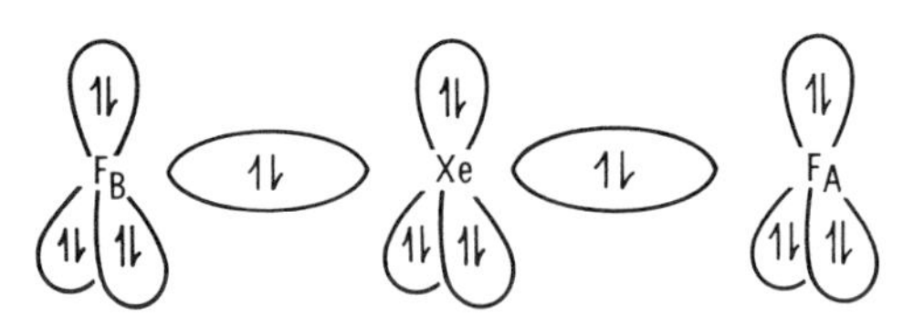

FIGURE 4.17 Localized lone pairs and bond pairs in FXeF, assuming (Xe5dz^2) participation in the bonding.

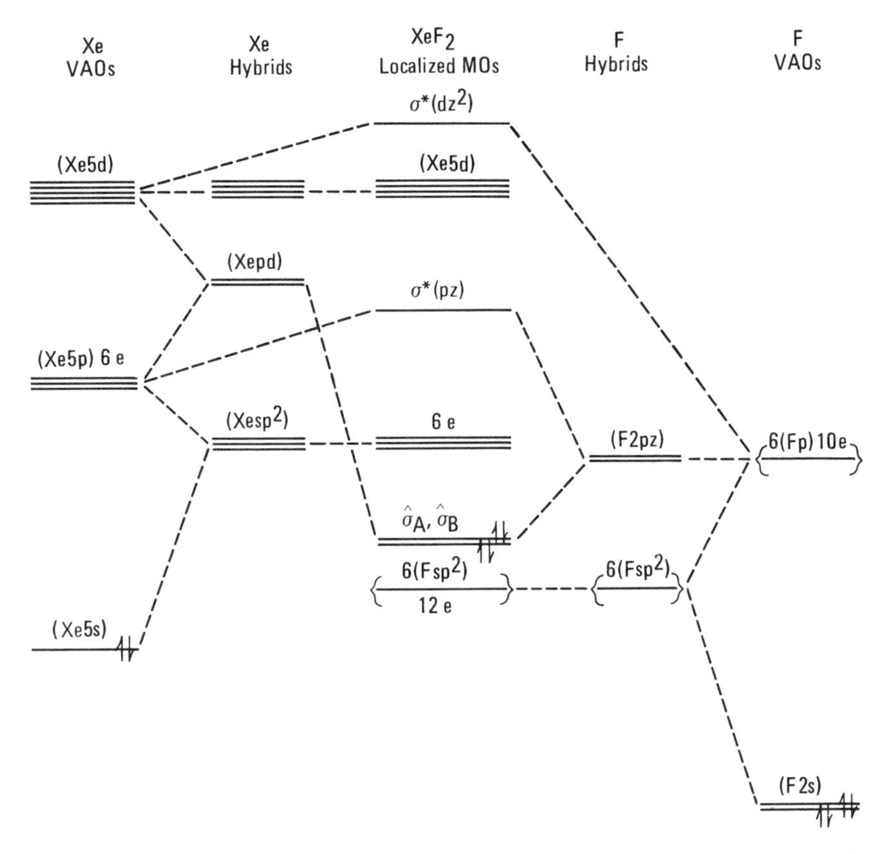

FIGURE 4.18 Localized MO energy level diagram for FXeF including (Xe5dz^2) participation.

Use the Node Game Homework Drawing Board to repeat our development of the delocalized and localized views for FXeF allowing (Xe$5d$) VAO participation. Follow the steps in the appropriate column of Table 4.7.

4.3. OCO

The electron dash structure normally drawn for this molecule is shown in Figure 4.19a. It implies that we have two lp's on each oxygen and a σ plus a π bond in each link. If we were to preassign an (O$2s$) and an (O$2p$) VAO on each oxygen to accommodate these lone pairs, a question regarding delocalization of the π bonds arises. Thus if the (p) VAO used for a lp on each oxygen is (O_A2px) and (O_B2px) (Figure 4.19b) we are restricted to forming *one* two-electron three-center delocalized π bond [i.e., with the ($2py$) VAOs on all three atoms] when we really might like two such delocalized π bonds [i.e., an additional one involving all three ($2py$) VAOs]. If we use (O_B2px) and (O_A2py) for one of the lone pairs on each O atom (Figure 4.19c) we can form no π bonds delocalized over all three centers; only two two-center two-electron π bonds [i.e., one in the *yz* plane on the right side and one in the *xz* plane on the left utilizing (O_A2py), (C$2py$) and (O_B2px), (C$2px$), respectively]. If we wish to delocalize the bonding as much as possible, we can preassign one lone pair on each oxygen to an (O$2s$) VAO. Then using the axis system for FHF^- in Figure 4.1 we can write Generator Table 4.5, wherein ten VAOs form five equivalence sets: Notice that we used an *s* GO with no spherical nodes to generate the ($2s$) VAO on carbon (cf., section 3.2). The ten SOs we have generated can be straightforwardly sketched and expressions written for them from the GO symmetries, as we have seen earlier. Since the *dxz* and *dyz* GOs generate only one SO each, both SOs are MOs, and since they have no contributions from the central neighboring atom, they are nonbonding:

$$\begin{aligned} \pi^0(dxz) &= \tilde{\pi}(dxz|\mathrm{O}p\pi) \\ \pi^0(dyz) &= \tilde{\pi}(dyz|\mathrm{O}p\pi). \end{aligned} \tag{4.19}$$

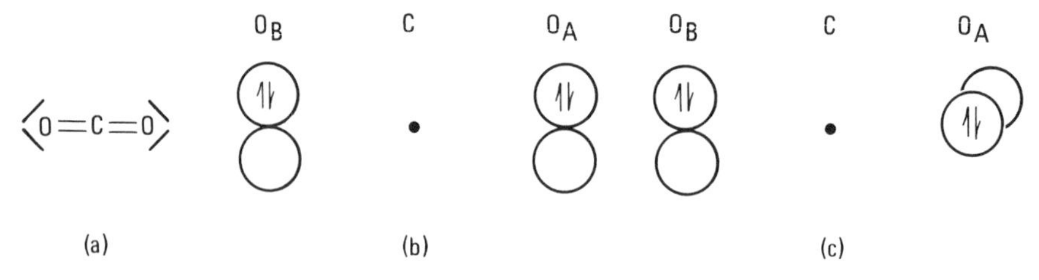

FIGURE 4.19 Electron dash structure for CO_2 (a) and two ways of preassigning its oxygen lone pairs [(b), (c)].

TABLE 4.5 Generator Table for CO_2

VAO Equivalence Sets	GOs					
	s	pz	px	py	dxz	dyz
$Cs = (C2s)$	n					
$Cp\sigma = (C2pz)$		n				
$Cp\pi = (C2px), (C2py)$			n	n		
$Op\sigma = (O_A 2pz), (O_B 2pz)$	n	n				
$Op\pi = (O_A 2px), (O_B 2px), (O_A 2py), (O_B 2py)$			n	n	n	n

The *px* and *py* GOs each generate two SOs: one on the central atom and one on the peripheral atoms. They give rise to two BMO–ABMO pairs, namely, $\pi(px)$, $\pi^*(px)$ and $\pi(py)$, $\pi^*(py)$, with the two BMOs degenerate and the two ABMOs also degenerate. Next we consider the *s*- and *pz*-generated SOs, of which there are two in each set in Table 4.5. These give BMO–ABMO pairs of the σ type, namely, $\sigma(s)$, $\sigma^*(s)$ and $\sigma(pz)$, $\sigma^*(pz)$. There are sixteen VAO electrons of which we have preassigned four to two (O2*s*) lone pair orbitals. In Table 4.5 we generated two NBMOs, four BMOs, and four ABMOs. Clearly the twelve remaining electrons occupy all of these MOs except the four ABMOs, which lie highest in energy. As seen in Figure 4.20, we have two occupied *three-center* σ BMOs and two occupied *three-center* π BMOs which together provide a bond order of 2.0 per link as expected from the electron dash structure. Notice, however, that our two occupied three-center π NBMOs are *not* implied by the electron dash structure. This situation reminds us of FXeF in which one of the electron dashes drawn between a pair of atoms had to be understood to symbolize a delocalized nonbonded electron pair. In the maximally delocalized view of CO_2 we have developed, one of the oxygen lone pairs on *each* oxygen must be viewed as residing in a π NBMO. These π NBMOs are *three-center* two-electron MOs, which lie at right angles to one another.

Without doing an accurate calculation, it is not possible to determine whether the best bonding picture of CO_2 is the maximally delocalized one we developed here or the one where each π bond is localized between C and O. A reliable rule to follow, however, is to delocalize as much as possible. In some molecules such as allene ($H_2C{=}C{=}CH_2$), this is not possible (See Exercise 8 at the end of the chapter.)

Use the Node Game Homework Drawing Board to draw the SOs and the MOs implied in Table 4.5 for CO_2. Follow the steps in the appropriate column of Table 4.7.

To generate the localized view for CO_2 corresponding to the maximally delocalized view, let us localize the two σ and the two π occupied BMOs in Figure 4.20 separately. The σ BMOs are generated by *s* and *pz* and so we place a pair of oppositely directed *spz* hybrid GOs at the molecular center as

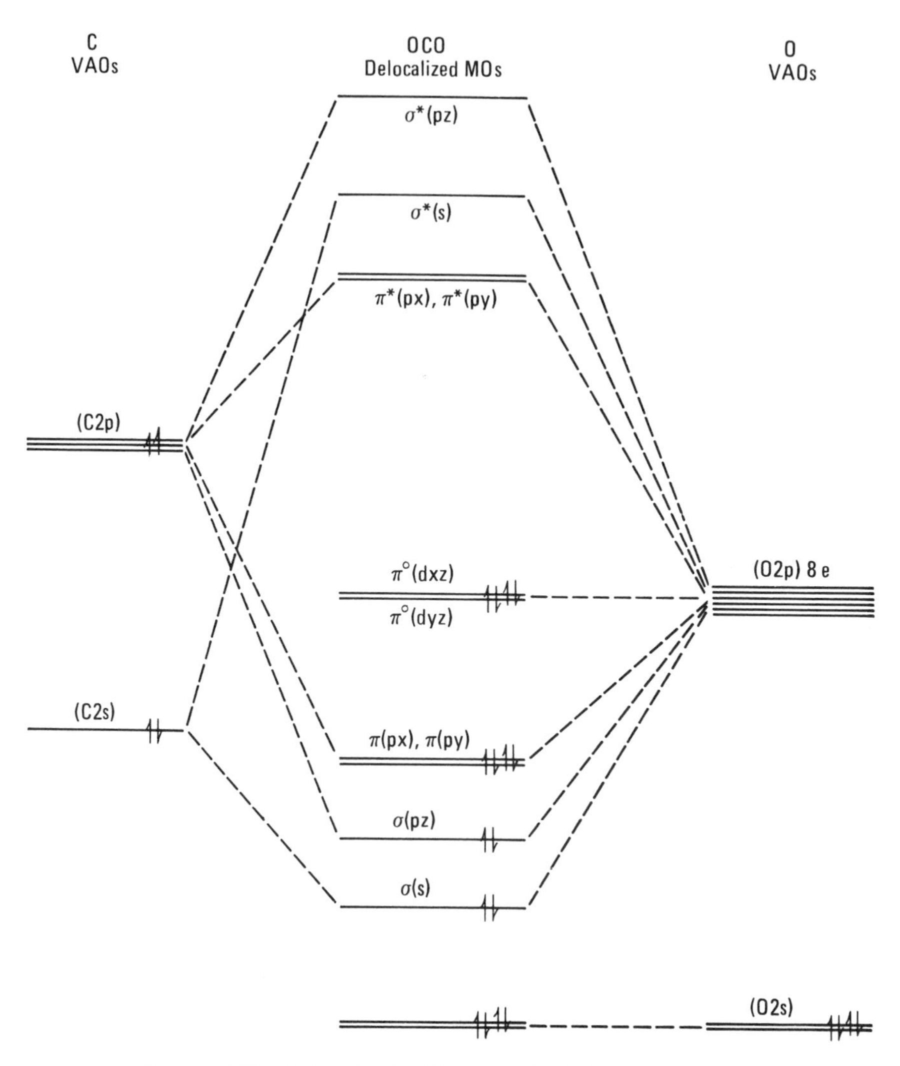

FIGURE 4.20 Delocalized MO energy level diagram for CO_2.

shown in Figure 4.21a. Each of these GOs leads to a localized BMO as shown in Figure 4.21b confined to two neighboring atoms with *no* contribution from the third atom. This assumes that the hybrid VAO on carbon is a *pure* (*sp*) hybrid. Actually, such a hybrid will probably be sp^{α}, where α is not quite 1. In that case the localized BMOs $\hat{\sigma}_A$ and $\hat{\sigma}_B$ will, of course, have to contain small antibonding contributions from ($O_B 2pz$) and ($O_A 2pz$), respectively, to preserve orthogonality.

Turning now to localization of the occupied π and π^0 MOs, their GOs are hybridized to give the $pxdxz_1$, $pxdxz_r$ and $pydyz_1$, $pydyz_r$ GOs discussed

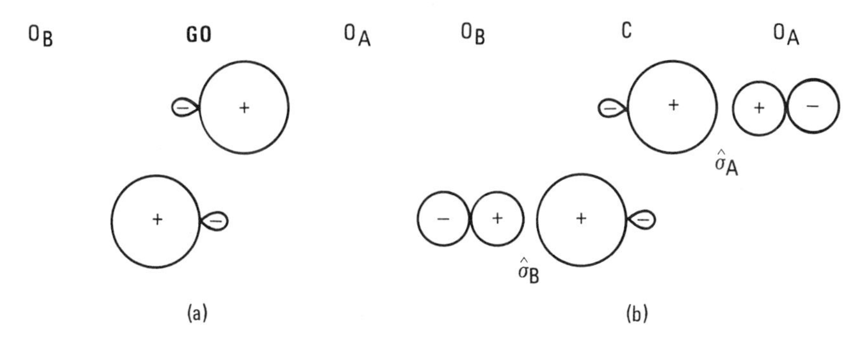

FIGURE 4.21 The *spz* hybrid GOs (a) and the localized MOs they generate (b) in CO_2.

earlier in connection with the π^0 MOs of FHF^-. The GOs in Figure 4.22a are seen in Figure 4.22b to call in only the $(2px)$ VAO on carbon since no (d) VAOs are available on this atom. Because the carbon cannot fully respond to the GO hybrid, there must be a small negative contribution from one of the oxygens, in addition to the substantial localization of bonding electron density in the opposite CO link stemming from the sizeable positive contribution of the other oxygen. These arguments are also applied to the localization of the occupied $\pi(py)$ and $\pi^0(dyz)$ MOs. Because of the antibonding contributions in $\hat{\pi}_A(pxdxz)$, $\hat{\pi}_B(pxdxz)$, $\hat{\pi}_A(pydyz)$, and $\hat{\pi}_B(pydyz)$, these localized MOs are only half-bonding between each link. Thus each link contains one full σ bond and two π half-bonds, yielding a total bond order of 1 +

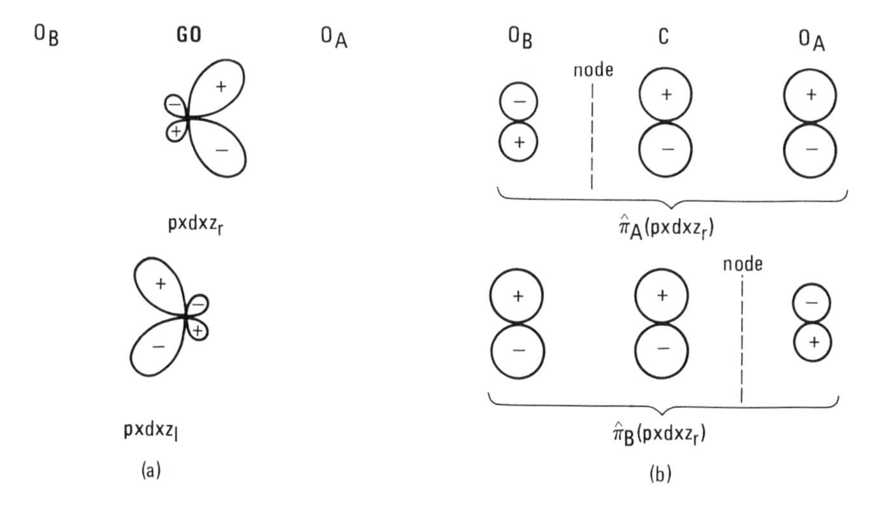

FIGURE 4.22 The *pxdxz* hybrid GOs (a) and the partially localized MOs they generate (b) in CO_2.

0.5 + 0.5 = 2.0 which is in accord with the electron dash structure in Figure 4.19a. The MO energy level diagram for this localized view is shown in Figure 4.23.

A second localized view of CO_2 is obtained by equivalent hybridization of the GOs that generate *all* the *bonding* MOs, namely, sp^3 hybridization. This procedure gives a tetrahedral GO hybrid set which calls in two lobes of an sp^2 VAO hybrid on each oxygen, as shown in Figure 4.24a. [Here each of the (O2*s*) lone pairs has been hybridized into the (sp^2) VAO set.] As seen in

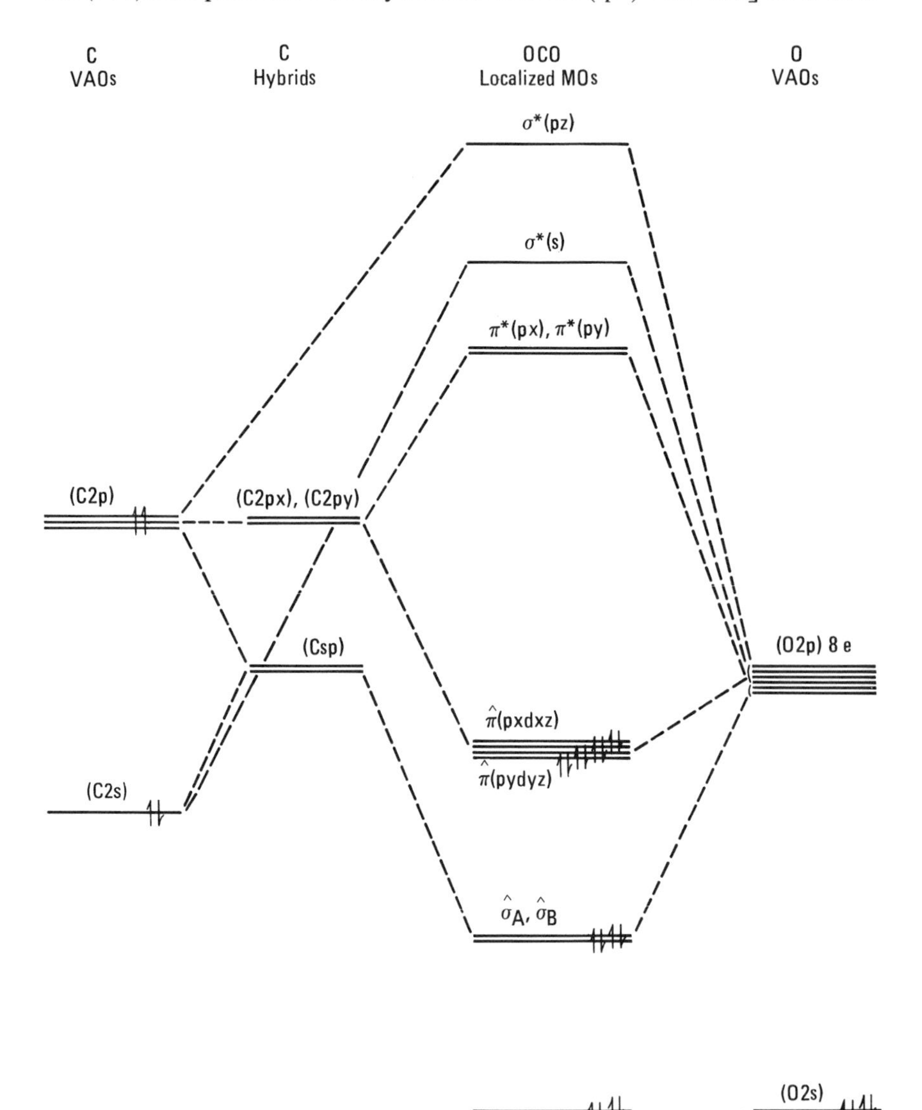

FIGURE 4.23 Localized MO energy level diagram for CO_2.

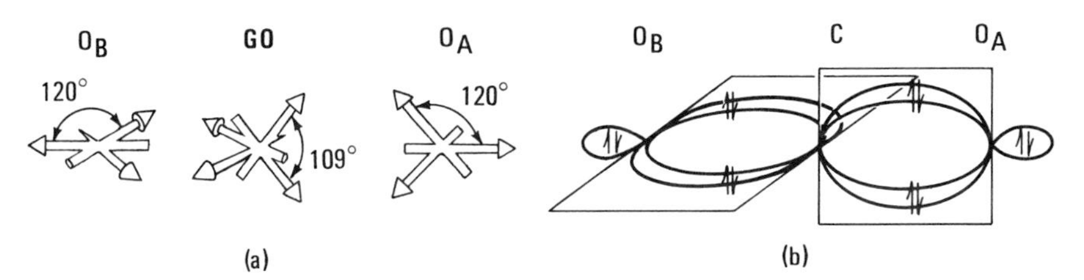

FIGURE 4.24 Oxygen VAO hybrids called in by sp^3 GOs for localized bonds (a) and the resulting localized banana bonds (b) in CO_2.

Figure 4.24b, this procedure leads to two pairs of mutually perpendicular banana bonds. The $\pi^0(dxz)$, $\pi^0(dyz)$ MOs now *cannot* be localized because hybridizing the *dxz*, *dyz* GOs merely gives us another identical VAO set rotated by some angle. Notice that the bond order from Figure 4.24 is still 2.0 as expected.

What if we were to hybridize the GOs which generated *all six* of the occupied MOs in Figure 4.20? The hybrid GO set is then a trigonal prismatically disposed sp^3d^2 set which, as shown in Figure 4.25a, calls in three of the lobes of an sp^3 VAO set on each oxygen atom. Notice that an $(sp^3dxzdyz)$ hybrid set has a trigonal prismatic geometry whereas an $(sp^3dx^2 - y^2dz^2)$ hybrid set is octahedral. In Figure 4.25b are shown the six banana bonds that arise by considering all six occupied MOs simultaneously. To see how the four VAOs on carbon contribute to these bonds, consider Figure 4.26a. Here we hybridize an *spxpy* carbon VAO set with a carbon *pz* AO to give one $(spxpy)^1(pz)^1$ set pointing three of its main lobes toward the right and the other set directing three such lobes toward the left. Each of these hybrid sets is 50% *spxpy* and 50% *pz* in character and each participates in three banana bonds with one oxygen, as shown in Figure 4.26b. Also shown in Figure 4.26b for both sets of banana bonds are the small antibonding contributions from the second oxygen which are required for orthogonality (because the carbon is unable to respond fully to the hybrid GOs) and which reduce the bonding order of each of the banana bonds to 2/3. The overall bond order

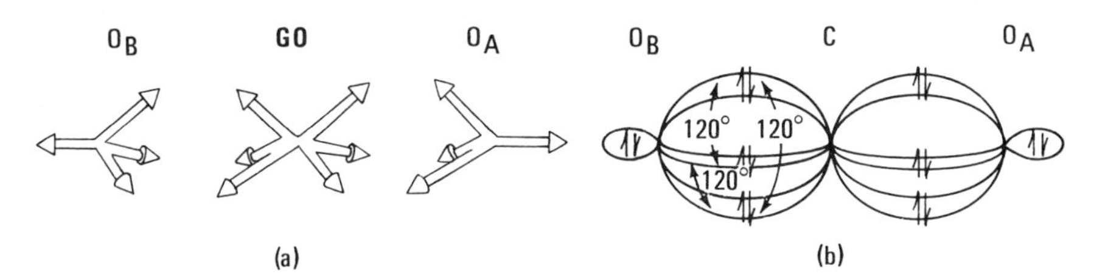

FIGURE 4.25 Oxygen VAO hybrids called in by $sp^3dxzdyz$ GO hybrids (a) and the resulting banana bonds in CO_2 (b). Note that only the main lobes of all the hybrid AOs are shown in (a). This procedure will also be followed in future diagrams.

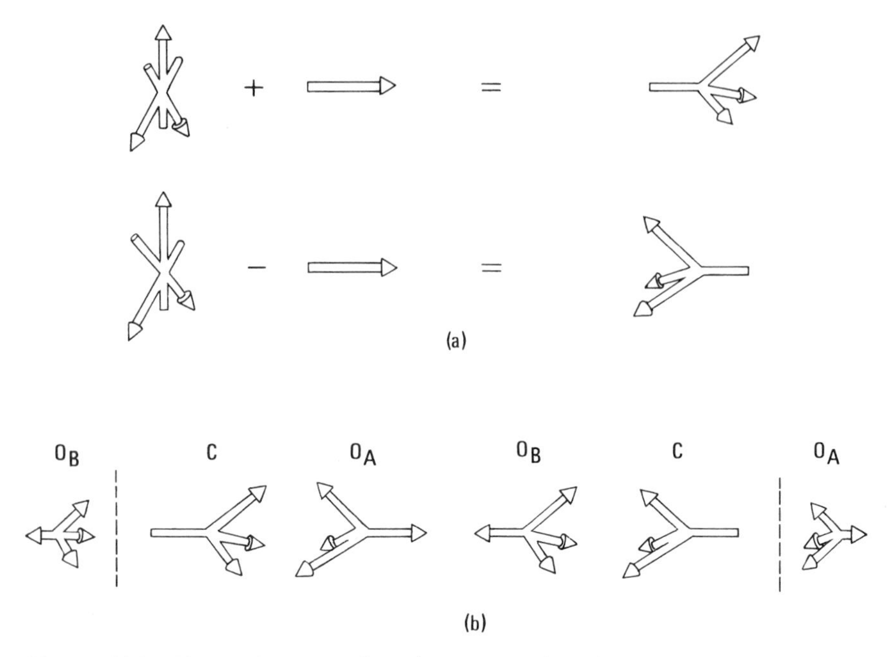

FIGURE 4.26 The hybrids $(spxpy)^1(pz)_r{}^1$ and $(spxpy)^1(pz)_l{}^1$ obtained by linear combination of *spxpy* and *pz* (a), and the six bonds in CO_2 (b), three of which are partially localized on the right side of the molecule and three on the left.

therefore remains 2.0 per link as expected from the delocalized view with which we started.

Use the Node Game Homework Drawing Board to draw the SOs associated with the three localized views we developed for CO_2. (See Figures 4.21 to 4.26.) Follow the steps in the appropriate column of Table 4.7.

From the preceding discussions it is clear that there is no one localized view which has two full bonding MOs in each link as well as two lone pairs on each oxygen. The electron dash structure in Figure 4.19a is therefore misleading. The point to be appreciated is that electron dash structures must be viewed with some suspicion since they are frequently too simplistic to relate much information of any significance concerning the delocalized or the localized bonding.

The valence electronic configuration and shape of a molecule is very often mimicked by another. This is true, for instance, for $NO_2{}^+$ and CO_2. These two molecules are said to be isoelectronic and isostructural. The vast majority of isoelectronic molecules are also isostructural. $I_3{}^-$ and XeF_2 are isostructural, although they are not strictly isoelectronic because the fluorines in XeF_2 use $n = 2$ valence orbitals while the outer iodines in $I_3{}^-$ use $n = 5$ valence orbitals. Some molecules are isostructural even though they are not isoelectronic. For example, SO_2 has a total of 18 valence electrons while NO_2 has 17. Yet both are bent according the VSEPR rules discussed later, though not with the same bond angle.

4.4. Vibrational Modes for Linear Triatomics

To pictorialize the vibrational modes for a linear molecule YZY, we proceed in a manner analogous to that described in Chapter 3.3 for the vibration of N_2. in Figure 4.27 are the nine atomic vectors (AVs) on the Y and Z atoms, which we must linearly combine into LCAVs permitted by GOs at the molecular center. As in the case of delocalized MOs, we can summarize the molecular motions information in a generator table. From Figure 4.27 we recognize that there are the four AV equivalence sets listed in the left-hand column of Generator Table 4.6. From these sets are generated nine symmetry motions (SMs) denoted by the symbol "m". Here "m" stands for the "motion" having the same symmetry as the GO heading the column. In other words, the GO indicates the direction of motion taken by an AV. The motions indicated in the first and the last two columns of Table 4.6 are depicted in Figure 4.28. In Figure 4.28a a symmetric vibration (v_1) generated by placing an *s* GO on the molecular center is represented. The motions in Figure 4.28b and c are rotations about the *y* and *x* axes of the center of mass (i.e., the Z atom), respectively. These rotations are generated by the *dxz* and *dyz* GOs. The *px*, *py*, and *pz* GOs each generate two SMs (one on the Y atoms and one on

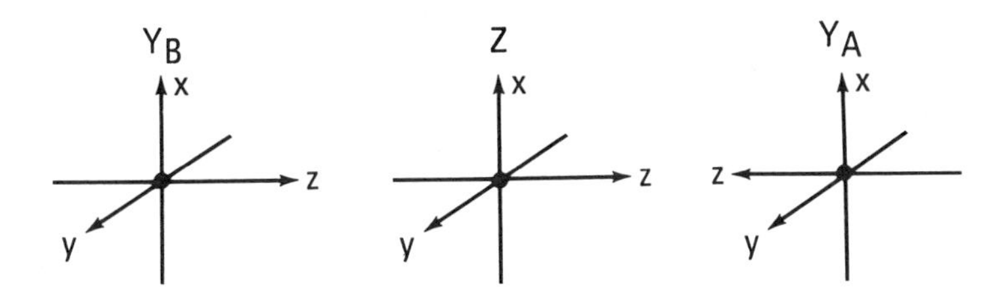

FIGURE 4.27 Atomic vector axis system for a linear YZY molecule.

TABLE 4.6 Generator Table for Linear YZY

	GOs					
AV Equivalence Sets	*s*	*pz*	*px*	*py*	*dxz*	*dyz*
Z(*z*)		m				
Z(*x*), Z(*y*)			m	m		
$Y_A(z)$, $Y_B(z)$	m	m				
$Y_A(x)$, $Y_B(x)$, $Y_A(y)$, $Y_B(y)$			m	m	m	m

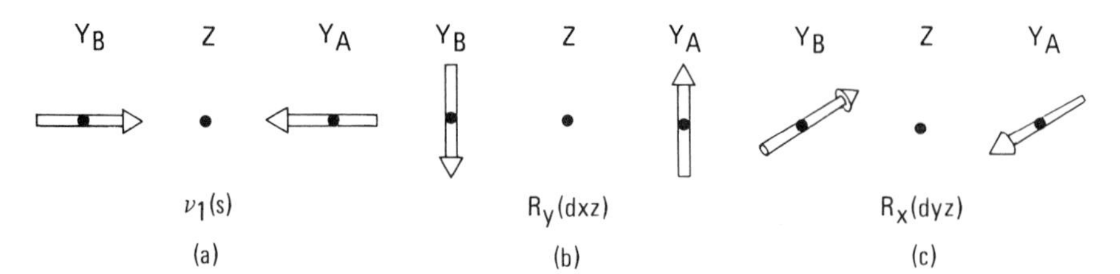

FIGURE 4.28 Symmetry and molecular motions generated by the (a) *s*, (b) *dxz*, and (c) *dyz* GOs.

Z) and so linear combinations of these SMs must be taken. These are shown schematically in Figure 4.29. The three positive linear combinations (sums) of these SMs clearly lead to translations (T) in the directions x, y, and z of the center of mass. The two negative combinations (differences) of the partner *px* and *py* SMs constitute a degenerate pair of bending vibrations of the molecule. The relationship $3n - 5 = 3(3) - 5$ tells us to expect four vibrational modes for a linear triatomic molecule, and we generated four: two bending and two stretching modes. Closely analogous modes arise in asymmetric linear triatomic XYZ. Although we generated the vibrations in Figures 4.28 and 4.29 in the order given by their subscripts, it should be noted that by convention in triatomic molecules, spectroscopists refer to the symmetric stretch (Figure 4.28a) as ν_1, the bending motion (Figure 4.29b) as ν_2, and the asymmetric stretch (Figure 4.29a) as ν_3.

Draw the symmetry motions for a linear YZY molecule using the Node Game Homework Drawing Board. (See Table 4.6 and Figures 4.28 and 4.29.) Follow the steps in the appropriate column of Table 4.7.

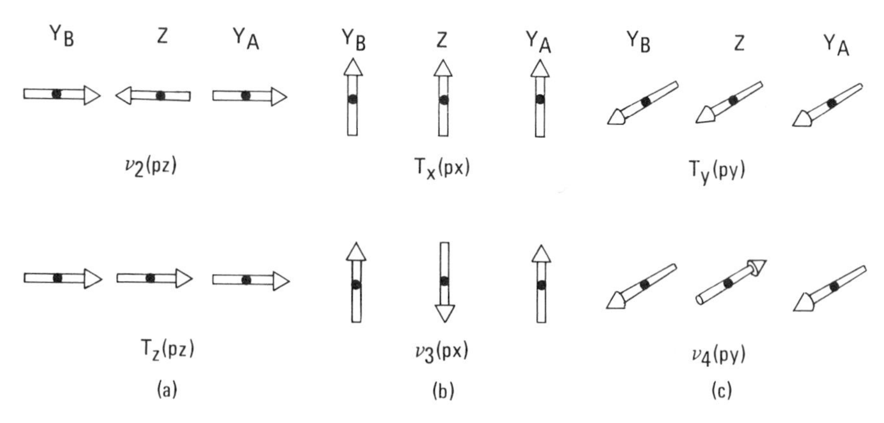

FIGURE 4.29 Symmetry and molecular motions obtained by linear combination of (a) *pz*-generated, (b) *px*-generated, and (c) *py*-generated symmetry motions.

Summary

By assuming linear geometries for FHF^-, FXeF, and CO_2, we saw from the delocalized views of these molecules that the sigma MOs of FHF^- and FXeF (with no *d* VAO participation on Xe) gave rise to a three-center four-electron MO system. In this scheme, two of the electrons are in a BMO and two are in an NBMO. In the corresponding localized views, these electron pairs could not be fully localized between two centers. By contrast, in CO_2 and FXeF (with *d* VAO participation on Xe) both electron pairs in the sigma framework are in BMOs, and when these occupied MOs are localized, doubly occupied MOs spanning only two centers were generated. In formulating bonding views of FXeF and CO_2, we found that electron dash structures, while very

TABLE 4.7 A Useful Summary of Procedures for Generating Pictorial Representations of Delocalized and Localized Bonding Views of a Molecule and for Visualizing its Vibrational Modes

Delocalized View	Localized View	Normal Modes
1. Decide on best electron dash structure.	1. Decide on which set(s) of doubly occupied delocalized MOs to localize.	1. Identify the motional atomic vector (AV) equivalence sets and list them in the generator table.
2. Establish the molecular geometry.	2. Hybridize the GOs which give rise to the set(s) in step 1.	2. Using GOs, draw all the symmetry motions (SMs).
3. After preassigning lps to (*s*) and suitably oriented (*p*) VAOs, identify all the VAO equivalence sets and list them in a generator table.	3. Using the hybridized GOs, draw the SOs.	3. Fill in the rest of the generator table (i.e., the GOs and m designations).
4. Using GOs, draw all the SOs, recalling that #SOs = #VAOs.	4. Draw the MOs, being aware of the consequences of a central atom on which the full set of SOs (i.e., hybridized VAOs) is not generated by the hybridized GO set.	4. Take linear combinations of SMs generated by the same GO and draw all the molecular motions.
5. Fill in rest of generator table (i.e., the GOs and the n, a, or b designations).	5. Draw the resulting MO energy level diagram and occupy the appropriate levels with electrons.	5. Identify the normal vibrational modes.
6. Take linear combinations of SOs generated by the same GO and draw all the MOs.	6. Verify that the bond order per link is the same as in the delocalized view.	
7. Draw the resulting MO energy level diagram and occupy the appropriate levels with electrons.		
8. Deduce the bond order per link.		

useful, are sometimes misleading regarding the number of lone pairs to be preassigned.

Using generator orbitals we saw that the AV equivalence sets of a linear triatomic molecule give rise to SMs which can be summarized in a generator table. From the permitted linear combinations of these SMs, the vibrational modes of these molecules were deduced.

Table 4.7 summarizes an abbreviated form of our recipes for developing the delocalized view, a localized view, and the molecular motions. These stepwise procedures, will be amplified in amended tables at the end of future chapters as the molecules we consider become gradually more complex. *In doing the exercises, be sure to follow these procedures closely.*

Exercises

1. For the linear species below, generate delocalized and localized views of the bonding. (a) BeH_2; (b) CH_2 (paramagnetic); (c) I_3^-; (d) NO_2^+.
2. Repeat Problem 1c including (*d*) VAOs.
3. Draw the MOs for FHF^- and sketch in any internuclear nodes as well as VAO nodes these MOs possess.
4. Generate the delocalized view of FHF^- after preassigning the lone pair electrons.
5. Draw the MO energy level diagrams corresponding to the localized views of CO_2 depicted in Figures 4.24b and 4.25b.
6. Using the GO approach, generate a delocalized view of $(CN)_2$ and also deduce its vibrational modes. [Hint: Treat the system as two symmetrical parts, namely, two carbons and two nitrogens, and take linear combinations of symmetry orbitals (and atomic vectors) that are generated by the same GO.]
7. Generate the delocalized view of N_3^+ in the linear geometry wherein it possesses two unpaired electrons.
8. Allene ($H_2C{=}C{=}CH_2$) exists as a nonplanar molecule in the ground state in which the (sp^2) frameworks of the CH_2 fragments are at 90° to one another. Draw the structure and comment on the degree of delocalization of the π bonding. Suggest a reason why the given conformation of allene is more stable than the planar one. (Hint: compare the number of electrons in bonding MOs for each conformation. Calculations show that this consideration is the overriding one.)
9. It appears that the CO molecules in linear $Ag(CO)_2^+$ bind by donating a pair of electrons to the metal. Generate the localized and delocalized view of the C—Ag—C bonding. Generate the delocalized view assuming that filled valence (*d*) orbitals on the metal can overlap the π^* MOs on the CO's.
10. The superconductor $LuNi_2B_2C$ contains linear BCB units, which to a reasonable approximation can be considered as B_2C^{3-} ions. Generate delocalized and localized bonding views of this anion. Is it diamagnetic? The linear BP_2^{3-} ion occurs in the yellow salt K_3BP_2. Is this ion diamagnetic?

11. The ion $[\{[(CH_3)_3C]_3C\}_2Na]^-$ has been isolated and shown by x-ray crystallography to contain a linear C—Na—C framework. Generate delocalized and localized bonding views for this framework.

12. The ion shown below sports a hydrogen midway between the two bridgehead carbons. Generate the delocalized and localized views of the C—H—C framework.

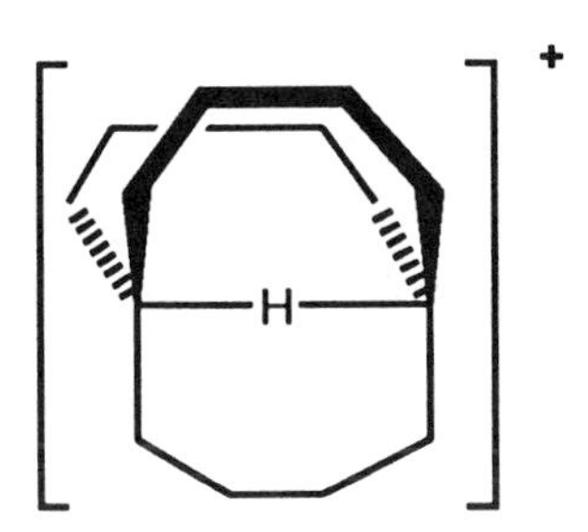

13. The dark blue anion $\{[(C_6F_5)_4Pt]_2Tl\}^{2-}$ has a linear PtTlPt framework:

Assume that each C—Pt bond is a localized two-center two-electron bond and that the platinum (formally in the 4+ oxidation state) uses an empty valence *p* orbital to interact with the thallium. (a) Develop the delocalized view of the Pt_2Tl^{2-} portion of the molecule. (b) How do you account for the fact that this anion is paramagnetic?

14. Organometallic compounds having opposing C≡O groups such as the one shown below display two characteristic C≡O stretching frequencies in their infrared spectra. These absorptions are approximated by motions of the oxygens back and forth against the carbons, and can be depicted as SMs of the oxygens along the OCMoCO axis. In other words, the oxygens can be viewed as vibrating against the relatively stationary $CMo(PF_3)_4C$ part of the molecule. (a) Draw the allowed CO vibrations using GOs. (b) Using GOs draw the allowed bending motions of oxygens against the carbons.

$$\mathrm{Mo(CO)_2(PF_3)_4}\ \text{(trans-CO; structure: } O\equiv C-Mo(PF_3)_4-C\equiv O)$$

15. Linear C_3 molecules have been observed in the spectra of comet tails and vaporized carbon. Develop a delocalized view of this molecule. Does it have any unpaired electrons?

16. Develop a delocalized and localized view for the linear C—Tl—C^+ portion of the $[H_3ClInCH_3]^+$ cation.

CHAPTER 5

Triangular and Related Molecules

In this chapter we will examine H_3^+, N_3^+, CH_3^+, and BF_3. The first two species are truly triangular in that they have an atom at each of the three corners of an equilateral triangle. Although the latter two molecules also have three identical atoms on the corners of a triangle, they possess a central atom as well.

5.1. H_3^+

This perhaps strange-looking ion was discovered by Thompson by mass spectrometry when he passed an electrical discharge through H_2 gas. This ion has also been found in space where it is quite stable. A triangular array of three H atoms bound to a cationic organometallic iridium compound has also been observed. Other triangular species that have been studied in the mass spectrometer are Li_3^+, Na_3^+, Na_2Li^+, $NaLi_2^+$, Li_2H^+, Na_2H^+, and $NaLiH^+$. Theoretical calculations show that the equilibrium H_3^+ configuration is an equilateral triangle with interatomic distances comparable to that in H_2. The equilateral triangular configuration was later confirmed in Coulomb explosion experiments.

To begin our discussion of the delocalized bonding in H_3^+, let us adopt the axis system in Figure 5.1. The dots denote atom centers and the triangle denotes our GO center. From the symmetry of H_3^+ we see that all of the (H1*s*) VAOs belong to the same equivalence set in Table 5.1. For example, rotation by 120° about the z axis brings ($H_A 1s$) into ($H_B 1s$), and ($H_B 1s$) into ($H_C 1s$). From Figure 5.2 we see that the three SOs expected from the three VAOs are generated by *s*, *px*, and *py* GOs. Note in Figure 5.2b that H_A is unable to contribute to $\tilde{\sigma}(py)$ since the *py* GO and ($H_A 1s$) behave differently with respect to reflection in the *xz* plane. Notice also in Figure 5.2c that the ($H_B 1s$) and ($H_C 1s$) contributions are only half as large as that of ($H_A 1s$). A way to visualize the reason for this result is to consider Figure 5.3. The (1*s*) VAO of H_A is seen to be called in as fully as possible by *px* and so we give this VAO a coefficient of 1. If we place an (H1*s*) VAO at point P directly

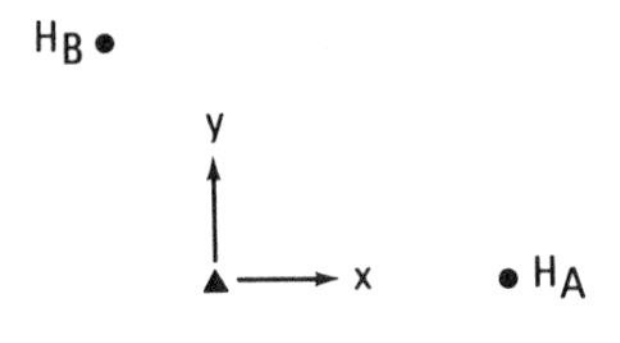

FIGURE 5.1 Axis system for H_3^+.

above *px*, then this GO is unable to call in a contribution from such a VAO because this VAO does not have a node where the GO does. Thus the coefficient of 1 for ($H_A 1s$) must decrease to zero if a (1*s*) orbital should move from the H_A position to point P. By the same token, this coefficient must rise negatively to -1 if the VAO at P moves to R. The path of a (1*s*) VAO around the circle back to H_A again is seen to reflect a cos function where the coefficient for the (1*s*) orbital at H_B is given by $\cos 120° = -1/2$. Similarly, the coefficient to ($H_C 1s$) is $-1/2$. Another way to rationalize the coefficients implied in Figure 5.2c is to recognize that they are required for orthogonality of this MO with the other two.

Since there is only one SO generated by each GO, the SOs are MOs and, because there is no central atom in this molecule, the atoms on the corners of the triangle must bind to one another. The bonding and antibonding nature of these MOs indicated in Table 5.1 can be pictorially recognized from Figure 5.2 by mentally eliminating the GOs. The MO $\sigma(s)$ in Figure 5.2a is bonding between all three pairs of atoms. The antibonding interaction between ($H_B 1s$) and ($H_C 1s$) in Figure 5.2b clearly marks this MO as antibonding. The antibonding interactions between ($H_A 1s$) and its two neighbors in $\sigma^*(px)$ (Figure 5.2c) exceed the single bonding interaction between ($H_B 1s$) and ($H_C 1s$), thus justifying the labeling of this MO as $\sigma^*(px)$.

Before we can draw an MO energy level diagram, however, we need to know whether the energies of $\sigma^*(py)$ and $\sigma^*(px)$ are the same or different. The orbitals certainly *look* different. What possible reason could there be for

TABLE 5.1 Generator Table for H_3^+

	GOs		
VAO Equivalence Sets	*s*	*px*	*py*
$Hs = (H_A 1s), (H_B 1s), (H_C 1s)$	b	a	a

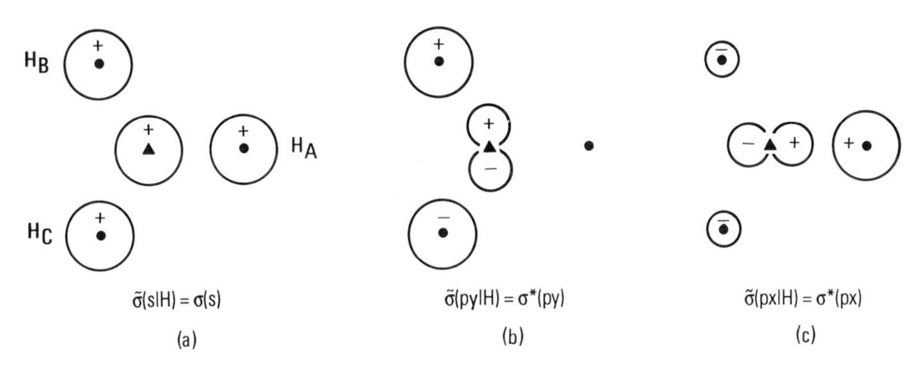

FIGURE 5.2 Molecular orbitals in H_3^+ generated by (a) s, (b) py, and (c) px.

their degeneracy? To appreciate the fact that these two ABMOs indeed are degenerate, we need to reconsider the triangular symmetry of the molecule. A triangle has a threefold axis of symmetry, which means that any point within the molecule when rotated by 120°, 240°, or 360° feels the same potential. The same is true if we reflect a point through the xy plane, for example. The $\sigma(s)$ MO and, of course, the s GO that generates it, have this full symmetry of the ion too. This is not true of the ABMOs *individually*, however, nor of the px and py GOs which generate them. But it is easy to see that rotation of a px GO by 120°, for example, creates a new p orbital, which must be a linear combination of px and py. In other words, the px and py GOs *taken together* do contain the full symmetry of our triangular molecule and this is also true for the two SOs they generate. This is easily visualized by squaring the wavefunctions of (px) and (py) and summing their electron densities. A three-dimensional plot of the sum resembles a perfectly circular donut lying in the

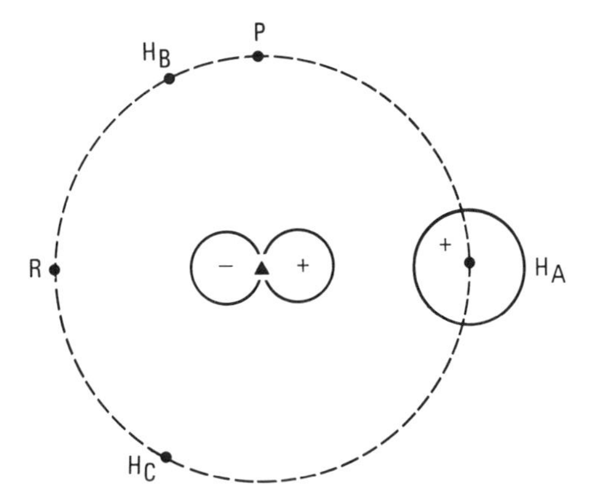

FIGURE 5.3 Diagram of points P and R in H_3^+ where the coefficients of (H1s) VAOs are 0 and -1 for an SO generated by px (see text).

xy plane. Such a shape has the full symmetry of a triangle or, for that matter, any regular polygon (Chapter 7). Because the *px*- and *py*-generated SOs are generated by GOs which are equal partners in containing the full symmetry of the H_3^+ ion, these SOs also have equal energies. It is always true that when a molecule has an SO that does not contain the full symmetry of the molecule, there is at least one other partner SO of equal energy, which also does not contain the full symmetry of the molecule. *Moreover, partner SOs are always generated from one VAO set by GOs that have the same l quantum number.* In atoms, any (*s*) orbital contains the full symmetry of the atom. Therefore, an *s*-generated SO cannot have a partner since a sphere already possesses the highest symmetry possible. A (*p*) orbital by itself obviously does not have the full symmetry of an atom but when all three (*p*) orbitals are taken together, their combined electron density generates a perfectly spherical charge density. A *p*-generated SO can have one or two partners a *d*-generated SO can have up to four partners, etc. In other words, the full set of AOs with a given l quantum number always contains the symmetry of a sphere. It turns out, however, that partner sets containing more than three members are rare.

The delocalized MO energy level diagram of H_3^+ shown in Figure 5.4 indicates that the bond order per link is 1/3. Since there is only one occupied MO in this molecule, no further localization is possible. The bond is a three-center two-electron bond, thus making it impossible to draw an electron dash

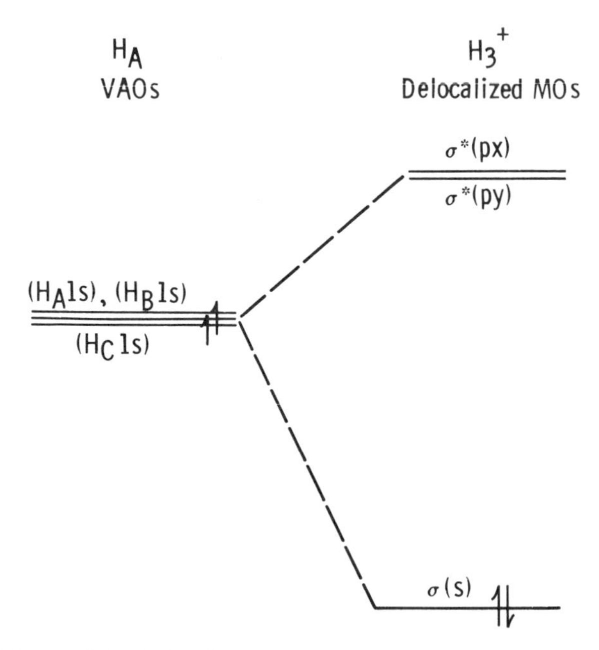

FIGURE 5.4 Delocalized MO energy level diagram for H_3^+.

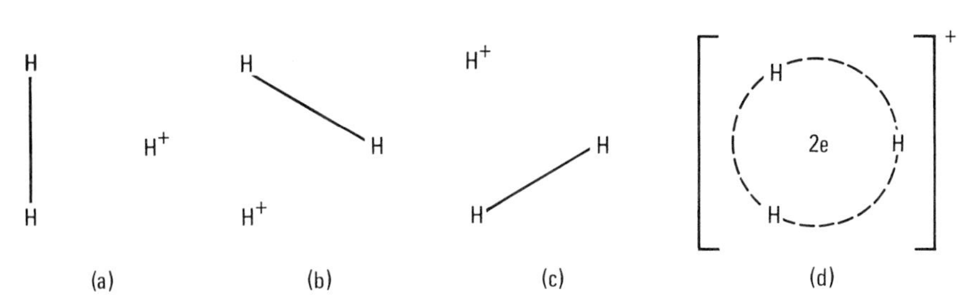

FIGURE 5.5 Electron dash structures for H_3^+ [(a), (b), (c)] and an average of these structures (d).

structure. However, we can draw the three so-called resonance structures shown in Figure 5.5a–c. Resonance structures are electron dash structures, which can be drawn by moving electrons around the structure while leaving all atoms stationary. Individually the resonance structures of H_3^+ are unsatisfactory, but taken together as an average structure (which is represented by Figure 5.5d) they confirm the experimental fact that all three atoms are chemically equivalent.

Use the Node Game Homework Drawing Board to develop the above delocalized view of H_3^+. Be sure to follow the steps in Table 4.7.

5.2. N_3^+

This ion formed by electron impact of N_2 at low pressures can also be studied in the mass spectrometer. Theoretical calculations reveal it to have all its electrons paired in the equilateral triangular geometry. Although a linear geometry (with two unpaired electrons) is calculated to be of lower energy for this ion, we will examine the metastable triangular arrangement since the bonding pattern in this geometry has important implications for more common molecules such as BF_3. The triangular form of the N_3^+ ion is isoelectronic with the cyclopropenium ion $(CH)_3^+$ (which is derived from cyclopropene by abstraction of a hydride ion). That is, N_3^+ and $(CH_3)^+$ each have the same number of valence electrons.

As with H_3^+, attempts to draw a single electron dash structure of N_3^+ and $(CH)_3^+$ are less than pleasing and we must draw resonance structures (e.g., Figure 5.6a, b, for each of which there are two additional similar structures achieved by moving the double bond and the charge around the rings). The $(CH)_3^+$ ion can be thought of as resulting from the movement of a proton from each nitrogen nucleus radially outward to the hydrogen position where it converts a nitrogen lone pair (lp) to a bonding electron pair (bp).

In developing the delocalized view of N_3^+, it is reasonable first to assign the lone pair on each nitrogen to an (N2*s*) VAO. Using the axis system in Figure 5.7 we construct the equivalence sets in generator Table 5.2. Note that

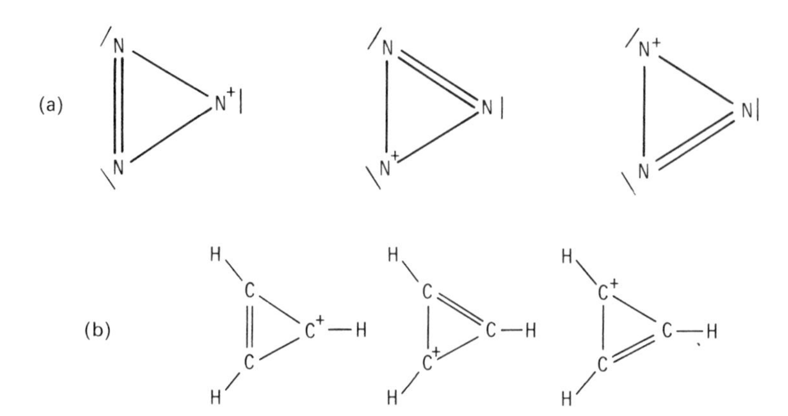

FIGURE 5.6 Electron dash resonance structures for N_3^+ (a) and $(CH)_3^+$ (b).

the N*p*r and N*p*t equivalence sets in this table denote *radially* and *tangentially* directed (*p*) VAOs, respectively. Generation of the SOs in the N*p*r equivalence set is done exactly in the same manner as for H_3^+ and you should do this using the Homework Drawing Board in Node Game. As shown in Figure 5.8, the *pz*, *dxz*, and *dyz* GOs generate three SOs among the VAOs of the N*pπ* set. Since these are the only SOs generated by these three GOs,

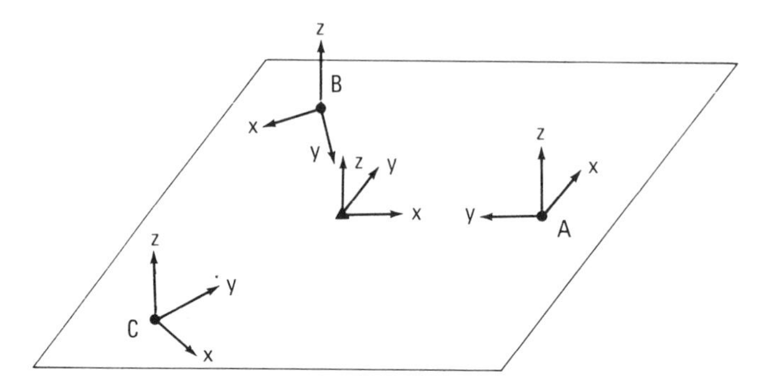

FIGURE 5.7 Axis system for N_3^+.

TABLE 5.2 Generator Table for N_3^+

	GOs						
VAO Equivalence Sets	*s*	*px*	*py*	*pz*	*dxz*	*dyz*	*f*3″
N*p*r = ($N_A 2py$), ($N_B 2py$), ($N_C 2py$)	b	a	a				
N*p*t = ($N_A 2px$), ($N_B 2px$), ($N_C 2px$)		b	b				a
N*pπ* = ($N_A 2pz$), ($N_B 2pz$), ($N_C 2pz$)				b	a	a	

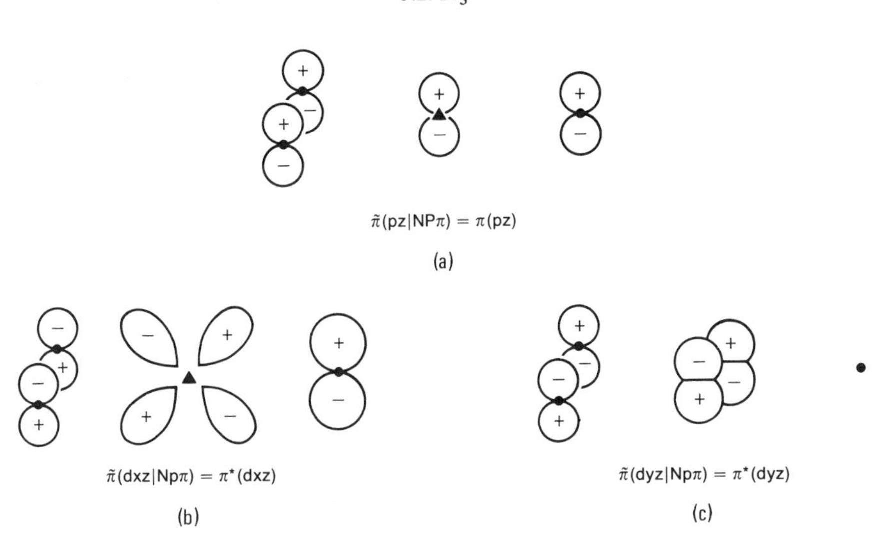

FIGURE 5.8 The π MOs generated in the N$p\pi$ equivalence set of N_3^+. Notice that the drawings for the (*dxz*) and (*dyz*) GOs in (b) and (c), respectively, are somewhat different.

these SOs are MOs. Using the same reasoning as for the MOs of H_3^+, the bonding and antibonding natures of the MOs in Figure 5.8, the coefficients (sizes) of the VAOs, and the degeneracy of the π^* *partner-set* MOs become clear. Notice that for clarity in drawing π^*(*dxz*), the size of ($N_A 2pz$) was doubled instead of reducing the sizes of ($N_B 2pz$) and ($N_C 2pz$) by half. This produces no difference in the overall results because the wave functions can be normalized later. From Figure 5.9 we see that *px* and *py* generate a partner set of MOs, which in contrast to the analogous situation in H_3^+, now gives a pair of bonding (rather than antibonding) MOs. That is, $\sigma(px)$ is bonding between the lobes directed toward one another between atoms B and C. While this is not true for $\sigma(py)$, the two bonding interactions between atoms A and B and A and C clearly outweigh the single antibonding interaction between B and C. The remaining SO in the Npt equivalence set is found to be generated by $f3''$. No GO simpler than $f3''$ gives this new SO. That $f3''$ is indeed correct here is seen by recognizing that since only one SO remains to be generated, it cannot be degenerated with another (i.e., have a partner) and indeed, $f3''$ is recognized from Figure 5.9c to contain the full symmetry of a triangle which it must if there is no partner SO. The MO $\sigma(f3'')$ is seen to be antibonding (rather than bonding as is the case with the *non*degenerate MO in the Npr and N$p\pi$ equivalence sets). Except for the *px*- and *py*-generated MOs in Table 5.2, all the MOs are generated by a single GO. By allowing the *px*- and *py*-generated antibonding and bonding MO pairs to interact (i.e., by linearly combining them) enhancement of their antibonding and bonding character will occur. This further splitting of their energies

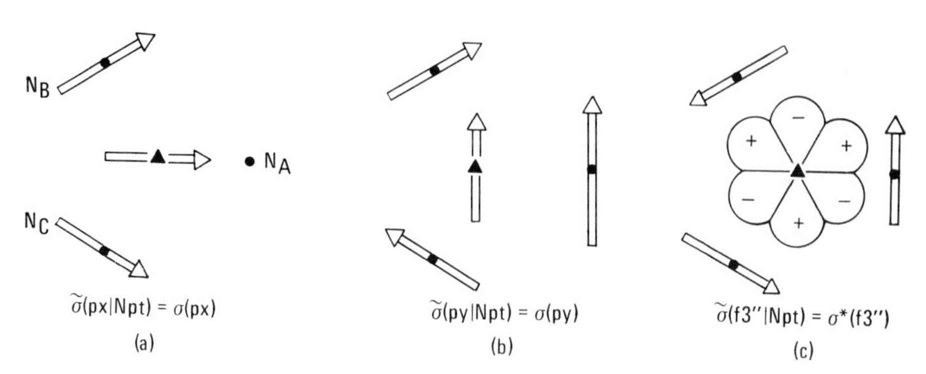

FIGURE 5.9 The σ MOs generated in the Npt equivalence set of N_3^+.

arises from the fact that these pairs of MOs are of similar energy and have identical symmetries since they are generated by the same GO in each case. Recall a similar occurrence in the MO energies of diatomics (Figure 3.15).

There is a point concerning our choice of the axis system for N_3^+, which needs further comment. If we had not chosen to point the GO x axis toward one of the nitrogen atoms, the $f3''$ GO could not have been used and would have been necessary to hybridize $f3'$ with $f3''$ to obtain a linear combination that would have a planar node intersecting each N atom. In future examples involving polygonal arrays of atoms in the xy plane, we will therefore generally point the GO x axis toward one of the atoms. The square array is an exception (Chapter 7).

In N_3^+ we have fourteen valence electrons, six of which we have preassigned to the three (N2s) VAOs. The remaining eight are expected to occupy the four BMOs we have generated. In Figure 5.10 a delocalized MO energy level diagram is given. In this diagram, all four BMOs are three-center two-electron bonds.

Before we go on to a localized view, let us modify our delocalized view somewhat to make our method for the treatment of future molecular ring systems more convenient to apply. If we had not preassigned our lone pair orbitals to (N2s) VAOs, we would have had an Ns equivalence set from which three SOs (MOs) would have been generated, namely, by s, px, and py GOs, as shown in Generator Table 5.3. Under these circumstances, the three linear combinations of the MOs generated by px, for example, would lead to three MOs in which $\sigma_s(px)$ would be essentially nonbonding compared to the other two [since the (N2s) VAO lies so low in energy], the tangential $\sigma_{px}(px)$ would still be quite bonding, and the radial $\sigma^*_{py}(px)$ would be quite antibonding. We could, of course, also find suitable orthogonal linear combinations of these MOs in which the s-dominated $\sigma^0_s(px)$ MO is entirely free of admixtures from Npt and thus contains Ns and Npr contributions only. That would mean that the predominant Npt contribution to $\sigma_{px}(px)$ would become even stronger. This hybridization scheme for the MOs generated by px

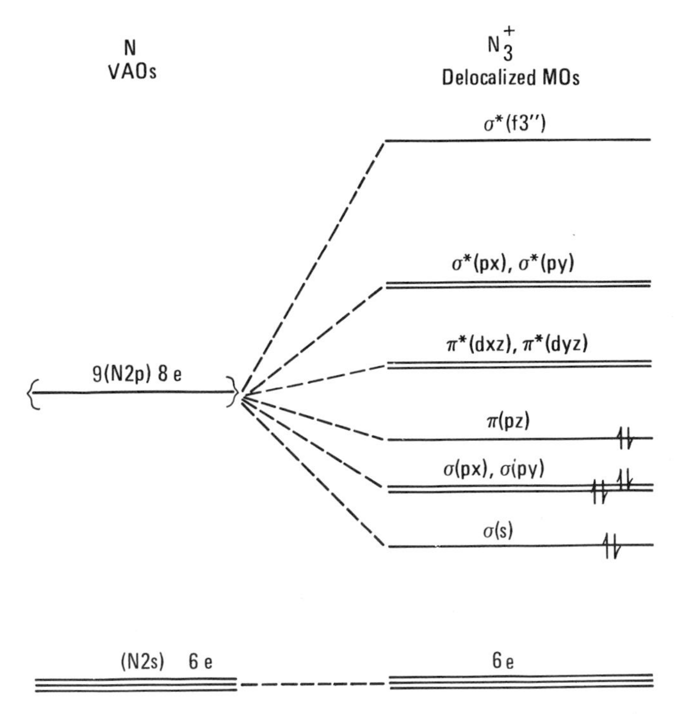

FIGURE 5.10 Delocalized MO energy level diagram for N_3^+.

(and similarly by *py*) is convenient since it creates a set of digonal (*s*)-dominated N*sp* hybrids pointed radially outward (N*spo*), and a set of digonal (*p*)-dominated N*sp* hybrids pointed radially inward (N*spi*). This would leave the N*pt* set intact. We could, of course, have arrived essentially at the same point by hybridizing the nitrogen atoms *at the start* so that each one has an (*s*)-dominated (N*spo*) and an (N*spi*) VAO *and preassigning a lone pair to each* (N*spo*). This procedure was useful in FHF^-, for example, and it will find great use in future applications. The modified generator table given as Table 5.4 and the MO energy level diagram in Figure 5.11 reflect this modification.

TABLE 5.3 Generator Table for N_3^+ (without lp Preassignment)

	GOs						
VAO Equivalence Sets	*s*	*px*	*py*	*pz*	*dxz*	*dyz*	*f*3″
N*s* = ($N_A 2s$), ($N_B 2s$), ($N_C 2s$)	b	a	a				
N*pr* = ($N_A 2py$), ($N_B 2py$), ($N_C 2py$)	b	a	a				
N*pt* = ($N_A 2px$), ($N_B 2px$), ($N_C 2px$)		b	b				a
N*pπ* = ($N_A 2pz$), ($N_B 2pz$), ($N_C 2pz$)				b	a	a	

TABLE 5.4 Generator Table for N_3^+ (with Preassignment of Lone Pairs in N*spo* Hybrids)

	GOs						
VAO Equivalence Sets	*s*	*px*	*py*	*pz*	*dxz*	*dyz*	*f*3″
N*spi* = (N_A2*spi*), (N_B2*spi*), (N_C2*spi*)	b	a	a				
N*pt* = (N_A2*px*), (N_B2*px*), (N_C2*px*)		b	b				a
N*pπ* = (N_A2*pz*), (N_B2*pz*), (N_C2*pz*)				b	a	a	

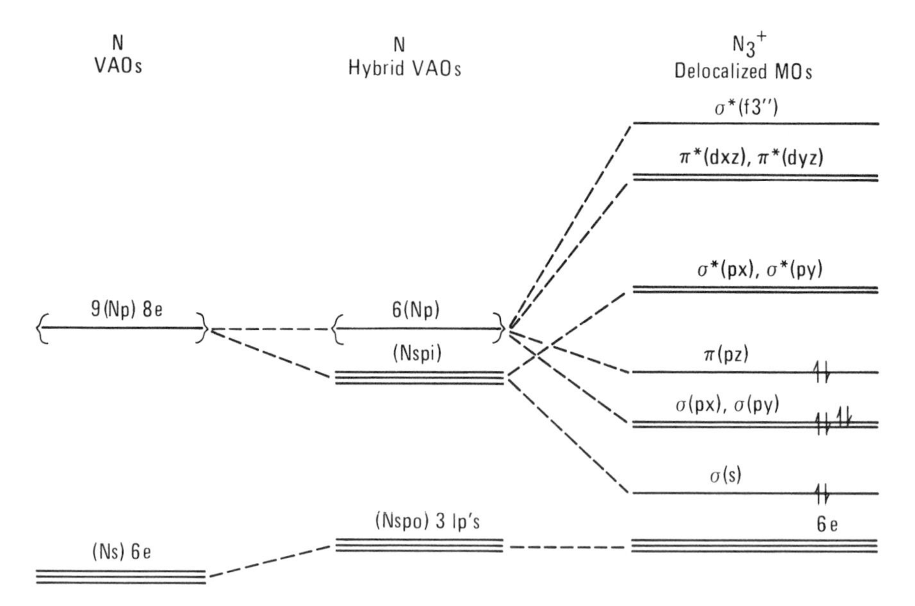

FIGURE 5.11 Delocalized energy level diagram for N_3^+ with preassignment of lone pairs in N*spo* hybrids.

Of course we could have included the N*spo* equivalence set in Table 5.4 in which case the *s* and the *px*, and *py* GOs would generate a bonding and two antibonding SOs, respectively. However, since all three of these SOs (MOs) are occupied anyway, we could always recast these MOs into the (N*spo*) VAOs (with small antibonding contributions from the other two atoms). For convenience we therefore choose to keep lone pairs localized in VAOs and to a first approximation this is valid. This procedure is reflected in the MO energy level diagram in Figure 5.11. In this diagram, the lone pairs are seen to have mainly (*s*) character in the (N*spo*) VAOs and the (*p*)-dominated $\sigma^*(px)$ and $\sigma^*(py)$ MOs have dropped in energy with the admixture of the (*s*) character, as has the $\sigma(s)$.

To obtain our localized picture of N_3^+, we begin by hybridizing the *s*, *px*, *py* GO set that generated the σ BMOs in Figure 5.11. This sp^2 hybrid set points its main lobes toward the three atom centers. Since we always prefer that the localized MOs be concentrated *between* atom centers, we take advantage of the concept developed in Chapter 2 that we can recast our present sp^2 GO set into one rotated by 60°. The main lobes are then directed at the bond midpoints. The effect of each lobe of this hybrid GO set is to call in an ($\mathrm{N}sp^\alpha$) hybrid from each nitrogen (Figure 5.12a) made up of an (N*spi*) and an (N*pt*) VAO (Figure 5.12b). The three localized MOs composed mainly of contributions from an ($\mathrm{N}sp^\alpha$) VAO on each pair of atoms are depicted in Figure 5.12c. It is clear that the ($\mathrm{N}sp^\alpha$) hybrids are not 180° apart (digonal, $\alpha = 1$) since an entire (N*pt*) VAO is involved plus the dominating (*p*) contribution in (N*spi*). Moreover, only part of the (N*s*) VAO contributes to the ($\mathrm{N}sp^\alpha$) VAO since most of the (N*s*) orbital set is already contributing to the (*s*)-dominated lp (N*spo*) orbital. Furthermore, the ($\mathrm{N}sp^\alpha$) orbitals cannot be facing directly toward each other along the internuclear axis since that would require an angle of 60° between each pair of hybrids on a nitrogen atom. (Even with pure *p* character, two orthogonal (*p*) VAOs must be 90° apart.) Since there is only one occupied π delocalized MO, no further localization of this MO is possible. The only difference between the delocalized MO energy level diagram in Figure 5.11 and its localized counterpart is that the occupied σ delocalized MOs are recast into three degenerate localized BMOs $\hat{\sigma}_A$, $\hat{\sigma}_B$, $\hat{\sigma}_C$ which lie two-thirds of the way up from $\sigma(s)$ to the degenerate $\sigma(px)$, $\sigma(py)$ pair.

Use the Node Game Homework Drawing Board to develop the preceding delocalized and localized views of N_3^+ wherein we localized the lone pairs. Follow the steps given in the approriate column of Table 5.8.

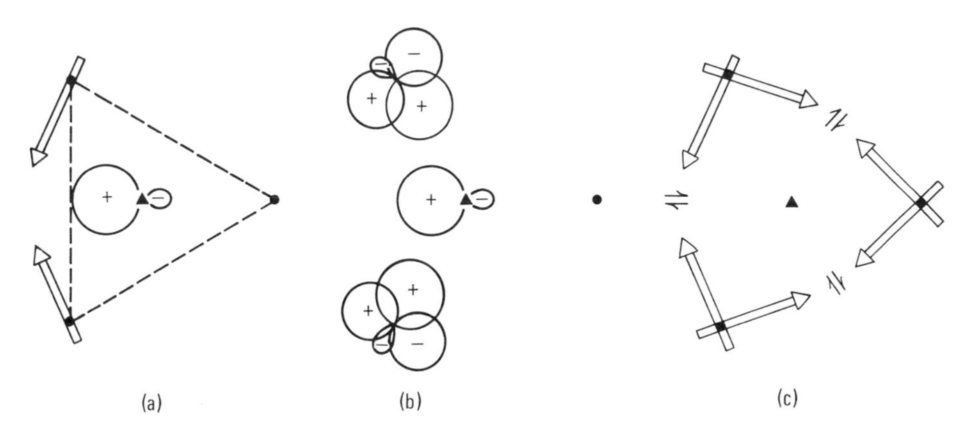

FIGURE 5.12 Generation by an sp^α ($\alpha \cong 2$) GO of a localized MO in N_3^+ between two nitrogens using a combination of their N*spi* and N*pt* VAOs (a), which is shown in more detail in (b). In (c) is shown a pictorial representation of all three localized MOs.

5.3. Electron Dash Structures and Molecular Geometry

Before discussing the bonding in more complex molecules, we will make some further observations about electron dash structures. If we can draw a reasonable electron dash structure, the geometrical structure can generally be inferred from it and the formulations of the delocalized and localized bonding patterns are greatly facilitated. How can we draw an electron dash structure, given a formula? Taking XeF_2 as an example, we could write the two electron dash structures in Figures 5.13a and b if we did not happen to know the experimentally determined structure. A simple calculation of the so-called formal charges shown above each atom permits us to make the correct choice. *The formal charge is given by the difference between the number of valence electrons in the neural atom and the sum of the lone-pair (lp) electrons and half the number of bonding pair (bp) electrons of the atom in the electron dash structure.* In general, the electron dash structure with the lowest set of formal charges is the one to choose (i.e., Figure 5.13a). If there is no alternative to having more than one formal charge of the same sign, then separating these charges by the largest number of atoms possible usually affords the best electron dash structure. How does the electron dash structure in Figure 5.13a allow us to deduce the geometry of the molecule? Electron dash structures generally reflect the localized MO picture. We have seen that in such a bonding view the localized valence MOs tend to avoid each other in space and that lone-pair localized MOs contain more s character and hence command a larger spherical angle around a nucleus than do localized BMOs. The recognition by Gillespie and Nyholm that actual bond angles adjust themselves to this behavior led to the "valence state electron pair repulsion" (VSEPR) rules, which state that the order of "repulsion" effects is lp–lp > lp–bp > bp–bp. Actually, electrostatic repulsions probably play a minor role and the mutual avoidance of localized MOs is largely governed by antibonding effects arising from the Pauli principle (which forces different electron pairs into different orbitals) and antibonding overlap effects between doubly filled orbitals.

In triatomic molecules, for example, it is possible to have zero (HBeH), one (ONO^-), two (H_2O), and three (FXeF) lone pairs on the central atom. The VSEPR rules easily predict a 180° bond angle for HBeH, <120° for ONO^-, and <109°28′ for H_2O. For five pairs of electrons around xenon in FXeF (three lone pairs and two bond pairs if (*d*) VAO participation on Xe is assumed), calculations show that if all "repelled" one another equally, they

0 0 0
|F—Xe—F|
(a)

0 1+ 1−
|F—F—Xe〉
(b)

FIGURE 5.13 Electron dash structures for two forms of XeF_2.

would arrange themselves on the surface of a sphere at the apices of a trigonal bipyramid or at the corners of a square pyramid, which is generally not much higher in energy than a trigonal bipyramid. In most examples, however, the trigonal bipyramidal geometry is adopted. Since lone pairs "repel" other lone pairs most strongly, they arrange themselves 120° apart from one another around the equatorial plane in the trigonal bipyramid while the bond pairs take up the remaining two positions on the axis, thus placing themselves 90° apart from the lone pairs. In H_3^+ the VSEPR rules apply in a negative sense; that is, there are no electron pair repulsions to prevent ring closure. Closing the ring is favorable here because it leads to three delocalized bonds instead of only two in a linear or bent arrangement.

Can lone pair densities postulated by the VSEPR rules be verified experimentally? This has been accomplished by careful low-temperature x-ray or neutron diffraction experiments. By substracting known free-atom or inner shell electron densities from the experimentally observed densities, electron difference maps are produced, which indicate locations of lone-pair densities. These difference maps are complemented by calculations, which though quite sophisticated, do contain approximations to keep computational costs within bounds.

5.4. BF_3

Examples of molecules with a central atom bonded to a triangular array of other atoms include BX_3, NO_3^-, CO_3^{2-}, BO_3^{3-}, CH_3, NH_3^+, BeH_3^-, CH_3^+, BH_3, BH_3^-, InX_3, and GaX_3. We will choose to examine BF_3 since it introduces another useful feature of electron dash structures. In Figure 5.14a is an electron dash structure which, although it does leave zero formal charges on the atoms, does not obey the Lewis octet rule. At a more fundamental level than we usually learn this rule, it says that when four VAOs are

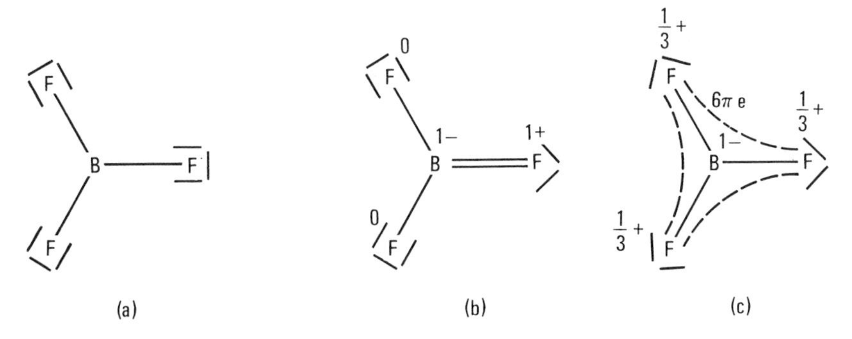

FIGURE 5.14 Electron dash structures for BF_3 in which the octet rule is only partially obeyed (a) and completely obeyed (b) but with the generation of formal charges. In (c) is an average electron dash structure for the three resonance structures implied by the one in (b).

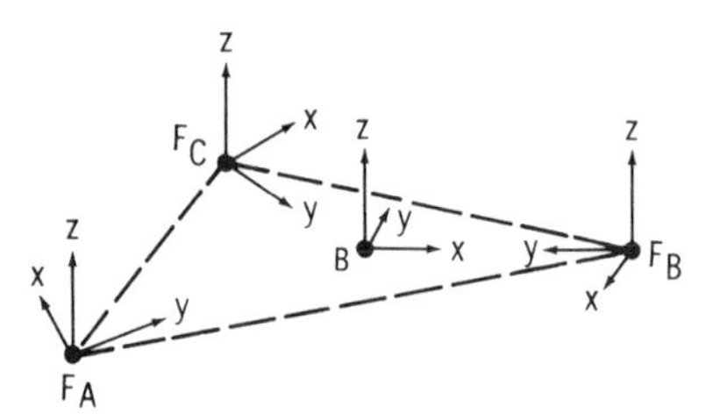

FIGURE 5.15 Axis system for BF_3.

available on an atom, they *all* become involved in the bonding scheme. Placing a double bond in one of the links (Figure 5.14b) by using one of the fluorine lone pairs shown in Figure 5.14a allows us to write in an analogous way two more equivalent structures with the double bonds in the other two links. The three resulting resonance structures involve a formal charge of -1 on the boron and a $+1/3$ charge on each fluorine. The presence of the double bond in these resonance structures implies that the second bond (which is in a π MO perpendicular to the molecular plane) is delocalized over all four atoms. The term *resonance* is taken to mean that if a single electron dash structure were to be drawn, it would be an average of the three resonance forms implied by Figure 5.14b. Such an average structure (Figure 5.14c) would have to involve a (2*pz*) VAO on each fluorine and the (B2*pz*) VAO, as indicated by the axis system in Figure 5.15. By assigning lp's to each of the (F2*pz*) VAOs in Figure 5.14a, there must be a total of six electrons in the delocalized π system since all three of the (F2*pz*) VAOs are involved in the delocalized π MOs. Because of the formal charges in Figure 5.14b and c, there is a polarization of the bonds in each of the links toward the fluorines. This is not unexpected because fluorine is more electronegative than boron. The presence of bond polarization means that we must also include the contribution of the three ionic resonance structures implied by Figure 5.16.

It is clear that the molecule must be planar since the σ bond pairs "repel" one another to 120°. Pi electron density also "repels" and its even distribution among all three links is expected to reinforce the trigonal planar geometry of the molecule. In Chapter 1 it was pointed out that double bonds are

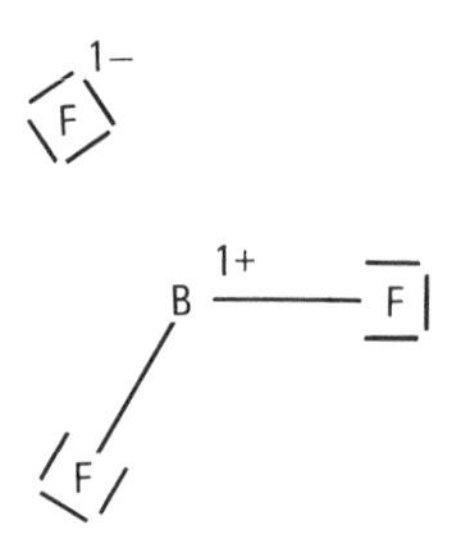

FIGURE 5.16 One of three ionic resonance structures for BF_3.

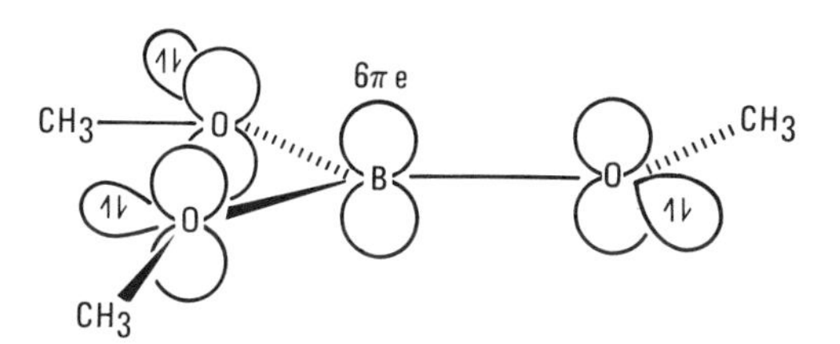

FIGURE 5.17 Structure of $B(OCH_3)_3$.

shorter than single bonds. Indeed the B—F bond in BF_3 is found experimentally to be shorter than expected for a B—F single bond. Could such bond shortening arise solely from coulombic attraction in the ionic resonance structures (Figure 5.16)? The answer is probably not, since π bonds are expected to experience restricted bond rotation while ionic interactions are not. Although restricted bond rotation is not possible to detect experimentally in BF_3, it is implied by the planar arrangement found in the BO_3C_3 portion of $B(OCH_3)_3$ (Figure 5.17), which is isoelectronic with BF_3 except that one of the lone pairs on the fluorines is converted to an O—CH_3 bond. If delocalized π bonding were not important in this molecule, a nonplanar conformation unfavorable to delocalization might be expected.

In developing the delocalized view of BF_3 we begin by assigning one lone pair on each fluorine to an (F2*s*) VAO and a second fluorine lone pair to an (F2*px*) VAO. In this way, the (F2*pz*) and (F2*py*) VAOs are reserved for the perpendicular π and the radial σ systems respectively, as shown in generator Table 5.5. For convenience, we will henceforth also indicate in the generator tables the number of VAO electrons to be accounted for over and above the preassigned lone pairs. Because the *s*-, *pz*-, *dxz*-, and *dyz*-generated MOs are analogous to those generated for N_3^+, their pictorial representations will not be repeated here. The new features in BF_3 are the pairs of SOs generated by *s*, *px*, *py*, and *pz* which we must linearly combine, and this is shown in Figure 5.18. In addition, there are the $\pi^0(dxz)$ and $\pi^0(dyz)$ MOs which are nonbonding owing to the lack of (*dxz*) and (*dyz*) VAOs on boron. The average bond

TABLE 5.5 Generator Table for BF_3 ($12e^-$)

	GOs					
VAO Equivalence Sets	*s*	*px*	*py*	*pz*	*dxz*	*dyz*
$Bs = (B2s)$	n					
$Bp\sigma = (B2px), (B2py)$		n	n			
$Bp\pi = (B2pz)$				n		
$Fpr = (F_A2py), (F_B2py), (F_C2py)$	n	n	n			
$Fp\pi = (F_A2pz), (F_B2pz), (F_C2pz)$				n	n	n

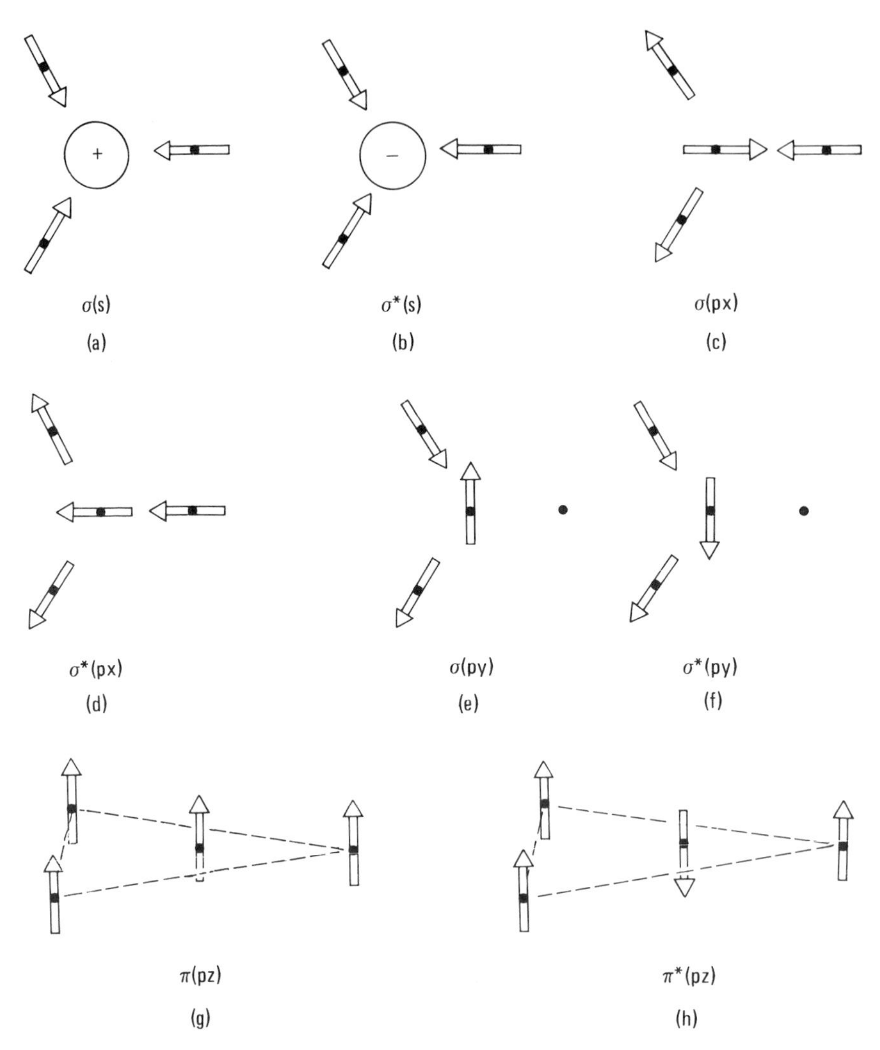

FIGURE 5.18 MOs generated by GOs in the plane [(a), (b), (c), (d), (e), (f)] and perpendicular to the plane [(g), (h)] of BF_3.

order from the average electron dash structure is expected to be 4/3 and this is confirmed by the MO energy level diagram in Figure 5.19.

Do Practice Exercise 2 in Node Game to review the steps we took in developing our delocalized view of BF_3.

For greater ease in visualizing the localized view, we choose to localize the σ and π MO sets separately. From our usual procedure, we conclude that the main lobes of an sp^2 GO set are directed toward the fluorines and they call in a set of localized MOs $\hat{\sigma}_A$, $\hat{\sigma}_B$, $\hat{\sigma}_C$, each composed mainly of an (sp^2) boron

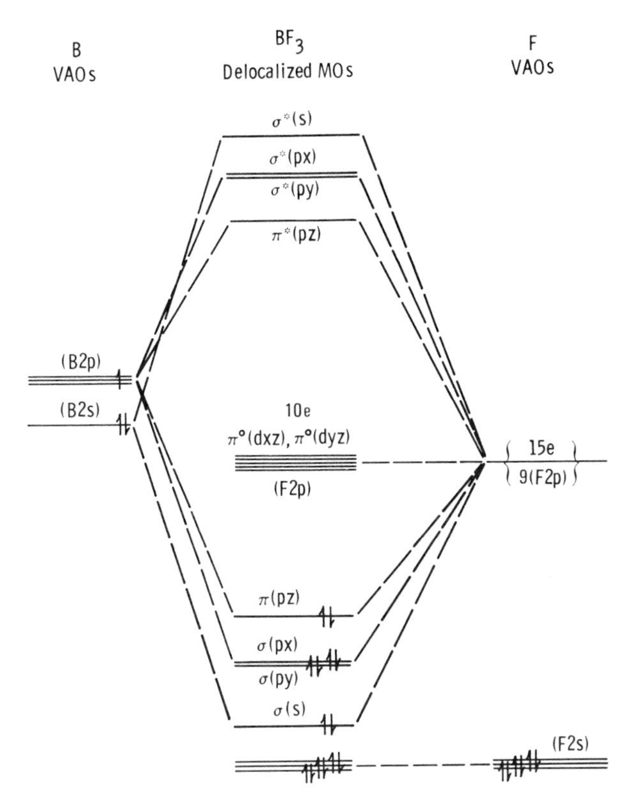

FIGURE 5.19 Delocalized MO energy level diagram for BF_3.

VAO and a fluorine (2*py*) VAO. The three π MOs in Figure 5.19 are similarly localized except that the *pzdxzdyz* GO set (of which one is depicted in Figure 5.20a) generates only the (2*pz*) VAO on boron (Figure 5.20b). As we have come to expect when the central atom does not have the full set of VAOs required by the GO hybrid set, the localized bond is partially cancelled by antibonding contributions from the other two atoms in Figure 5.20b in order to preserve orthogonality. Each localized π BMO contributes a bond order of $+1$ in one B—F link and a bond order of $-1/3$ in each of the two others, resulting in an overall π bond order of 1/3 in each link, as was found in the delocalized view. In Figure 5.21 is shown the localized MO energy level diagram. In it the fluorine lone pairs are depicted as hybrid VAOs. Such hybrids are, of course, more localized than *s* and *p* lone pairs in canonical VAOs.

Use the Homework Drawing Board in Node Game to develop the delocalized view we have just seen for BF_3. Use the steps in the appropriate column of Table 5.8 as a guide.

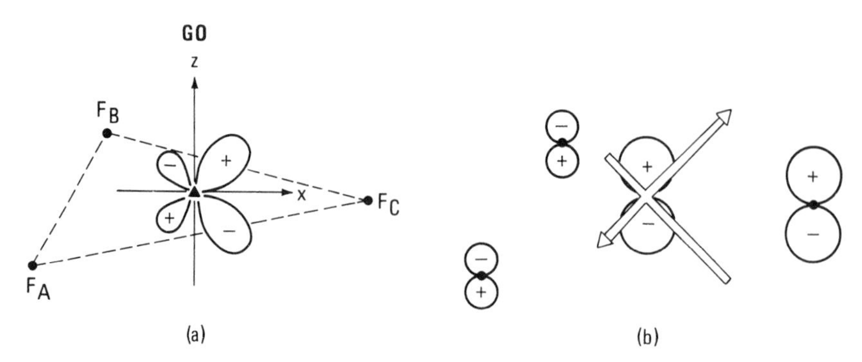

FIGURE 5.20 The *pzdxzdyz* GO in BF_3 (a) which generates a partially localized π BMO (b) from the (B2*pz*) and the (F2*pz*) VAOs. Note the "double headed" and "double tailed" notation for the pd^2 GO.

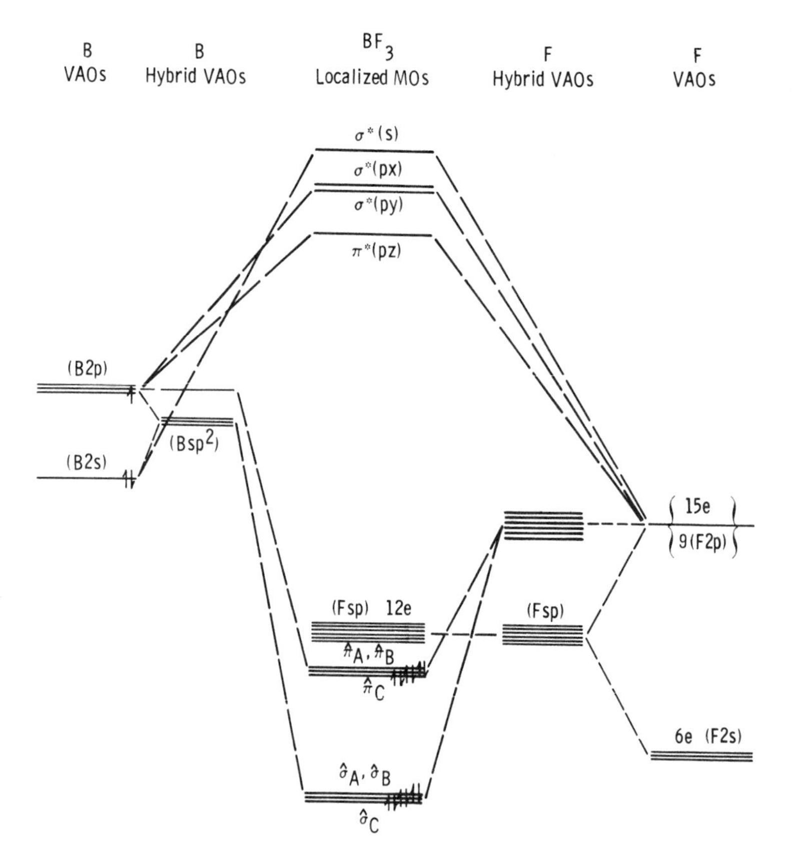

FIGURE 5.21 Localized MO energy bond diagram for BF_3.

In closing this chapter, we briefly consider the molecular motions available to a triangular molecule with and without a central atom. For the latter case these motions are summarized in Figure 5.22 and generator Table 5.6. In this and future generator tables we will frequently list only the generic symbols for the equivalence sets and not their individual members. In Figures 5.22a–c and d–f are shown the top views of GO-induced symmetry motions (SMs) among the radial and tangential AVs, respectively. In Figures 5.22g–i are shown perspective views of the SMs among the "π" AVs. It is clear from Figure 5.22 that the *s* GO leads to a vibration, that *pz* generates a translation, and that *dxz*, *dyz*, and $f3''$ each provide a rotation. Note that the larger AV contribution on atom A of Z_3 (by a factor of 2) generated by *dxz*, compared to the two smaller ones on atoms B and C in Figure 5.22h, is in agreement with the fact that the momentum of the first atom must be twice that of the other two in order for the angular momentum in the positive z direction to be equal to the sum of the angular momenta of the two in the negative z direction. Since *px* and *py* each generate two SMs, linear combinations must be taken. In Figure 5.23a is shown a superposition of the two AVs of Figures 5.22b and d as a sum and in Figure 5.23b it is shown as a difference. By taking the linear combination in Figure 5.23b with a somewhat higher (negative) contribution of the SM in Figure 5.22d, the vectorial sums of the motions on atoms B and C of Z_3 would be parallel with the movement of atom A (Figure 5.24) and we would have a translation along the $-x$ direction of the center of mass. The other linear combination (Figure 5.23a) represents a vibration that stretches the Z_B—Z_C bond while the Z_B—Z_A and Z_C—Z_A bonds are compressed. In a similar manner the linear combinations of the SMs in Figures 5.22c and e shown in Figures 5.25a and b, respectively, lead to a translation in the y direction and a vibration. The total of three vibrations we have generated is in accord with the $3n - 6$ rule. Since ν_2 and ν_3 are generated by partner GOs, their frequencies are degenerate.

If we now introduce a central atom Y into the molecule Z_3, Generator Table 5.6 acquires the added AV sets denoted by Y in Generator Table 5.7. Figure 5.22a depicts ν_1. Easily visualized (see Figures 5.18g and h) are the *pz*-generated SMs, which give rise to T_z and the vibration ν_2, respectively. Let us now focus on the three *px*- and *py*-generated SMs. From our discussion on N_3^+, we have already deduced the appearance of the SMs in the Z*y* and Z*x* AV equivalence sets generated by *px* and *py*. Figure 5.26 shows the SMs of the *py*-generated AVs in the Y*xy*, Z*y*, and Z*x* equivalence sets. From generator Table 5.7 we note that one of the linear combinations to be obtained from the (*px*) GO column must be a non-vibrational motion, namely T_x. Similarly, a T_y translation arises from the (*py*) GO. In the case of T_y, for example, it is clear that the appropriate linear combination of the AVs for this translation involves the sum in equal proportions of the SMs in Figures 5.26a and c plus a larger proportion of the negative of the SM in Figure 5.26b which serves to point the resultant motions of atoms B and C of YZ_3 precisely along the y direction. For the two vibrations of *py* symmetry, imagine

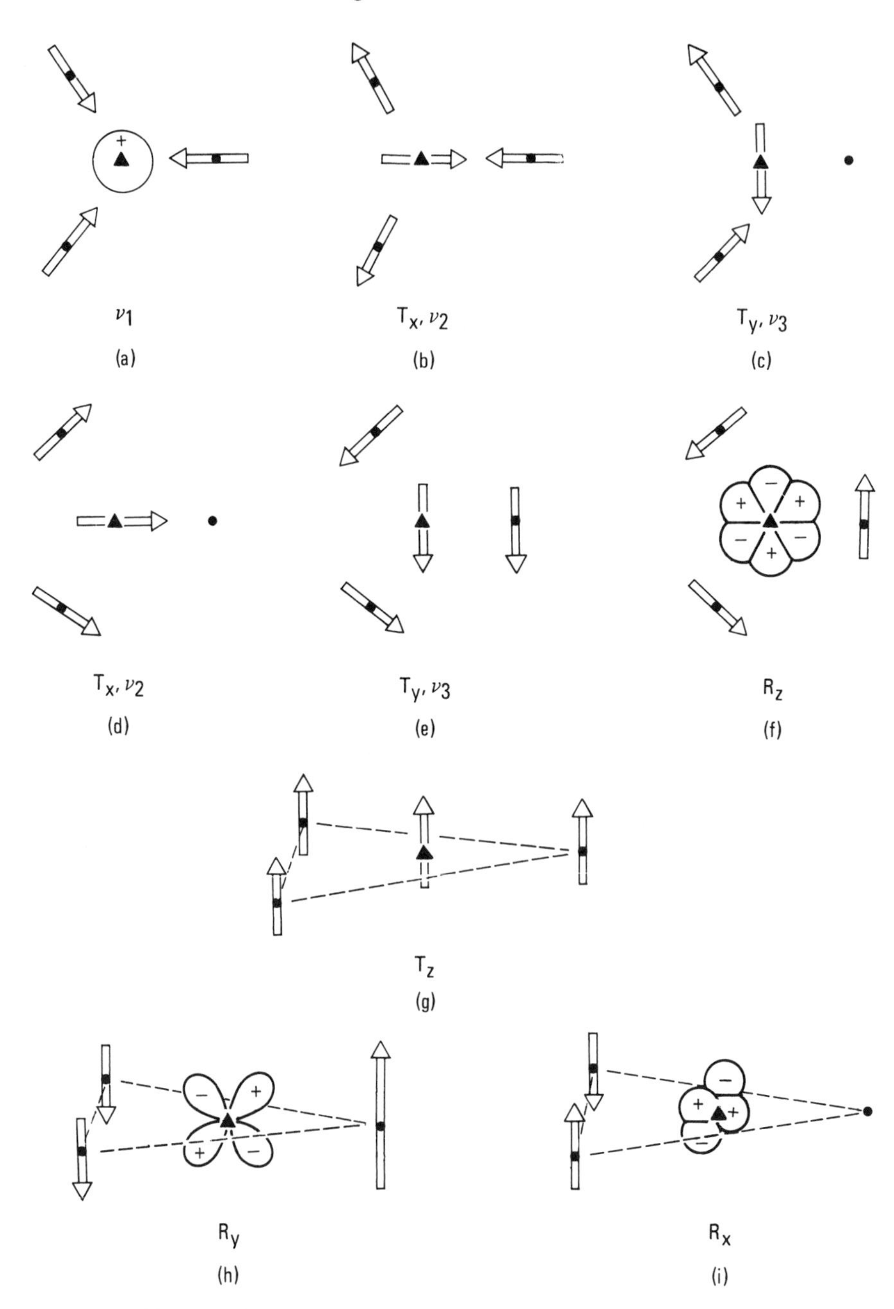

FIGURE 5.22 Molecular motions for a triangular Z_3 molecule.

TABLE 5.6 Generator Table for Triangular Z_3

	GOs						
AV Equivalence Sets	*s*	*px*	*py*	*pz*	*dxz*	*dyz*	*f*3″
Z*y*	m	m	m				
Z*x*		m	m				m
Z*z*				m	m	m	

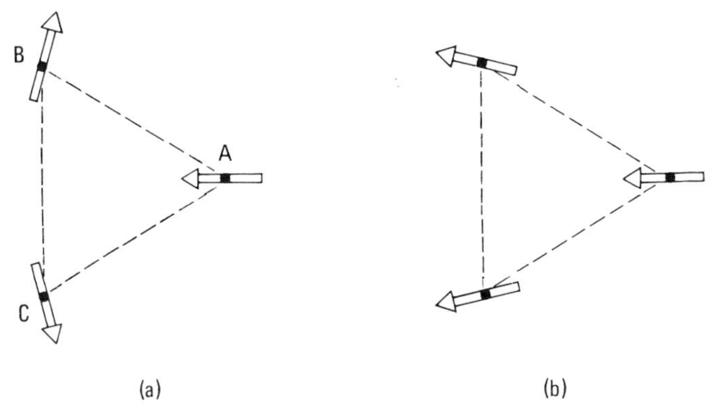

FIGURE 5.23 Linear combinations of the atomic vectors of Figure 5.22b and d. In (a) is shown the sum of the vectors and in (b) is shown the difference.

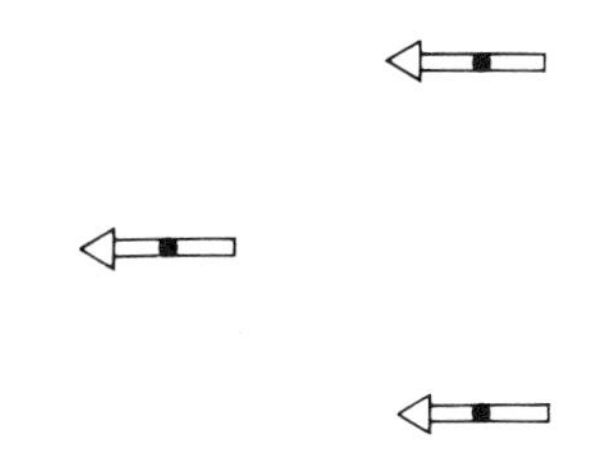

FIGURE 5.24 Linear combination shown in Figure 5.23b with a more negative contribution of the SM in Figure 5.22d.

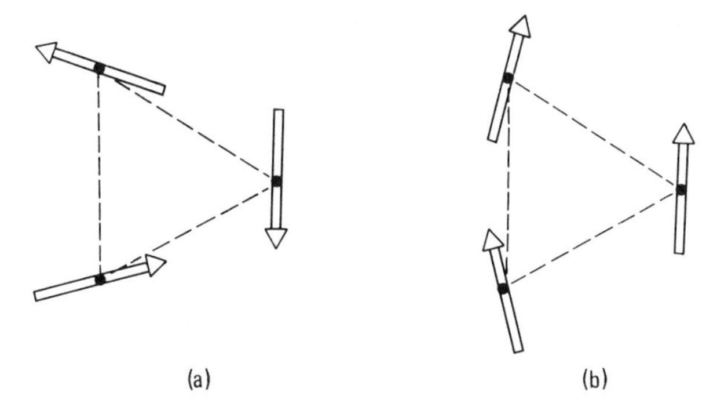

FIGURE 5.25 Linear combinations of the atomic vectors in Figures 5.22c and e. The sum is shown in (a), and the difference is shown in (b).

TABLE 5.7 Generator Table for Trigonal Planar YZ_3

	GOs						
AV Equivalence Sets	*s*	*px*	*py*	*pz*	*dxz*	*dyz*	*f*3″
Y*xy*		m	m				
Y*z*				m			
Z*y*	m	m	m				
Z*x*		m	m				m
Z*z*				m	m	m	

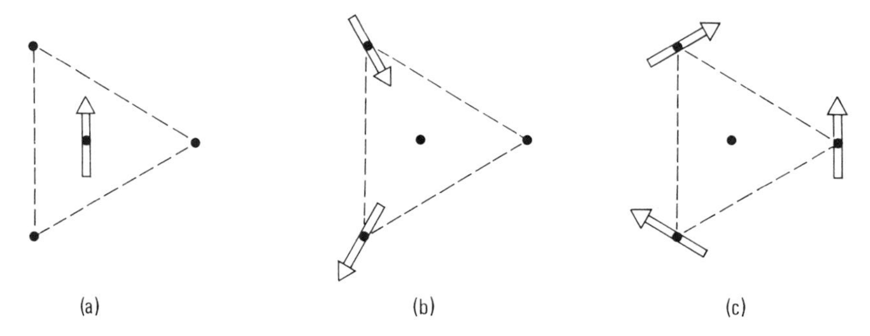

FIGURE 5.26 Symmetry motions of the *py*-generated atomic vectors in the (a) Y, (b) Z*y*, and (c) Z*x* equivalence sets of trigonal planar YZ_3.

the Y atom moving as shown in Figure 5.26a. A counteracting vibratory motion would be obtained by adding either the peripheral atom motion shown in Figure 5.26b or the negative of the one in Figure 5.26c. The precise admixture of the motions in Figures 5.26b and c with that in Figure 5.26a is determined by the relative masses and force constants. These motions give rise to ν_3 and ν_4 depicted in Figure 5.27.

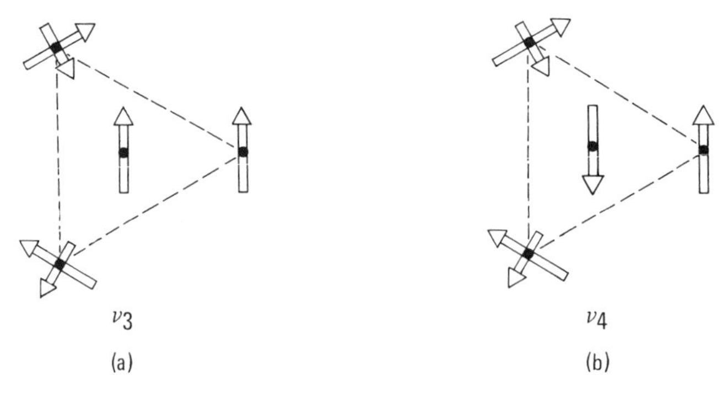

FIGURE 5.27 Linear combinations of the SMs shown in Figures 5.26a and c in which a small positive contribution of Figure 5.26b has been added.

Similar considerations of the px-generated SMs in Table 5.7 give rise to T_x, ν_5, and ν_6. The total of six vibrational nodes is expected on the basis of the $3n - 6$ formula.

Employing the steps in the appropriate column in Table 5.8 as a guide, draw the SMs of a triangular Z_3 molecule and of a trigonal YZ_3 molecule using the Homework Drawing Board of Node Game.

Summary

The new concepts introduced during our examination of the bonding in triangular and related molecules included: the determination of the coefficients of VAOs in SOs (e.g., in H_3^+) by means of simple trigonometric relationships, the recognition of partner sets of GOs and MOs, the preassignment of *sp*-out and *sp*-in hybrid VAOs on peripheral atoms to accommodate lone pairs (e.g., in N_3^+), the rehybridization of peripheral atom VAOs in order to localize bonding electron pairs between pairs of atoms (as in N_3^+), and the calculation of formal charges as an aid in deciding on the atom connections in molecules. In our discussion of BF_3, the importance of resonance and bond polarity was stressed as well as the role of π bonding in molecular planarity [e.g., $B(OCH_3)_3$]. In visualizing the molecular motions of triangular planar YZ_3 species by means of GOs, a method was developed for determining the dominant SM contributions to the molecular motions when three SMs are generated by the same GO.

Table 5.8 is an amplified version of Table 4.7. Additions are in italics.

Exercises

1. For the species below, draw electron dash structures indicating any resonance forms and formal charges. Then generate delocalized and localized views of the bonding for each. (a) Na_3^+; (b) CH_3; (c) NO_3^-; (d) $InCl_3$; (e) cyclic $C_3H_3^{2+}$ dication (Is it paramagnetic or diamagnetic?)

2. Verify pictorially that no simpler GO than $f3''$ gives an SO in Figure 5.9c.

3. Repeat Problem 1d including (d) VAOs on In for the π system of $InCl_3$.

4. Repeat Problem 1 for the BO_3 portion of $B(OCH_3)_3$. Assume that the OC and CH bonds are localized.

5. Sketch the SMs in the px GO column of Table 5.7. Show how the ν_5 and ν_6 vibrations and the T_x translation arise from these SMs.

6. $Fe(CO)_5$ is trigonal bipyramidal. Using sketches and an appropriate GO table, describe the motions of the trigonal planar $Fe(CO)_3$ moiety. (The FeCO bonds are all linear.)

TABLE 5.8 A Useful Summary of Procedures for Generating Pictorial Representations of Delocalized and Localized Bonding Views of a Molecule and for Visualizing its Vibrational Modes

Delocalized View	Localized View	Normal Modes
1. Decide on best electron dash structure *using lowest possible formal charges and drawing resonance forms. In such forms some lone pairs may be involved in a π system.*	1. Decide on which set(s) of doubly occupied delocalized MOs to localize.	1. Identify the motional atomic vector (AV) equivalence sets and list them in the generator table.
2. Establish the molecular geometry *from VSEPR theory if possible.*	2. Hybridize the GOs that give rise to the sets(s) in step 1.	2. Using GOs, draw all the symmetry motions (SMs).
3. After preassigning lone pairs to (*s*) and suitably oriented (*p*) VAOs, identify all the VAO equivalence sets (*hybrids may also be used*) and list them in the generator table.	3. Using the hybridized GOs, draw the SOs.	3. Fill in the rest of the generator table (i.e., the GOs and m designations).
4. Using GOs, draw all the SOs, recalling that #SOs = #VAOs.	4. Draw the MOs, being aware of the consequences of a central atom on which the full set of SOs (i.e., hybridized VAOs) is not generated by the hybridized GO set. *If necessary, rehybridize peripheral atom VAOs to locate bond pairs between atoms.*	4. Take *appropriate* linear combinations of SMs generated by the same GO and draw all the molecular motions.
5. Fill in rest of generator table (i.e., the necessary GOs and the n, a, or b designations). *Establish partner GO and SO sets.*	5. Draw the resulting MO energy level diagram and occupy the appropriate levels with electrons.	5. Identify the normal vibrational modes.
6. Take linear combinations of SOs generated by the same GO and draw all the MOs.	6. Verify that the bond order per link is the same as in the delocalized view.	
7. Draw the resulting MO energy level diagram and occupy the appropriate levels with electrons.		
8. Deduce the bond order per link.		

7. Predict the structures of the following and justify your answer on bonding considerations. (a) H_2CCHBF_2; (b) $(C_6H_5)_2BOCH_3$; (c) $N(SCF_3)_3$ (Hint: Consider (*d*) VAOs on sulfur which as a set are sufficiently reduced in energy by the very electronegative CF_3 groups.)

8. Repeat Problem 1 for the following. In generating the bonding views, treat the cyclic moiety in two parts. Assume localized bonds for any peripheral atoms. (a) Li_2H^+; (b) the hypothetical HCCH(BH) species [obtained by removing H^- from the known $HCCH(BH_2)^-$ ion].

9. The compound shown below is approximately planar [except for the R = $(CH_3)_3C$ groups] and the NCC portions of the NCR molecules are linear, lying on lines radiating away from the center of the hexagonal plane. The connecting

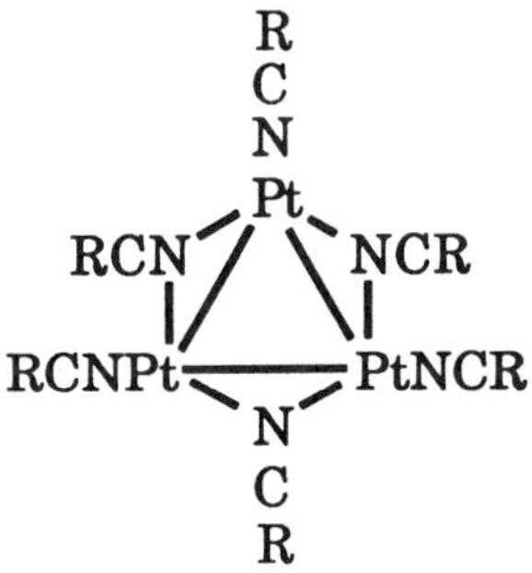

R
C
N
Pt
RCN NCR
RCNPt PtNCR
N
C
R

lines shown are meant to indicate bonding but not necessarily bond pairs. Assume the bonds in the RCN moieties are localized. Our objective is to generate the delocalized bonding pattern in the ring system. To reach this objective we will take the following steps. (a) Draw the Lewis structure of the $(CH_3)_3CCN$ molecule. (b) Assume that the electrons binding the nitrogen atoms into the ring system and to the Pt atoms come from the nitrogen lone pair. What localized orbital will you choose for this lone pair? (c) Assume that all ten valence electrons on each platinum remain in the (*d*) VAO set. Identify the empty VAOs available on each platinum. (d) Which of these VAOs can become involved in MOs lying in the plane of the ring? (e) Assume that the N lone pair donated to Pt uses a (6*sp*) hybrid Pt VAO. (f) Pictorially generate the delocalized MOs arising from the three oppositely directed (6*sp*) VAOs. (g) Pictorially generate the MOs arising from the three N lone-pair orbitals and the three tangential Pt VAOs. (h) Sketch a delocalized MO energy level diagram for the ring system and occupy the MOs appropriately with the electrons available. (i) How many electrons bind the three-membered Pt_3 ring? (j) How many electrons bind the six-membered Pt_3N_3 ring?

10. The number of CO stretching frequencies are often used to characterize organometallic complexes of the type shown below. (See Problem 14 in Chapter 4 for more information.) Draw the permitted CO stretching modes.

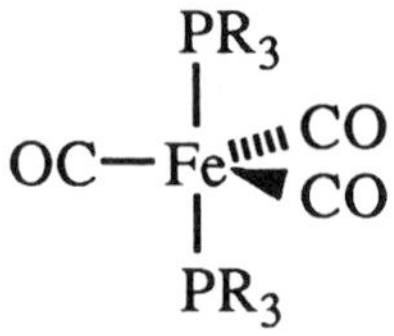

11. The unique oxygen atom, the three metals and the oxygens of the water molecules in $[M_3(O_2CCH_3)_6(O)(OH_2)_3]^{1+}$ (M = trivalent V, Cr, Mn, Fe, Co, Ru, Rh, Ir) form a trigonal plane as shown below. Each acetate ion (shown schematically in the structure of the complex, and separately as an average resonance form) bridges two metals via lone pairs from their oxygens, so that each metal is surrounded octahedrally by a total of six oxygens. Develop the delocalized and localized views of the central M_3O portion of the ion. Show the atom in which the formal positive charge resides.

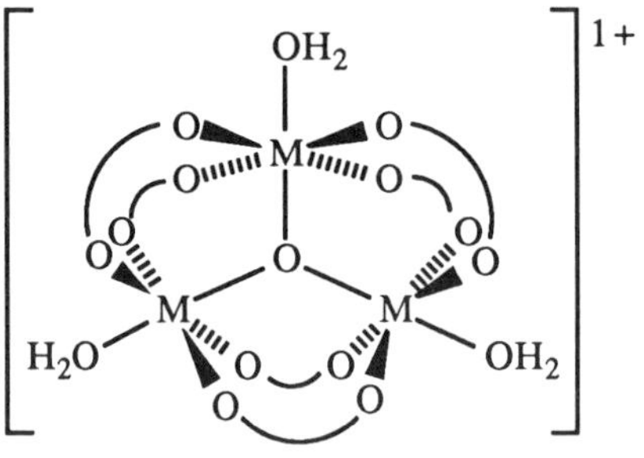

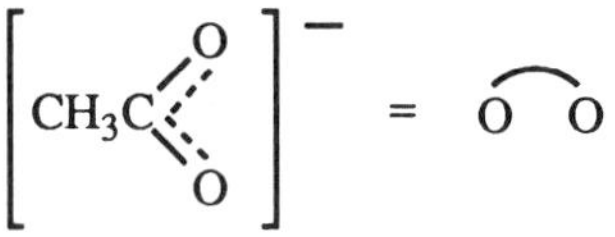

12. Generate the localized and delocalized view of the CuC_3 bonding in trigonal planar $Cu(CO)_3{}^+$. Include valence (d) orbital participation from the metal.

13. Develop a delocalized view for the planar molecule $C(CH_2)_3$ that possesses two unpaired electrons.

Chapter 6

Bent Triatomic Molecules

Among the more common bent triatomic molecules are HCH, HOH, HSH, HSeH, HTeH, ONO, ONO^-, O_3, FOF, $I_3{}^+$, FBrF, and $OClO^-$. Some molecules are bent for reasons that are not obvious since they disobey VSEPR theory. In this category we have BaH_2 (whereas BeH_2 is linear), Na_2O (whereas Li_2O is linear) and TiF_2 (whereas $CrCl_2$, $MnCl_2$, $FeCl_2$, $CoCl_2$, and $NiCl_2$ are linear). Furthermore, MgF_2, CeO_2, ZrO_2, and TiO_2 are also bent. In this chapter we treat two well understood examples: one without π bonding (H_2O) and one with π bonding ($NO_2{}^-$).

6.1. H_2O

The electron dash structure of H_2O in Figure 6.1 tells us that there are four electron pairs around oxygen in a localized view. Four electron pairs can maximally repel one another by arranging themselves at the corners of a tetrahedron in which the angles are 109°28′. As mentioned in the previous chapter, we expect the bond angle in the chalcogen hydrides HChH (Ch = O, S, Se, Te) to be less than the tetrahedral angle since according to the VSEPR concept, lp–lp repulsions exceed bp–bp repulsions.

To develop the delocalized view of H_2O we begin by assigning the two oxygen lone pairs to the (O2*s*) and (O2*px*) VAOs in the axis system shown in Figure 6.2. The reason for choosing (O2*px*) here is that of the three (*p*) orbitals on oxygen, only (O2*px*) can be hybridized with (O2*s*) to localize the oxygen lone pairs in (*sp*) hybrid VAOs that do not lie in the same plane as the O—H links. Another rationale for choosing (O2*px*) for an oxygen lone pair is that of the three oxygen (2*p*) VAOs, only (O2*px*) does not have the proper symmetry to interact with the (H1*s*) VAOs. Mixing (O2*py*) or (O2*pz*) with (O2*s*) would force the axis of the digonal lone pairs to be in the molecular plane and this is not expected on the basis of VSEPR theory. In Figure 6.2 it will be noted that the GO center (indicated by the triangle) is not coincident with the oxygen. This is because the oxygen is not the center of the molecule, since a point in the molecular plane directly below the oxygen on the *z* axis

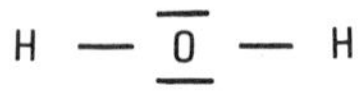

H — O — H

FIGURE 6.1 Electron dash structure for H_2O.

in Figure 6.2, for example, does not feel the same potential as it would at the same distance directly above the oxygen along the same axis in the molecular plane. It is clear that any GO location we decide upon should be equidistant from the two hydrogens and any point along the z axis meets this requirement. We also want the GO center placed such that an as yet undetermined set of GO hybrids on it (for the localized view) can point between the O and H atoms. This can be conveniently accomplished by locating the GO anywhere on the z axis between the oxygen and a line connecting the two hydrogens. It is this type of reasoning we will always use to obtain our GO center in future examples. In generator Table 6.1 are listed the three VAO equivalence sets we have available for bonding. The SOs are shown in Figure 6.3 and the linear combinations of SOs generated by the same GOs are depicted in Figure 6.4. The delocalized MO energy level diagram in Figure 6.5 tells us that the bond order is 1.0, as is expected on the basis of the electron dash structure.

Do Practice Exercise 3 in Node Game to do a "hands on" review of the delocalized view of H_2O we have developed. Use the steps in the appropriate column of Table 6.4 as a guide.

The localized view of H_2O is obtained as usual by using hybrids of the GOs associated with the occupied delocalized BMOs. Thus spy_l and spy_r on the GO center are directed toward the O—H interatomic regions where bond pair concentration is expected. If we were to hybridize the (O2py) and (O2pz) VAOs such that they are rotated as shown in Figure 6.6, spy_l would call in (H_B1s) and (O2pz), and spy_r would call in (H_A1s) and (O2py). Each (H1s) VAO would interact with one (O2p) VAO to form a localized two-center bond. Since we have no s-character in such a pair of rotated (O2p) VAOs, they would still be 90° apart and the HOH bond angle would also be 90°. While such a view is quite useful for H_2S, H_2Se, and H_2Te wherein the

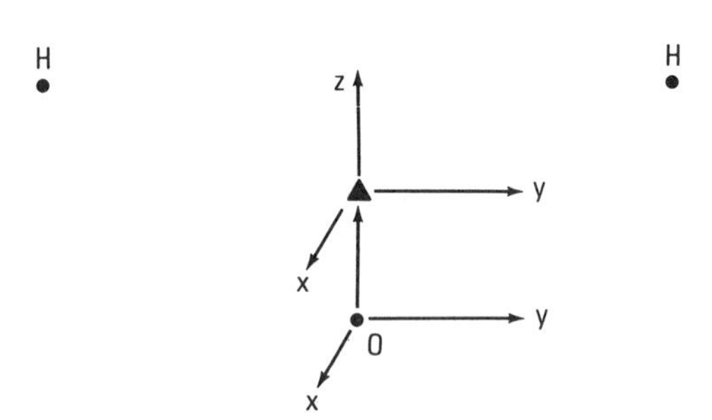

FIGURE 6.2 Axis system for H_2O. The GO center is denoted by the triangle.

TABLE 6.1 Generator Table for H_2O ($4\ e^-$)

	GOs	
VAO Equivalence Sets	s	py
$Opy = (O2py)$		n
$Opz = (O2pz)$	n	
$Hs = (H_A 1s), (H_B 1s)$	n	n

bond angles are 92°, 90°, and 90°, respectively, it leaves something to be desired in the case of H_2O for which the bond angle is 104°. By requiring less p-character in the oxygen lp's and more in the bp's this objection can be overcome, however. For example, if the oxygen VAOs are nearly sp^3 hybridized, the HOH bond angle can be nicely accommodated. That is, the hybrids could be created with somewhat more s-character in the lone pairs and a bit less in the bone pairs. This means, of course, that our original assumption that the ($O2s$) and ($O2px$) canonical atomic orbitals are reserved for lone pairs is modified somewhat. These conclusions are reflected in the localized MO energy level diagram in Figure 6.7 in which the oxygen VAOs are crudely assumed to be equivalent sp^3 orbitals.

An interesting point arises from a consideration of the relative energies of the lone pairs and the bond pairs in Figures 6.5 and 6.7. Photoelectron spectroscopy (PES) experiments confirm calculations that the first electron ionized from water is from an oxygen lone pair. These results appear to conflict with the statement made in Chapter 1 that the delocalized view accommodates spectroscopic results better than the localized view. Recall,

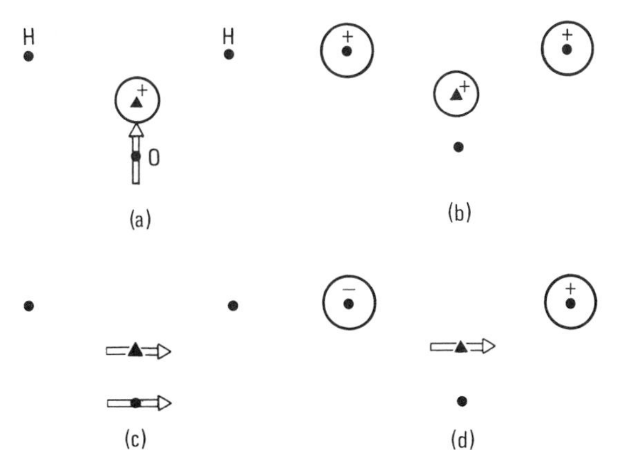

FIGURE 6.3 Sketches of the SOs of H_2O.

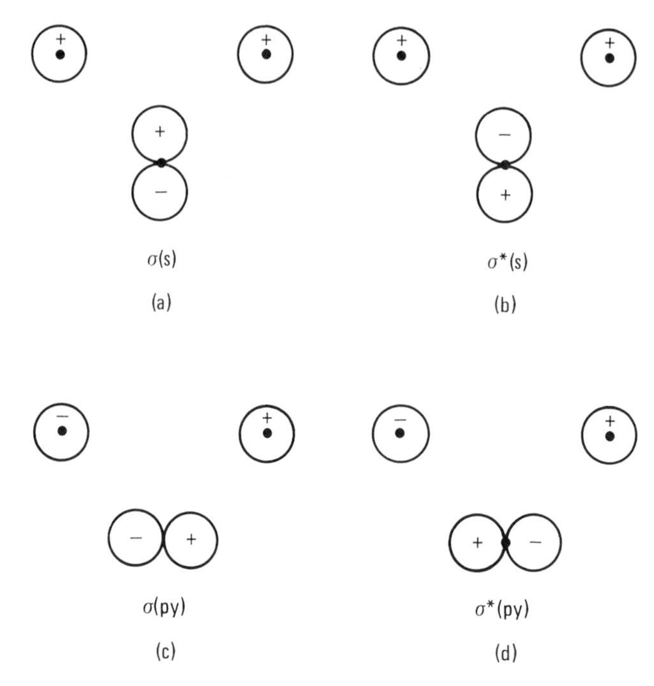

FIGURE 6.4 Sketches of MOs obtained by the linear combinations of SOs in Figure 6.3.

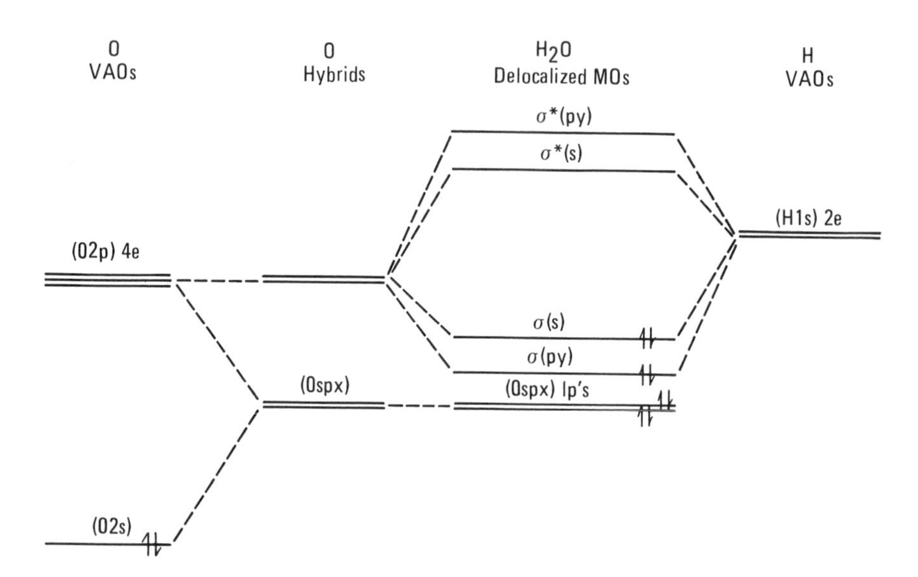

FIGURE 6.5 Delocalized energy level diagram for H_2O.

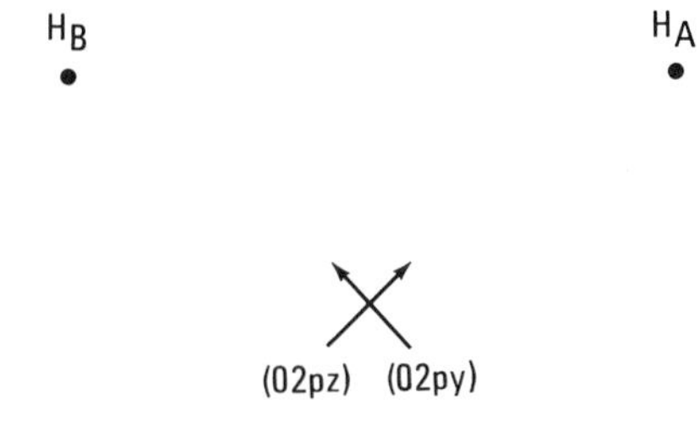

FIGURE 6.6 The result of rotating the (O2*py*) and (O2*pz*) VAOs from their positions as given in Figure 6.2 by means of hybridizing them.

however, that in Figure 6.5 we localized the lone pairs. If we leave them in the canonical VAO's, the higher-energy lone pair is in (O2*px*), which indeed lies higher in energy than the two BMOs. The (O2*s*) lone pair then lies below the two BMOs, and this has also been verified by calculations and PES.

Use the Node Game Homework Drawing Board and Table 6.4 to review this development of a localized view for H_2O.

6.2. NO_2^-

From the electron dash structure of this molecular ion in Figure 6.8a we see that two resonance structures can be drawn, of which the average is represented by the electron dash structure in Figure 6.8b. It should be noted that

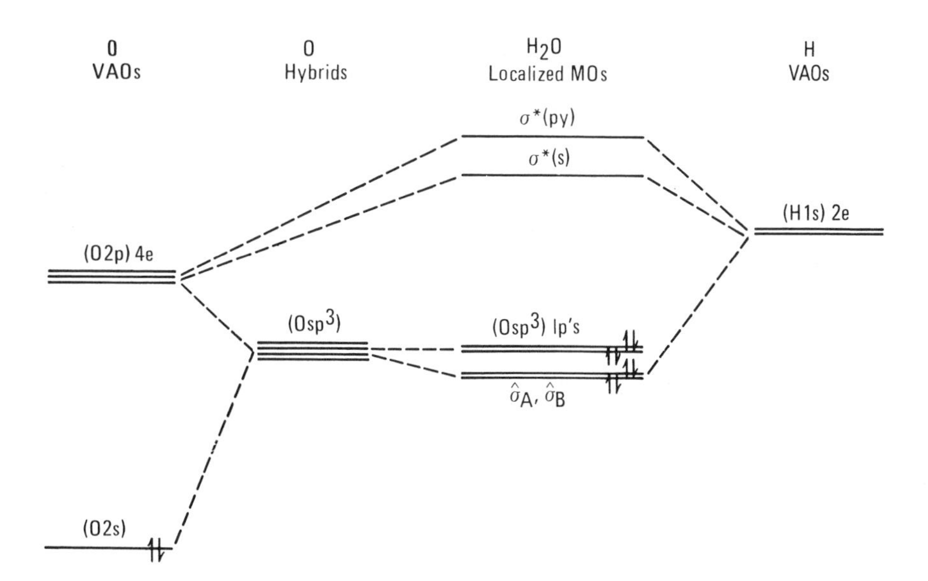

FIGURE 6.7 Localized MO energy level diagram for H_2O.

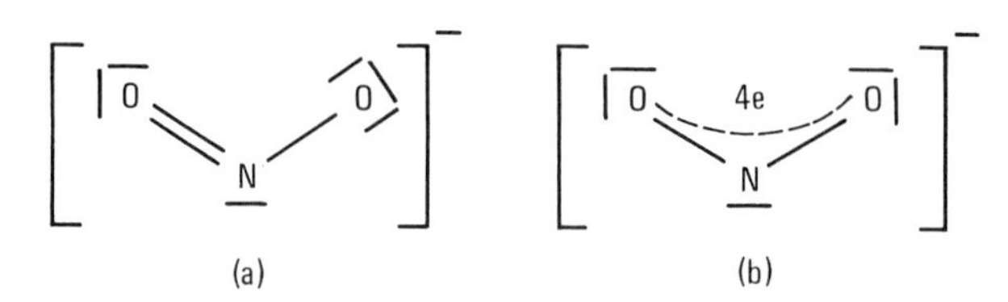

FIGURE 6.8 One resonance structure for NO_2^- (a) and a representation of an average electron dash structure (b).

in the axis system of Figure 6.9, the maximum delocalization of the π bond in NO_2^- suggested by Figure 6.8b requires the participation of the (O2*px*) VAO on each oxygen as well as the (N2*px*) VAO. This means that in addition to the π bonding electron pair in Figure 6.8a, a lone pair from the oxygen associated with the other N—O link is also engaged in the delocalized π system of Figure 6.8b. From these electron dash formulations we conclude that there is present a delocalized three-center two-electron π bond and also a three-center two-electron nonbonding π MO. We can further conclude from these electron dash structures and VSEPR considerations that NO_2^- is bent. The experimentally determined bond angle of 115° suggests that the repulsion of the nitrogen lone pair by the combination of the two σ bonds and the two half π bonds is greater than the mutual repulsion of the σ plus π bonding electron density in the two links. As we will see shortly, however, the delocalized π nonbonding electron pair is actually quite localized on the oxygens and therefore these electrons are not repelled strongly by the nitrogen lone pair.

We begin our discussion of the delocalized view of the bonding in NO_2^- by assigning lone pairs to VAOs. The oxygen lone pairs are assigned to the (O2*s*) and (O2*py*) VAOs on each oxygen in Figure 6.9 and the nitrogen lp is assigned to (N2*s*). We could, of course, localize these lp's further by placing them in a pair of digonal *spy* VAOs on each oxygen. This leaves eight electrons for the delocalized σ and π systems. The n entries under the GOs in generator Table 6.2 are obtained from the sketches in Figure 6.10. By reversing the sign of the nitrogen (or oxygen) SOs in Figures 6.10a–c, we obtain the

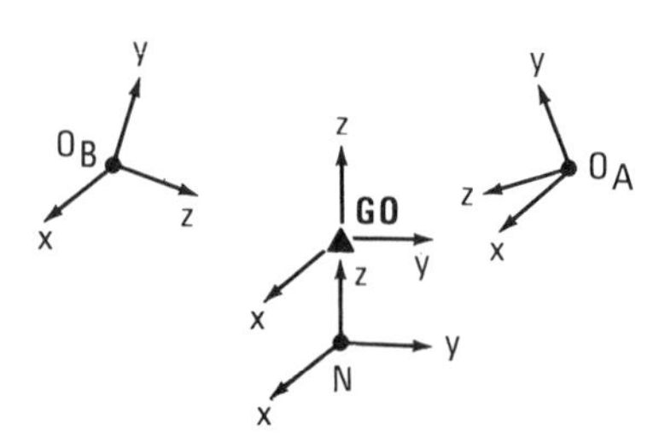

FIGURE 6.9 Axis system for NO_2^-.

TABLE 6.2 Generator Table for NO_2^- (8 e^-)

	GOs			
VAO Equivalence Sets	*s*	*px*	*py*	*dxy*
N*pr* = (N2*pz*)	n			
N*pt* = (N2*py*)			n	
N*p*π = (N2*px*)		n		
O*pr* = ($O_B 2pz$), ($O_A 2pz$)	n		n	
O*p*π = ($O_B 2px$), ($O_A 2px$)		n		n

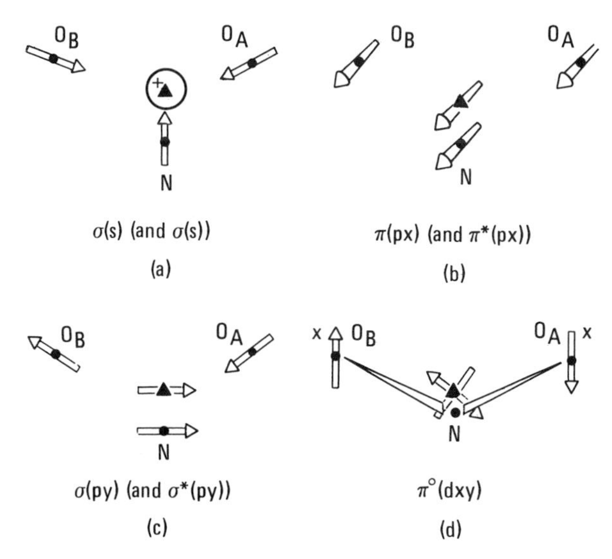

FIGURE 6.10 MOs for NO_2^- obtained by linear combination of the SOs given in Table 6.2 (a)–(c). The GOs giving rise to these MOs are also shown. The SO in (d) is a NBMO. Notice that orientation of the molecules' geometry in (d) is shown with the nitrogen coming out of the plane of the paper (as indicated by the long wedges).

antibonding counterparts of the BMOs shown in this figure. The delocalized MO energy level diagram is given in Figure 6.11 and the bond order of 1.5 agrees with that suggested by the electron dash structure. Notice that because the (*dxy*) GO is unable to call in a VAO on the nitrogen (Figure 6.10d), the electron pair density in $\pi^0(dxy)$ is localized on the oxygens (Figure 6.11).

Use Table 6.4 and the Node Game Homework Drawing Board to review the delocalized view of NO_2^- we have just developed.

The localized view of NO_2^- is conveniently obtained by using *spy* hybrid GOs to localize the two σ bonding MOs, and *pxdxy* hybrids to localize the occupied π MOs. The digonal *spy* hybrids call in contributions from the

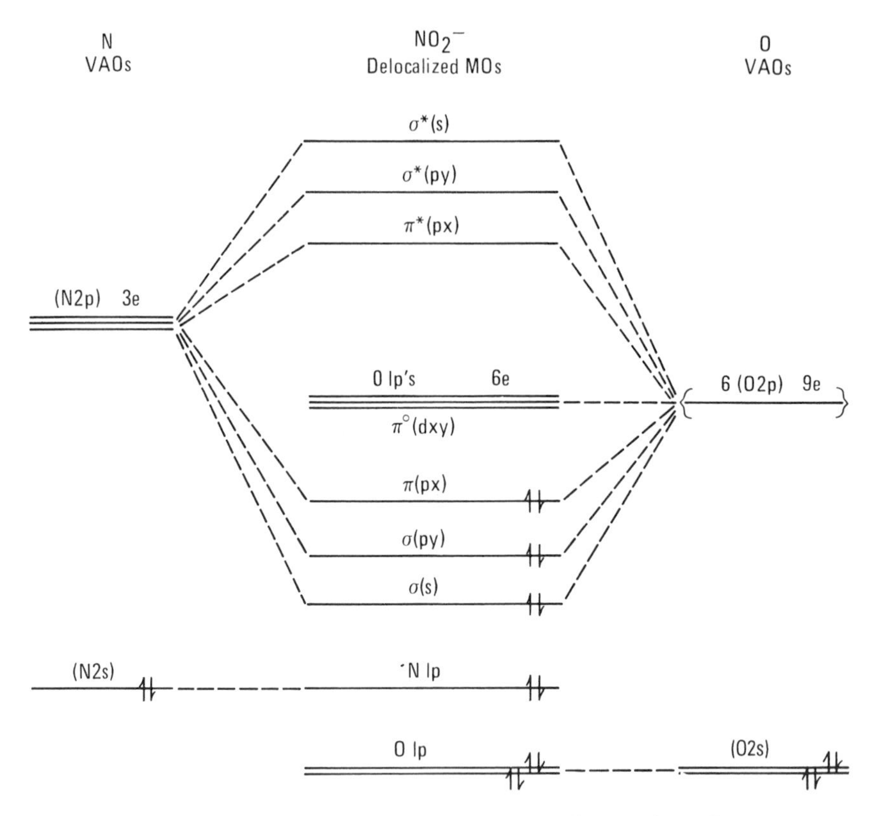

FIGURE 6.11 Delocalized MO energy level diagram for NO_2^-.

corresponding (2*pz*) VAOs on the oxygens and (N2*py*) and (N2*pz*) VAOs, which are suitably rotated via hybridization to localize electron density within the links (Figure 6.12). Increased localization of these MOs as well as of the nitrogen lone pair can be achieved by hybridizing the (N2*s*), (N2*py*), (N2*pz*) set to nearly sp^2 VAOs. The *pxdxy* hybrid GOs calls in the ($O_A 2px$) VAOs as shown in Figure 6.13, but the nitrogen VAOs respond only with (N2*px*). As

FIGURE 6.12 Localized σ MOs of NO_2^- called in by digonal *spy* GOs.

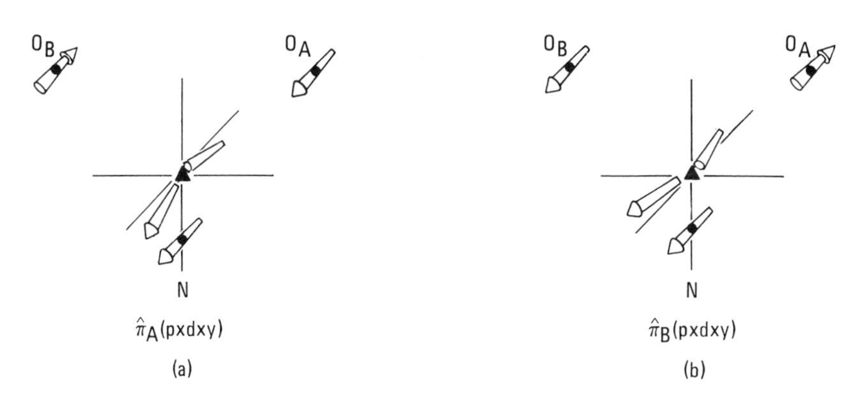

FIGURE 6.13 Localized π MOs of NO_2^- called in by the $pxdxy_r$ (a) and $pxdxy_l$ (b) hybrid GOs.

we have come to appreciate in such cases, $\hat{\pi}_A$ $(pxdxy)$ and $\hat{\pi}_B$ $(pxdxy)$ include small antibonding contributions from the ($O_B 2px$) and ($O_A 2px$) VAOs, respectively, in order to preserve orthogonality. These antibonding contributions also reduce the localized bond order of 1.0 to 0.5. In Figure 6.14 is depicted a localized MO energy level diagram.

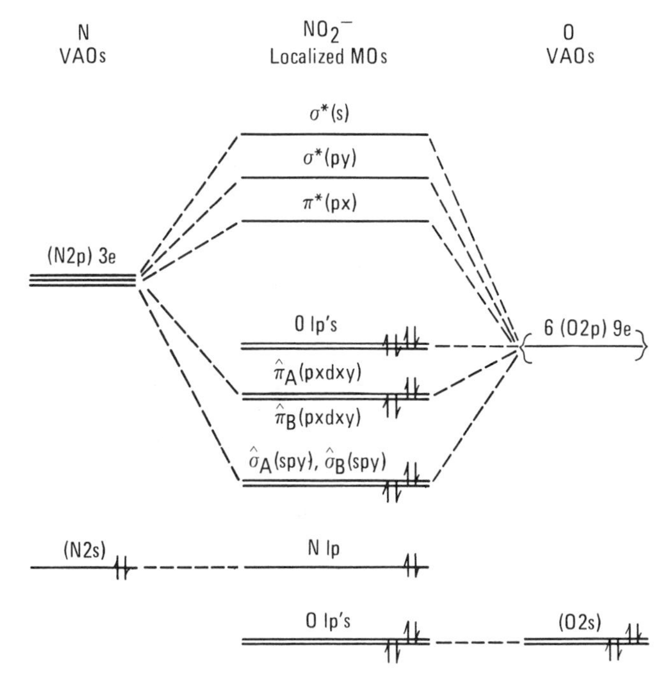

FIGURE 6.14 Localized MO energy level diagram for NO_2^-.

TABLE 6.3 Generator Table for Bent YZ_2

	GOs			
VAO Equivalence Sets	*px*	*py*	*pz*	*dxy*
Yr = Y*z*			m	
Yt = Y*y*		m		
Yπ = Y*x*	m			
Zr = Z_Az, Z_Bz		m	m	
Zt = Z_Ay, Z_By		m	m	
Zπ = Z_Ax, Z_Bx	m			m

Use the Homework Drawing Board in Node Game and the steps in the appropriate column of Table 6.4 to develop our localized view of the bonding in NO_2^-.

The molecular motions of a bent YZ_2 molecule are summarized in generator Table 6.3. For these motions, the GO center represents the center of mass of the molecule. It should be noted that while generator Table 6.2 contains an *s* but not a *pz* GO, the opposite is the case in generator Table 6.3. This is because in the latter table the Z tangential AVs are included, whereas the oxygen tangential AOs in NO_2^- contained preassigned lone pairs. Since *s* generates SMs with the Yr and Zr AV equivalence sets, but *s* does not generate an SM with Zt (whereas *pz* does), we use *pz* instead of *s*. Thus since *pz* generates more SMs than *s*, we use *pz*. All the SMs are drawn in Figure 6.15 wherein we mentally substitute Y and Z atoms for N and O, respectively. In Figure 6.15a we see that the *px* GO calls in the Y*x* AV on nitrogen and $+Z_Ax + Z_Bx$ in the partner Z_Ax, Z_Bx AV equivalence set on the Z atoms (see generator Table 6.3). This linear combination, which is a sum of these two symmetry motions, gives rise to translation along *x*. By taking the difference between these symmetry motions (e.g., by reversing the sign of Y*x*), a rotation around the *y* axis of the center of mass is obtained. In Figure 6.3b, *py* is seen to call in $+Yy$ and the two partner sets on the Z atoms as $-Z_Ay + Z_By$ and $Z_Az - Z_Bz$. Since we have three symmetry motions here, three molecular motions must be produced. One of these will shortly be seen to be a rotation and another a translation. Because one GO cannot give rise to more than one rotation or one translation, the third motion must be a vibration. The molecular rotation around *x* can be seen to arise from the linear combination $+Yy + Z_Ay - Z_By$, which represents the counterclockwise rotation of each atom at the same speed around the center of mass. In this molecular motion there is no contribution from Z_Az, Z_Bz. The translation along the GO direction is easily visualized by sketching $Yy + a[-Z_Az + Z_Bz] + b[-Z_Ay + Z_By]$, where the contribution of the middle term (given by *a*) exceeds that of *b* in the third term so that the vector resultants are parallel to the GO (and to Y*y*). The precise ratio of the contributions (coefficients *a* and *b*) of the terms associated with the Z_Ay, Z_By and

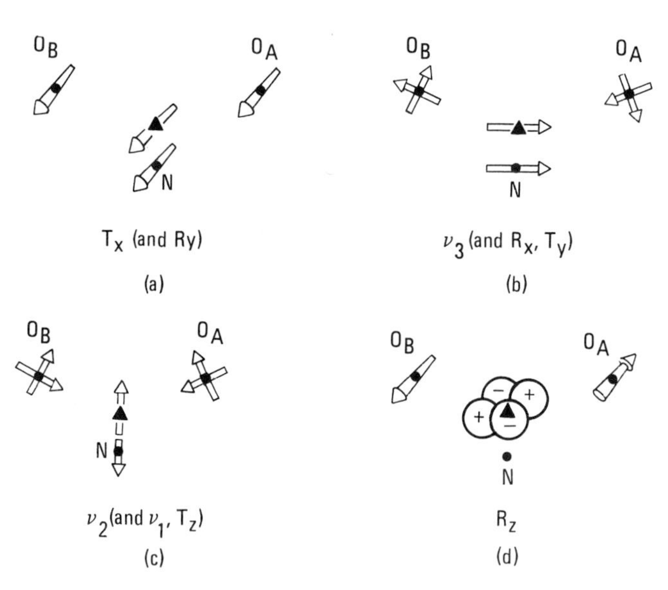

FIGURE 6.15 Sketches of the SMs given in Table 6.3 for NO_2, an example of a bent YZ_2 molecule.

$Z_A z$, $Z_B z$ AV equivalence sets in this molecular motion depends on the bond angle. Only for an angle of 90° will it be 1 : 1. The third molecular motion is some linear combination of the symmetry motions that represent a vibration. A requirement of all vibrations is that the center of mass remain stationary and that there be no net angular momentum of the molecule (i.e., no net rotational motion) during any part of the vibration. Thus if N moves in the direction shown in Figure 6.15b, then we must have a contribution from $Z_A z - Z_B z$ to counteract the motion of Y in the y direction. Depending on the relative masses and force constants there can be a contribution from $Z_A y$ and $Z_B y$ that directs the motions of these atoms either downward toward or upward from the mass center. A contribution from $+Z_A y - Z_B y$ is ruled out, however, by the fact that it would impart a rotation around x. The resulting vibration, which can now be sketched, is an unsymmetrical stretching mode (ν_3 in the parlance of the spectroscopist).

In Figure 6.15c it is easy to see what combinations of the three symmetry motions depicted there give the translation along the z direction of the center of mass. Since no combination gives a rotation, the other two combinations must be vibrations. Allowing Y to move in the direction shown in Figure 6.15c demands counteracting movements from the Z atoms with differing contributions of their two symmetry motions that depend upon the relative masses and force constants. Generally, the relative contributions are such that one of the vibrations is a nearly pure bond bending mode, ν_2, and the other a nearly pure bond stretching mode, ν_3 (i.e., the symmetrical stretch called ν_1 by spectroscopic convention). The rotation around z of the center of mass is shown in Figure 6.15d. From the $3n - 6$ formula we expect three vibrational modes, as indeed we have found. Note that since none of the p GOs are partners in the bent triatomic geometry, all the vibrations generated by them are nondegenerate and will have different frequencies.

Use the Homework Drawing Board in Node Game and the steps in the appropriate column of Table 6.4 to generate the SMs of a bent YZ_2 molecule.

Summary

In examining H_2O and NO_2^- we became aware of the convenience of placing our GO center between the middle atom and the pair of atoms bound to it on a line that bisects the angle in the molecular plane. In the localized view of H_2O we saw how VAOs on the oxygen could be hybridized to form highly localized two-center two-electron bonds to the hydrogens. A feature of NO_2^- is the possibility of resonance, which leads to a delocalized π system. Finally, the molecular motions of bent YZ_2 molecules were clarified by the GO method.

An updated recipe for using the GO approach to bonding and molecular motions appears in Table 6.4.

Exercises

1. Draw electron dash structures (indicating formal charges) and develop the delocalized and localized bonding views of (a) O_3; (b) H_2Te (90° angle), (c) OF_2; (d) ClO_2^- and ICl_2^+.
2. From their delocalized MO energy level diagrams, rationalize why the bond distance in NO decreases when it is ionized to NO^+ but the bond lengths are not significantly changed on going from NO_2^- to NO_2.
3. Rationalize why the bond angle in NO_2^- (115°) increases to 134° in NO_2.
4. When a positive species, such as a proton, reacts with water, it attacks the middle atom in the molecule, but when such a reaction occurs with NO_2^-, one of the end atoms is attacked. Account for these phenomena by considering the delocalized bonding views of these substrates.
5. The reaction of Na + S_8 in $[(CH_3)_2N]P{=}O$ solvent gives the blue S_3^{2-} ion. Draw the electron dash structure for this ion, draw its structure, and develop a delocalized and localized view for it.
6. Develop the delocalized and localized views of the SSeS framework of $Se(SCN)_2$. Treat the bonding in the SCN moieties as localized.
7. Develop the delocalized and localized views of bent NH_2^-, NH_2 and NH_2^+, all of which are known.
8. Thermal decomposition of cyclopropane $[(CH_2)_3]$ leads to the diradical $H_2CCH_2CH_2$, which has two unpaired electrons. Using VSEPR theory, draw its

Table 6.4 A Useful Summary of Procedures for Generating Pictorial Representations of Delocalized and Localized Bonding Views of a Molecule and for Visualizing its Vibrational Modes

Delocalized View	Localized View	Normal Modes
1. Decide on best electron dash structure using lowest possible formal charges and drawing resonance forms. In such forms, some lone pairs may be involved in a π system.	1. Decide on which set(s) of doubly occupied delocalized MOs to localize.	1. Identify the motional atomic vector (AV) equivalence sets and list them in the generator table.
2. Establish the molecular geometry from VSEPR theory if possible.	2. Hybridize the GOs which give rise to the sets(s) in step 1.	2. Using GOs draw all the symmetry motions (SMs).
3. After preassigning lone pairs to (*s*) and suitably oriented (*p*) VAOs, identify all the VAO equivalence sets (hybrids may also be used), and list them in the generator table.	3. Using the hybridized GOs, draw the SOs.	3. Fill in the rest of the generator table (i.e., the GOs and m designations).
4. Using GOs *located at a molecular center or centroid*, draw all the SOs, recalling that #SOs = #VAOs.	4. Draw MOs, being aware of the consequences of a central atom on which the full set of SOs (i.e., hybridized VAOs) is not generated by the hybridized GO set. If necessary, rehybridize peripheral atom VAOs to locate bp's between atoms.	4. Take appropriate linear combinations of SMs generated by the same GO and draw all the molecular motions.
5. Fill in rest of generator table (i.e., the necessary GOs and the n, a, or b designations). Establish partner GO and SO sets.	5. Draw the resulting MO energy level diagram and occupy the appropriate levels with electrons.	5. Identify the normal vibrational modes.
6. Take linear combinations of SOs generated by the same GO and draw all the MOs.	6. Verify that the bond order per link is the same as in the delocalized view.	
7. Draw the resulting MO energy level diagram and occupy the appropriate levels with electrons.		
8. Deduce the bond order per link.		

structure and indicate the geometry of each carbon. In what type of orbital(s) would you expect to find each unpaired electron?

9. Calculations show that LiH_2 and LiH_2^+ are bent. Why is a bent configuration preferred to a linear one? (Recall H_3^+.) Develop delocalized views of these molecules.

10. Draw the permitted CO stretching modes of the complex below. (See Problem 14 in Chapter 4 for more information.)

$P(C_6H_5)_3$
$(C_6H_5)_3P$ W CO
$(C_6H_5)_3P$ CO
$P(C_6H_5)_3$

11. Water molecules lose an electron in an electrical discharge giving H_2O^+, which also probably exists in comet tails and interstellar space. The bond angle in H_2O^+ has been calculated to be 5.4° larger than in H_2O. Comment on this observation and develop the delocalized view of H_2O^+.

12. When $KAsF_6$ is subjected to gamma rays, bent F_3^{2-} ions are detected. Develop a delocalized view of this ion and comment on its magnetism (i.e., whether it is diamagnetic or paramagnetic). (Hint: You will want to utilize a σ^* ABMO.)

13. When $F_2C{=}CF_2$ is photolyzed, singlet state (no unpaired electrons) CF_2 is formed whereas when CF_4 is subjected to a glow discharge, triplet (two unpaired electrons) CF_2 is formed. In one of these experiments the FCF bond angle is 134.8°, while in the other it is 104.9°. Develop delocalized views for these two geometries of CF_2 and suggest which geometry corresponds to the singlet and which to triplet state.

CHAPTER 7

Polygonal Molecules

Examples of ring compounds in organic and inorganic chemistry are legion. Rings that have more than three atoms and are saturated (i.e., have no double bonds) generally possess a puckered ring structure with roughly tetrahedral bond angles, in accord with VSEPR considerations. The members of such rings are connected by two σ bonds only and each member has in addition two lone pairs (e.g., S_8), a lone pair and a bond pair (e.g., *cyclo*-$(PPh)_5$), or two bond pairs [e.g., *cyclo*-C_6H_{12}]. The bonding in saturated systems can be qualitatively understood by realizing that each three-atom fragment is a bent triatomic molecule. An important class of ring molecules contains planar species of which we have already examined two members, namely, H_3^+ and N_3^+. Several members of the polygonal class are particularly interesting because they possess delocalized π bonds. We have already examined one example of such a molecule, namely, N_3^+. Now we address ourselves to ring systems with more than three members, which lie in the same plane *only* because of delocalized π bonding.

7.1. Te_4^{2+}

Four different electron dash structures (Figure 7.1) can be drawn for this deep purple cation. The acyclic structure in Figure 7.1a is unreasonable in view of the adjacent charges. As is implied by one of the four resonance structures in Figure 7.1b, however, both the charges and the π bonding are more effectively delocalized around a cyclic four-membered ring. The average electron dash structure for these resonance forms is shown in Figure 7.1c. Although the structure in Figure 7.1d is also cyclic, the π bond in it is restricted to the position shown since migration of the double bond to any other position would bring the nonplanar trisubstituted Te atom into a planar configuration with its three substituents (Figure 7.1e). *It is a cardinal rule in drawing resonance structures that only electrons may be shifted and not atoms.* Therefore, Figures 7.1d and 7.1e depict two electron dash structures that are not resonance structures even though both of them display the same atom

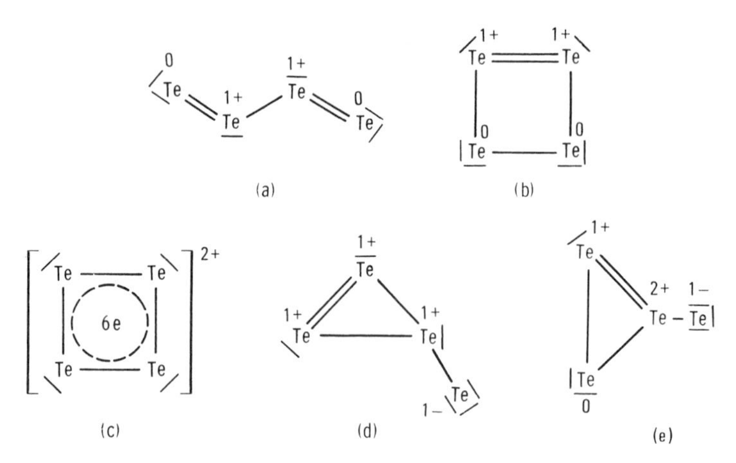

FIGURE 7.1 Possible electron dash structures for Te_4^{2+} [(a), (b), (d) (e)]. In (c) is shown an average electron dash structure of the resonance forms implied by the structure in (b).

connectivity pattern. Both structures also suffer from the destabilizing presence of adjacent charges that are the same. From Figures 7.1b and c, we see that for delocalization of the π bond to occur, two lone pairs from the lower two Te atoms in the resonance structure in Figure 7.1b must be housed in nonbonding π MOs. The Te_4^{2+} ring is classified as *aromatic* since it obeys the $4n + 2$ rule. This rule says that a ring is aromatic if the total number of π electrons is equal to $4n + 2$ where n is an integer. In Te_4^{2+}, $n = 1$.

For the delocalized view of Te_4^{2+} we first assign the lone pairs to (Te5s) VAOs. Using the axis system in Figure 7.2, we construct generator Table 7.1 from the SOs generated from the various GOs depicted in Figure 7.3. These SOs are the MOs of the ring and their bonding, antibonding, and nonbonding natures become evident by examining their nodal patterns. The nodal patterns for the MOs in the LCAOs implied in Figures 7.3b and c are shown in Figures 7.4a, b, and Figures 7.4c, d, respectively. Here we assume in determining the overall bonding nature of an MO that σ interactions outweigh the in-plane interactions of the Tept equivalence set. The MOs designated as being of the π type are oriented perpendicular to the plane of the

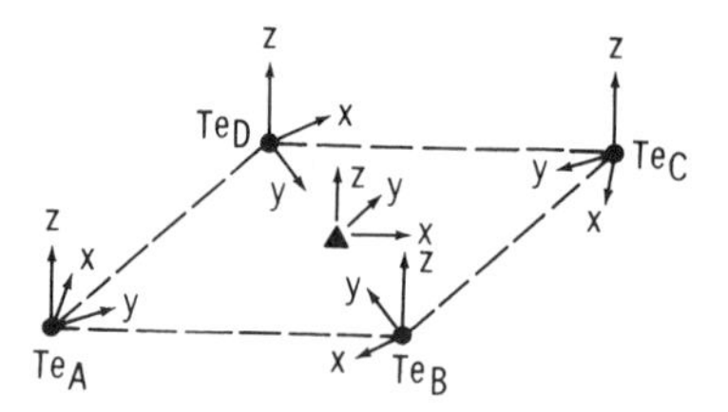

FIGURE 7.2 Axis system for Te_4^{2+}.

TABLE 7.1 Generator Table for Te_4^{2+} (14 e^-)

	GOs									
VAO Equivalence Sets	s	px	py	pz	dxy	$dx^2 - y^2$	dyz	dxz	$f2''$	$g4''$
Tepr = $(Te_A 5py)$, $(Te_B 5py)$, $(Te_C 5py)$, $(Te_D 5py)$	b	n	n		a					
Tepπ = $(Te_A 5pz)$, $(Te_B 5pz)$, $(Te_C 5pz)$, $(Te_D 5pz)$				b			n	n	a	
Tept = $(Te_A 5px)$, $(Te_B 5px)$, $(Te_C 5px)$, $(Te_D 5px)$		n	n			b				a

molecule as in N_3^+ (Figure 5.8). The MOs labeled as σ MOs are in the plane of the molecule as they were in N_3^+ (Figure 5.9). The delocalized MO energy level diagram in Figure 7.5 implies that the average bond order is 5/4 per link. Moreover, since there are six electrons in the π MOs, we have agreement with the $4n + 2$ aromaticity rule (i.e., $4n + 2 = 6$ if $n = 1$) and this π occupation is also reflected in the average electron dash diagram in Figure 7.1c.

The axis system chosen for Te_4^{2+} appears unusual in that contrary to the axis system used for N_3^+ (Chapter 5) and also for further examples of polygonal species in this chapter, neither the x nor the y axis points to an atom. Because of the fourfold symmetry of a square as well as of a pair of orthogonal axes, there is no overriding reason to point the axes toward an atom. This is because the GOs are easy to identify (i.e., no hybrid GOs are necessary) whether the x and y axes point to atoms or to the midpoints between them. The reason we chose to point these axes between the atoms is that the MOs are more delocalized in the delocalized view than if we had chosen to direct to direct x and y toward the atoms (see Problem 8).

Using the Node Game Homework Drawing Board and the steps in the appropriate column in Table 7.7, develop the above delocalized view of the bonding in Te_4^{2+}.

For the localized view let us hybridize the GOs of the σ MO set separately from those of the π MO set. The main lobes of an $spxpydx^2 - y^2$ hybrid set for the σ MOs point to the edges of the square in Figure 7.2. These lobes call in two-electron σ bonds between each pair of tellurium atoms, which involve linear combinations of the in-plane radial and tangential VAOs. Neglecting neighbor overlap, we can reorient these orthogonal in-plane (tellurium p) VAOs as shown in Figure 7.6. For the localized π bonds a $pzdxzdyz$ GO set is required. The main lobes of each member of such a set point as shown in Figure 7.7a. Together, the three members point to the corners of a trigonal prism. The orientation of these hybrids with respect to rotation about the z axis of Te_4^{2+} is arbitrary and two possible directions for the main lobes of these hybrids will be discussed. In one of these orientations, one member hybrid points its main lobes over and under an edge of the molecule (Figure 7.7b), and in the other these two lobes are pointed over and under an atom

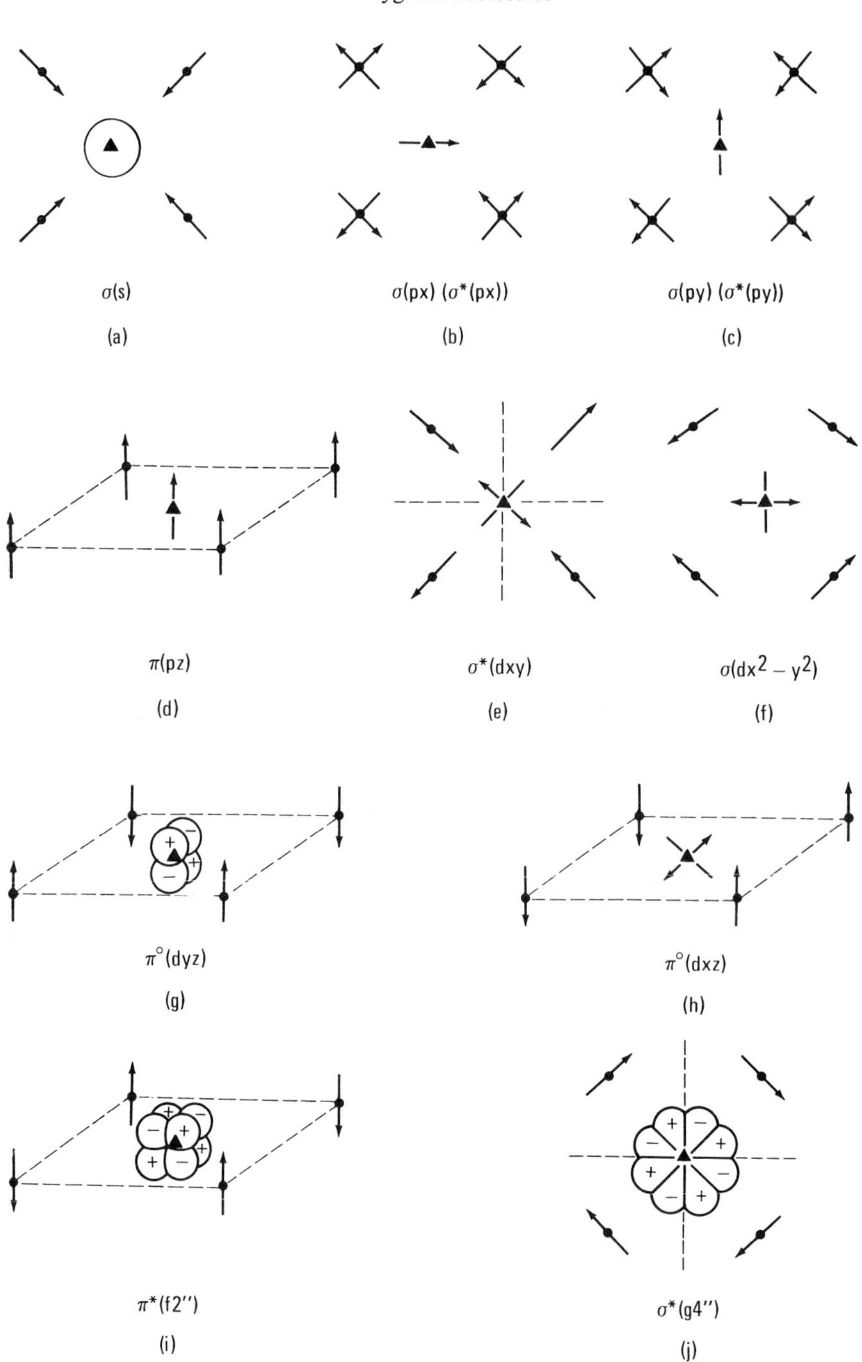

FIGURE 7.3 Drawings of the SOs of Te_4^{2+} in Table 7.1. In this and subsequent figures, the pictorial representations of *p* and *d* orbitals will often be given by two-dimensional drawings instead of three-dimensional shafts.

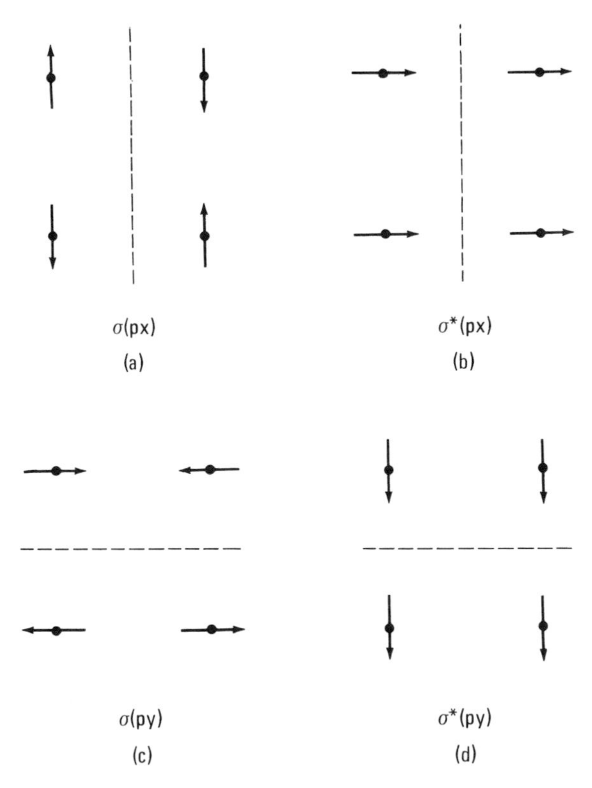

FIGURE 7.4 Drawings of the MOs obtained upon linearly combining the SOs in Figure 7.3b [(a), (b)] and in Figure 7.3c [(c), (d)].

center (Figure 7.7c). Although the localized π MOs in these two orientations are all of equal energy, they do not all have identical shapes (Figures 7.7b and c) and hence different atom contributions are involved. The localized MO energy level diagram is easy to draw and this is left as an exercise at the end of the chapter.

Use the Homework Drawing Board in Node Game and the appropriate columns of Table 7.7 to develop our localized view of the bonding in Te_4^{2+} and its SMs.

The molecular motions of a square tetra-atomic molecule such as Te_4^{2+} is summarized in generator Table 7.2. Note that these SMs look like the MOs in generator Table 7.1 and in Figures 7.3 and 7.4. The SMs corresponding to these figures are ν_1 (Figure 7.3a), ν_2 (Figure 7.4a), T_x (Figure 7.4b), ν_3 (Figure 7.4c), T_y (Figure 7.4d), T_z (Figure 7.3d), ν_4 (Figure 7.3e), ν_5 (Figure 7.3f), R_x (Figure 7.3g), R_y (Figure 7.3h), ν_6 (Figure 7.3i), and R_z (Figure 7.3j). The six frequencies found in our GO analysis are in agreement with the $3n - 6$ rule discussed earlier.

Use the Homework Drawing Board in Node Game to generate the molecular motions of Te_4^{2+}.

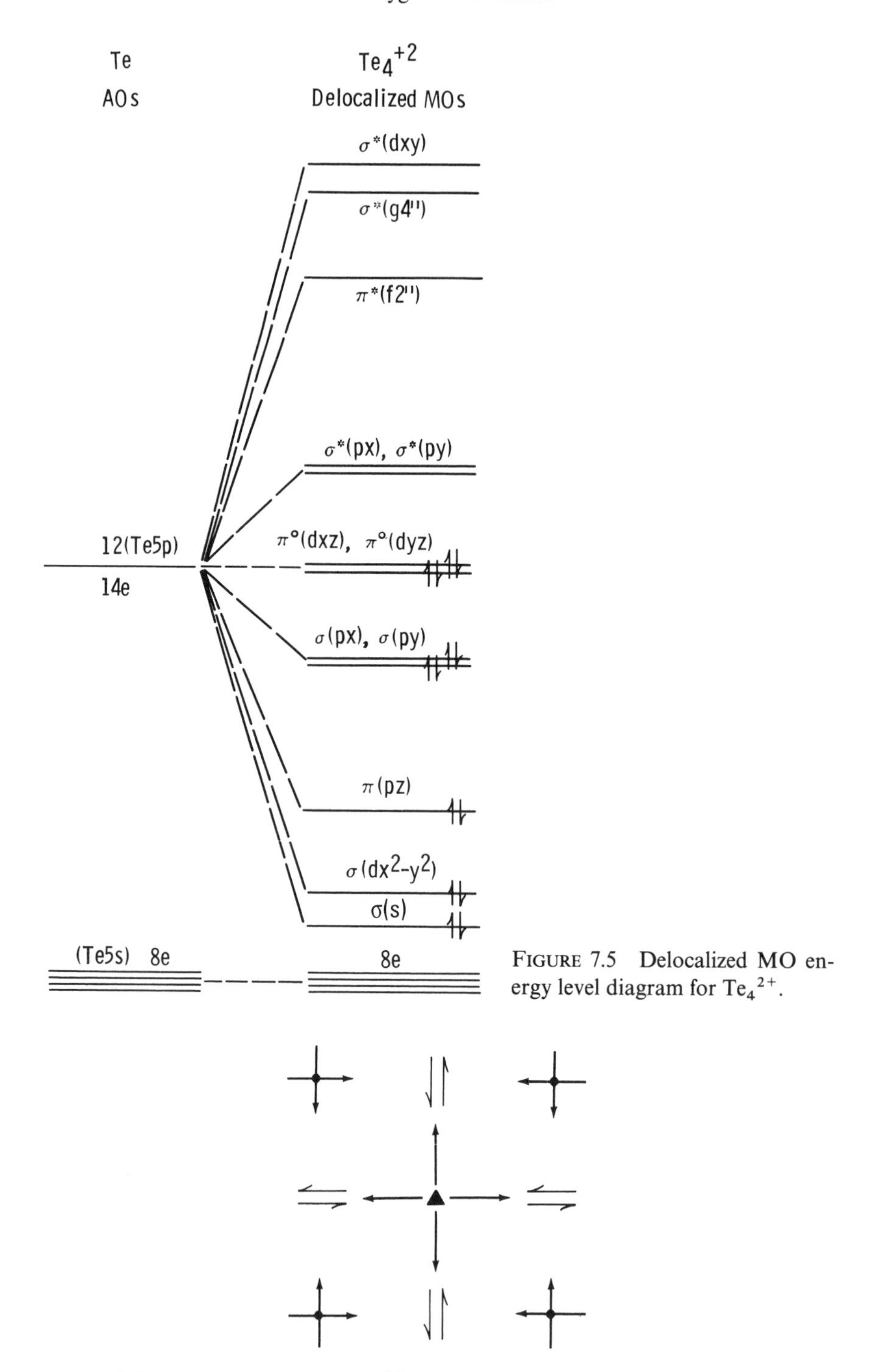

FIGURE 7.5 Delocalized MO energy level diagram for Te_4^{2+}.

FIGURE 7.6 Localized bonds in Te_4^{2+} called in among the reoriented Te (p) VAOs by the $s\ pxpydx^2 - y^2$ GO lobes.

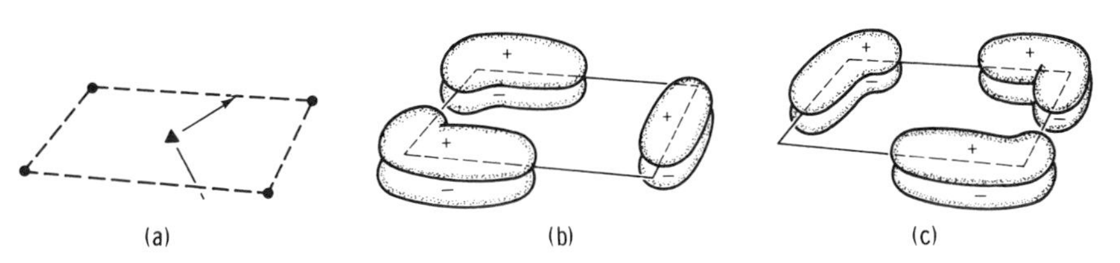

FIGURE 7.7 Main lobes of one member of a trigonal *pzdxzdyz* GO set (a). In (b) and (c) are shown two possible orientations of the three localized π MOs in Te_4^{2+}.

7.2. *cyclo*-C_4H_4

Cyclobutadiene is an unstable four-membered ring, which has been detected in a low-temperature photochemical reaction. The π delocalization in this molecule is similar to that encountered in Te_4^{2+}, but the present molecule is not aromatic. As we will see, the electronic structure of the ground state depends upon whether the molecule is square or rectangular. For the square geometry, the ground state is a triplet (two unpaired electrons) and, hence, paramagnetic. For the rectangular geometry the ground state can be a singlet (all spins paired) and, hence, diamagnetic. We first discuss the square geometry.

The σ bonding system of *cyclo*-C_4H_4 is analogous to that of Te_4^{2+}. The two species are related to each other in the same manner as the σ systems of N_3^+ and *cyclo*-$C_3H_3^+$ (Chapter 5) are related. In the present discussion the π system is the feature of interest. (We will discuss the σ system of a planar *cyclo*-$C_nH_n^x$ system in our examination of *cyclo*-$C_5H_5^-$ in the next section). In our discussion of the π MOs of square *cyclo*-C_4H_4, we will see that these MOs are qualitatively the same as those of Te_4^{2+}. The essential difference between the two molecules is that the π system of *cyclo*-C_4H_4 has two electrons less than that of Te_4^{2+} (Figure 7.8). Consequently there are only two electrons available for the degenerate NBMOs $\pi^0(dxz)$, $\pi^0(dyz)$. Hence

TABLE 7.2 Generator Table for Z_4

	GOs									
AV Equivalence Sets	*s*	*px*	*py*	*pz*	*dxy*	dx^2-y^2	*dxz*	*dyz*	*f*2″	*g*4″
Zr	m	m	m		m					
Zt		m	m			m				m
$Z\pi$				m			m	m	m	

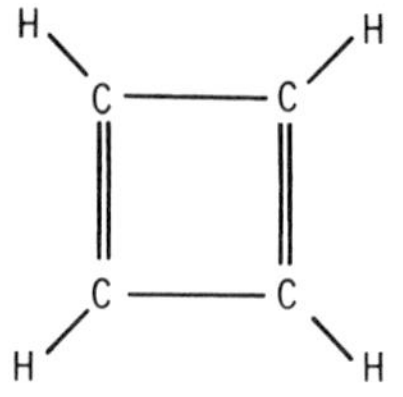

FIGURE 7.8 Electron dash structure for cyclobutadiene.

one electron will go in each NBMO and the two spins will be parallel to give a triplet ground state.

To localize our π system, we recall that only MOs with the same occupation numbers can be mixed to obtain localized MOs. Hence no further localization is possible for the one doubly occupied π BMO. The two singly occupied π NBMOs, on the other hand, are degenerate partners, and mixing them will not produce localized MOs, but only a different set of degenerate partners. Thus no really localized view is possible.

Thinking in terms of resonance structures, one would be led to a description of the π-electronic structure of *cyclo*-C_4H_4 as an equal mixture of the two structures implied by Figure 7.8. However, each of these resonance structures also implies doubly occupied orbitals only, and hence each corresponds to a singlet (diamagnetic) state. It follows that this is also true of any linear combination of the two structures. Hence the resonance description is appropriate only for the lowest-lying singlet excited state and not the triplet ground state. This argument holds for all molecules with resonance structures. In general, the resonance description corresponds to the lowest singlet state of the molecule and not to its ground state if the ground state happens to be of higher multiplicity than a singlet.

The rectangular geometry of *cyclo*-C_4H_4 can be obtained by deforming the square geometry into a rectangle through elongation along the x axis of the GO center. (It is of course arbitrary whether the elongation occurs along the x or the y axis.) As a result of this distortion, $\pi(pz)$ becomes less bonding, $\pi^0(dxz)$ gains bonding character to become $\pi(dxz)$, $\pi^0(dyz)$ gains antibonding character to become $\pi^*(dyz)$, and $\pi^*(f2'')$ reduces its antibonding character. These changes can be seen from the alterations in the pairwise interactions among the MO lobes. If the distortion and, hence, the removal of the degeneracy in the nonbonding π MOs is sufficiently large, the ground state will correspond to double occupancy of $\pi(dxz)$ with $\pi^*(dyz)$ being empty, so that a singlet state results. Molecules frequently distort geometrically to remove an orbital degeneracy, and a sufficient orbital energy separation in such a distortion leads to electron pairing in the lower-lying MO. This process is called the *Jahn-Teller effect*, in honor of its discoverers.

The singlet π MO state lends itself to localization since there are two fully occupied delocalized π MOs and they arise from a *pz* and a *dxz* GO. The *pd* hybrid GO set constructed from these AOs consists of an oppositely directed pair of hybrid GOs (such as were used for one of the localized forms of CO_2 in Chapter 4) pointing toward the midpoints of the shorter two sides of the rectangle. They generate localized MOs that provide two-center two-electron bonds between the corresponding atoms. Thus the localized bond picture corresponds to the electron dash structure of Figure 7.8 having the two double bonds between those atoms which, *in the rectangular geometry*, lie closer to each other. Molecules such as *cyclo*-C_4H_4 having $4n$ π-electrons (rather than $4n + 2$) are said to be *anti-aromatic*.

It has recently been observed that at 25K, *cyclo*-C_4H_4 rapidly interconverts between the two rectangular structures that can be drawn (i.e., one elongated along the x axis, the other along y). These are not resonance forms because the shifting of the double bonds also involves movement of atoms. When this happens, the structures are called *valence tautomers*.

The vibrational modes of square *c*-C_4H_4 can be obtained by generating the molecular motions of the C_4 square and the H_4 square separately, using the GO center to also represent the center of mass, and then taking linear combinations of those motions generated by the same GO. Of course, the motions of the C_4 and H_4 squares duplicate those of the Z_4 square discussed in the previous section. By following along in Figures 7.3 and 7.4 it can be seen that linear combinations of duplicates of each of the SMs shown in these figures lead to ν_1, ν_2 (Figure 7.3a); ν_3, ν_4 (Figure 7.4a); ν_5, T_x (Figure 7.4b); ν_6, ν_7 (Figure 7.4c); ν_8, T_y (Figure 7.4d); ν_9, T_z (Figure 7.3d); ν_{10}, ν_{11} (Figure 7.3e); ν_{12}, ν_{13} (Figure 7.3f); ν_{14}, R_x (Figure 7.3g); ν_{15}, R_y (Figure 7.3h); ν_{16}, ν_{17} (Figure 7.3i) and ν_{18}, R_z (Figure 7.3j). The $3n - 6$ formula tells us that we should have eighteen vibrations, and indeed our GO analysis verifies this.

Use the Homework Drawing Board in Node Game to generate the SMs *cyclo*-C_4H_4, following the steps in the appropriate column of Table 7.7.

7.3. *cyclo*-$C_5H_5^-$

The cyclopentadienide ion, made by proton abstraction from cyclopentadiene with a strong base, is an important intermediate in the synthesis of a wide range of organometallic compounds. From the electron dash structure in Figure 7.9, it can be seen that part of the stability of this anion arises from its aromatic character. Thus each carbon contributes one π electron, and the carbon bearing the negative charge in a resonance structure (such as is seen in Figure 7.9a) contributes one more. Again $n = 1$ in the $4n + 2$ rule. In discussing the bonding of *cyclo*-$C_5H_5^-$, as well as planar aromatic species of the general formula *cyclo*-$C_nH_n^x$, it is convenient to assign the H—C electron

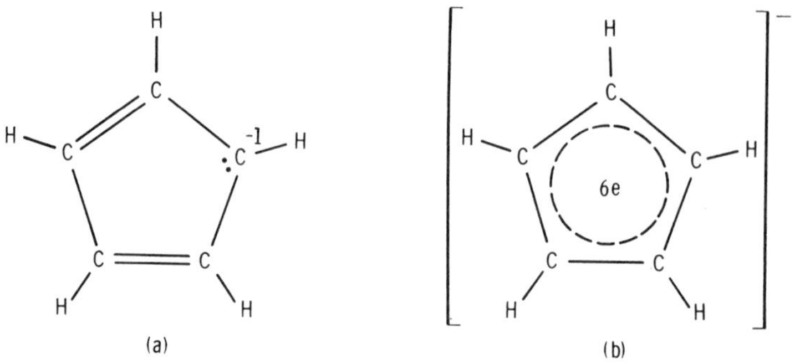

FIGURE 7.9 An electron dash structure of *cyclo*-$C_5H_5^-$ (a). In (b) is shown an average electron dash structure of the resonance forms exemplified by (a).

pairs in the Lewis structure to localized MOs. It should be recalled from our discussion of the delocalized view of N_3^+ that the nitrogen lone pairs could be localized in (N*spo*) VAOs. Thus it is not necessary to choose canonical AOs to form the VAO equivalence sets. Any hybrid AO set could serve equally well. In N_3^+ we used an inwardly directed (N*spi*) VAO having less *s*-character along with (N2*px*) and (N2*pz*) for the delocalized bonding. In *cyclo*-$C_nH_n^x$ [or any cyclic planar *cyclo*-$(ZY)_n^x$ species for that matter] we can follow a similar path. In the case at hand, we use (C*spo*) VAOs and corresponding (H1*s*) VAOs for localization of the C—H bonds and we employ (C*spi*) VAOs for the radial σ bonding. Although the outward- and inward-pointing hybrids have different admixtures of (*s*) and (*p*) character, we will not be concerned with these differences here.

Employing the axis system in Figures 7.10 and 7.11 and the SOs in Figure 7.12, we can construct generator Table 7.3. In arriving at the proper GOs in this table, it is of great importance to appreciate the fact that *p*, *d*, and *f* GOs contained in it have "dumbbell," "four-leaf clover," and cube-like symmetries, respectively, and that these symmetries are superimposed on the pentagonal geometry of the five carbons of *cyclo*-$C_5H_5^-$. Using the schematic top view of the system shown in Figure 7.11 as an aid, it is seen, for example, that a *px* GO calls in ($C_C spi$), ($C_D spi$), and ($C_E spi$) with positive signs in the hybrid main lobes while the same VAOs of C_A and C_B are called in with negative signs. Moreover, the coefficients for these VAOS will be in the order $C_D > C_A = C_B > C_C = C_E$. As a second example, it is easy to see from Figure 7.11 that the $dx^2 - y^2$ GO calls in (sp_r) lobes on C_A, C_B, and C_D with the same sign but that these VAOs on C_C and C_E are called in with opposite signs. From a consideration of each SO in Figure 7.12, their overall bonding or antibonding nature indicated in generator Table 7.3 can be determined. Because of the relative complexity of the molecule, no attempt has been made in Figure 7.12 to indicate the VAO coefficients by relative sizes of the arrows.

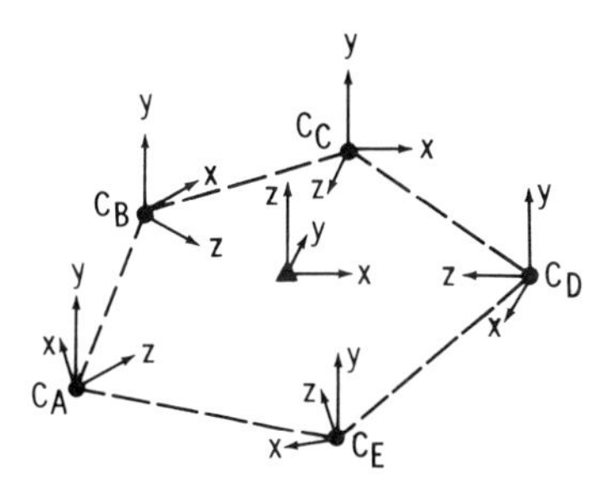

FIGURE 7.10 Axis system for *cyclo*-$C_5H_5^-$.

TABLE 7.3 Generator Table for C_5 framework in *cyclo*-$C_5H_5^-$ ($16e^-$)

	GOs										
VAO Equivalence Sets	*s*	*px*	*py*	*pz*	*dxy*	dx^2-y^2	*dxz*	*dyz*	*f*2′	*f*2″	*h*5′
C*sp*-in = (C_A*spi*), (C_B*spi*), (C_C*spi*), (C_D*spi*), (C_E*spi*)	b	b	b		a	a					
C*px*σ = (C_A2px), (C_B2px), (C_C2px), (C_D2px), (C_E2px)		a	a		b	b					a
C*py*π = (C_A2py), (C_B2py), (C_C2py), (C_D2py), (C_E2py)				b			b	b	a	a	

This less rigorous but qualitatively correct procedure will be followed generally throughout the remainder of this book.

Linear combinations of the pairs of SOs generated by the same GOs give rise to pairs of MOs which are split apart in energy. Because of the fivefold symmetry of pentagonal $C_5H_5^-$, the determination of which linear combination in Figure 7.13 corresponds to the BMO and the ABMO is difficult. We would expect that the BMO consists mainly of the dominantly bonding SO generated in the C*sp*-in equivalence set and that the ABMO would consist

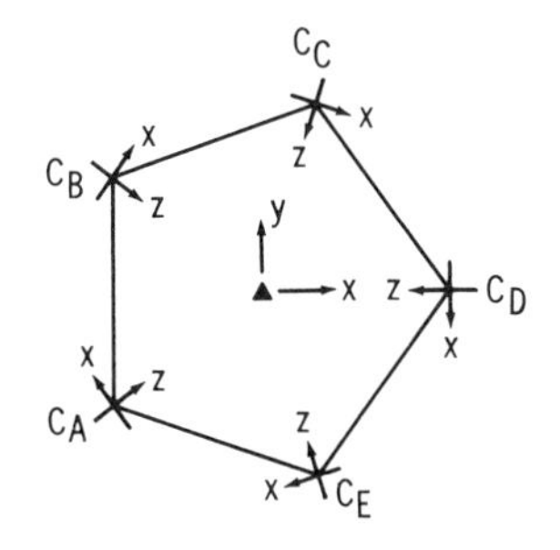

FIGURE 7.11 Top view of the orbital and GO in-plane axes for *cyclo*-$C_5H_5^-$.

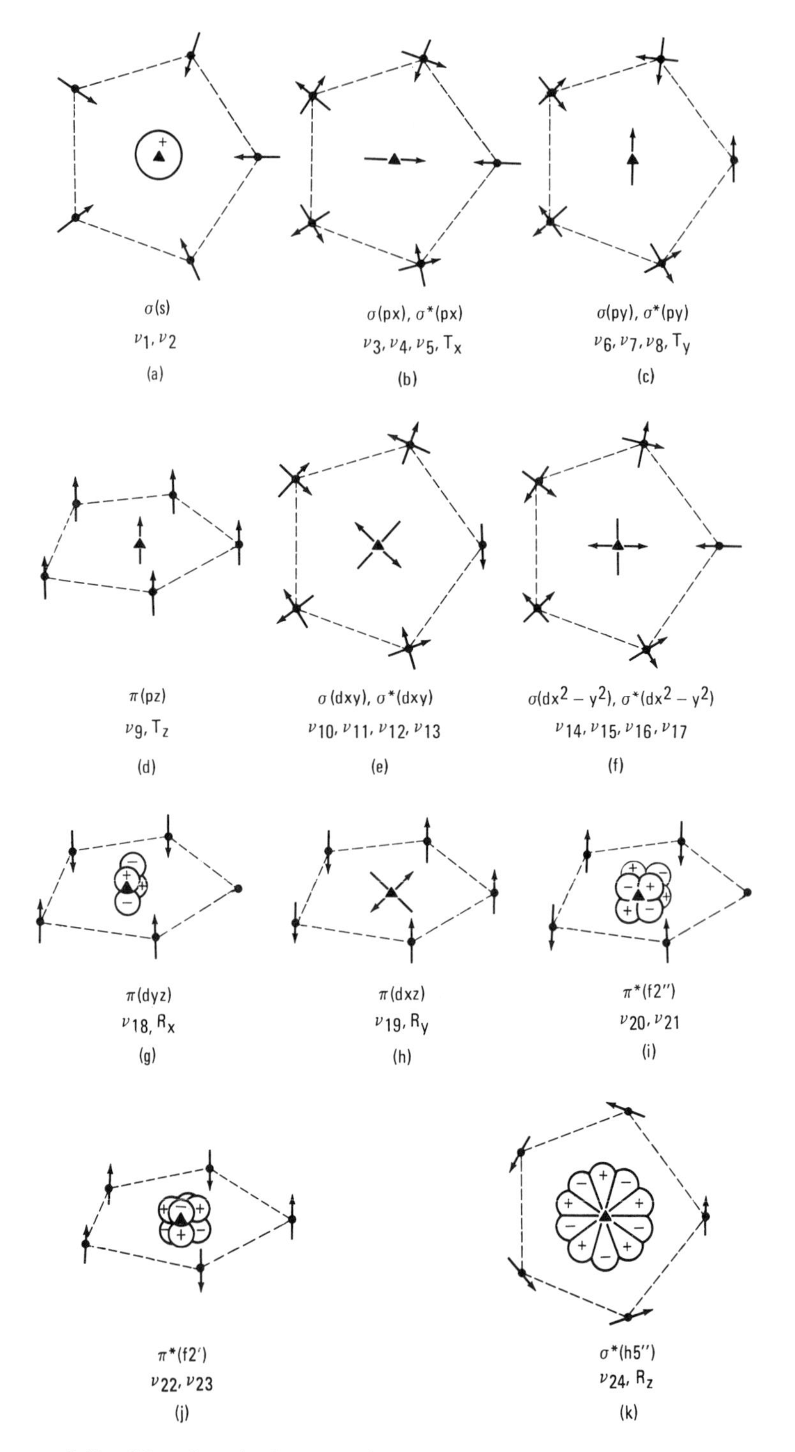

FIGURE 7.12 SOs of *cyclo*-$C_5H_5^-$. These drawings also represent the SMs of a pentagonal array of atoms.

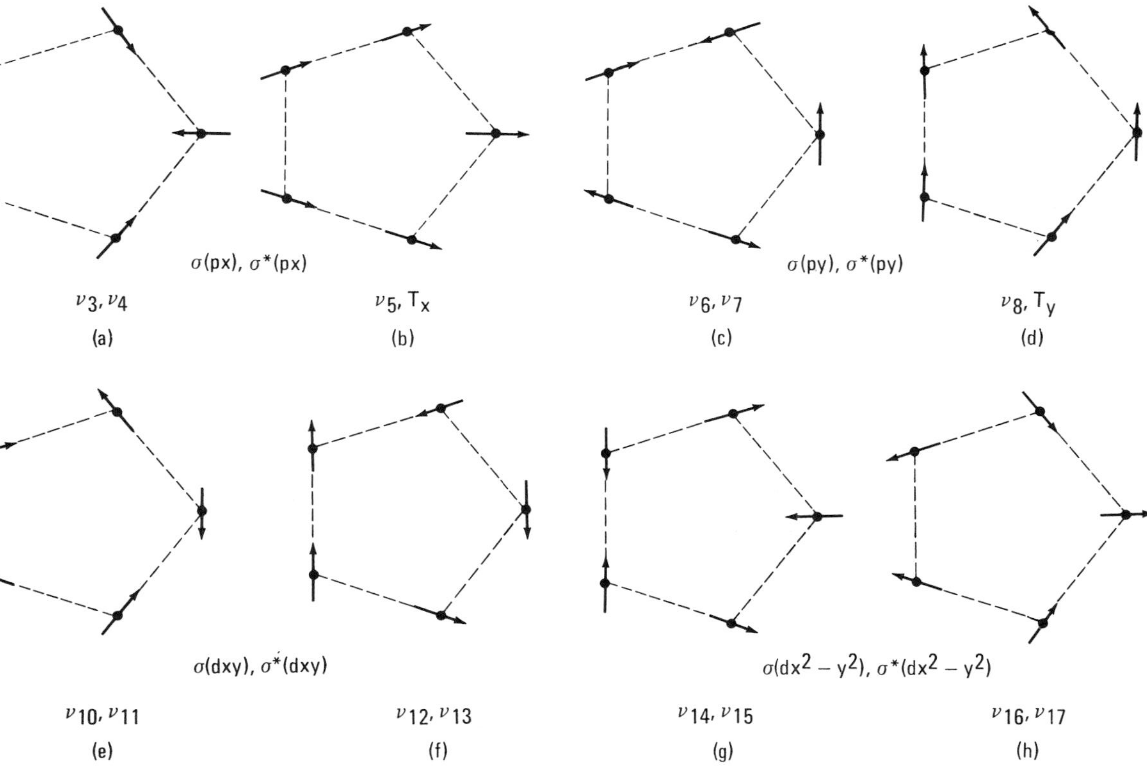

FIGURE 7.13 Linear combinations of SOs for *cyclo*-$C_5H_5^-$ given in Figure 7.12. These drawings also depict the linear combinations of SMs for a pentagonal array of atoms.

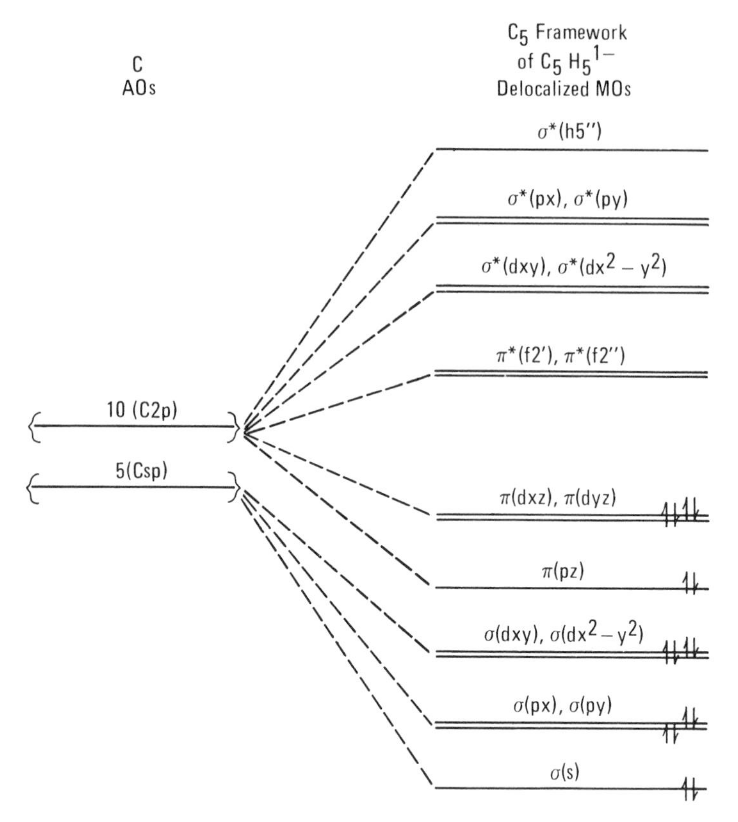

FIGURE 7.14 Delocalized MO energy level diagram for C_5 framework of *cyclo*-$C_5H_5^-$.

mainly of the dominantly antibonding SO generated in the $Cpx\sigma$ equivalence set. Similar considerations hold for the remaining MOs. Since ten of the twenty-six valence electrons are accounted for in the localized C—H bonds, sixteen electrons remain for filling the eight BMOs in Figure 7.14. Of course, the order of the BMOs and ABMOs in this diagram is not possible to ascertain without an accurate calculation. The average σ bond order is 1.0 and the average π bond order is 3/5, giving an overall average bond order of 8/5 per link. As expected from the $4n + 2$ aromaticity rule ($n = 1$), there are six electrons in the π MO system.

It should be noted that the tangential (px) VAOs are given the set label $Cpx\sigma$ because their overlap resembles σ interactions, even though they are at an angle. In higher-order polygons such overlaps become increasingly "head on" and hence more "σ-like" in their appearance. In Te_4^{2+} these VAOs were given the set label Tept (where t denotes tangential) since in a square system such orbitals are exactly halfway between σ- and π-type overlap. Note also that we consistently refer to MOs composed of inwardly directed VAO lobes in polygonal molecules as σ-type MOs even though they also involve angular overlap. Actually, these MO labels apply in a strict sense only to diatomics and in more complex systems these designations are more arbitrary. We will adopt the convention that MOs composed of (p) VAOs that interact in a truly "side-on" manner will be labeled π MOs, whereas less than strictly side-on overlap of these VAOs will give MOs labeled σ.

Do Practice Exercise 4 in Node Game to review the process for developing our delocalized π MOs for *cyclo*-$C_5H_5^-$. Use the Homework Drawing Board in Node Game and the appropriate column in Table 7.7 to do the same for our delocalized σ MOs for this ion.

The σ bonding MOs in *cyclo*-$C_5H_5^-$ stem from an sp^2d^2 hybrid GO set whose members point their main lobes toward the corners of a regular pentagon. By directing these lobes midway between the carbon atoms, we conclude that the five localized σ bond pairs can be housed in five two-center bonds, each of which could be made up of C2p VAOs from neighboring carbon atoms, as shown in Figure 7.15. Taking advantage of the partially hybridized carbon VAOs we already have, however, we could hybridize C2s with *both* the C2px and the C2pz VAOs so that we have a new outward-pointing Cspo, and two C2sp^γ hybrids directed along the edges of the pentagon. Localizing the occupied π MOs of Figure 7.14 involves the use of a pd^2 GO hybrid which we have already encountered in Te_4^{2+}. The directions we choose for the trigonally directed members of this set are arbitrary and in Figure 7.16 are shown two orientations of the localized π bonding electron pair regions that result from two convenient orientations of the pd^2 hybrid GOs. The top view in Figure 7.16a shows that two of these localized MOs are equivalently positioned with respect to atom centers and consequently have

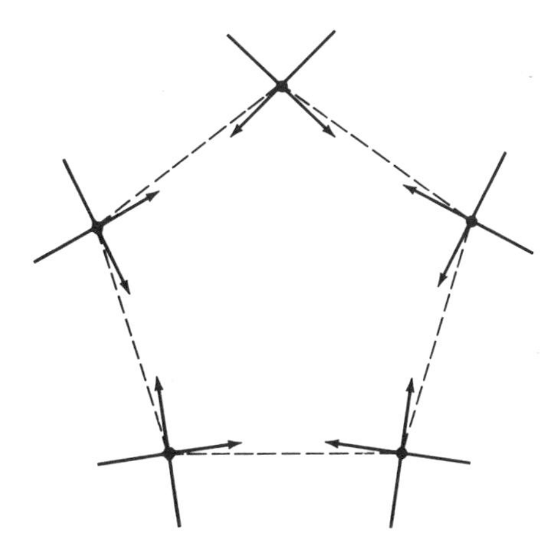

FIGURE 7.15 Carbon VAOs in *cyclo*-$C_5H_5^-$ reoriented for localized bonds between C atoms.

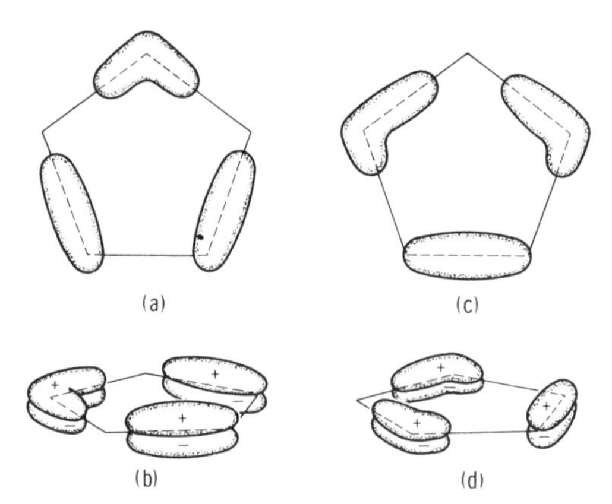

FIGURE 7.16 Two views of one orientation of the localized π MOs in *cyclo*-$C_5H_5^-$ [(a), (b)]. Two views of a second orientation are shown in (c) and (d).

the same shape. The third localized MO, being differently placed with respect to atom centers, has a different shape. This figure shows that the Lewis structure in Figure 7.9a, which implies two π bonds and a doubly occupied (p) AO perpendicular to the molecular plane, is a somewhat misleading oversimplification in that the unique π-type localized MO is almost a three-center π bond and the other two π bonds are correspondingly polarized.

Since the choice of the carbon toward which we pointed the first hybrid GO is arbitrary, there are five different localized MO *sets*. These five MO representations are not to be superimposed (as resonance structures would be). Rather, they are all equivalent and each by itself is an adequate representation of the molecular state.

Another useful localized description for *cyclo*-$C_5H_5^-$ is obtained from a different orientation of the hybrid GOs. Pointing one of these hybrid GOs toward the midpoint of a bond gives rise to the localized MOs shown in Figure 7.16c and d. It is apparent that while all these GOs point between carbon atom pairs, two are off-center. The localized MO energy level diagram is easy to draw and this is left for you to do as an exercise.

We thus find that the three π electron pairs in *cyclo*-$C_5H_5^-$ can be housed in three localized regions toward which the pd^2 hybrid GOs point. It may be mentioned that the geometrical shape of the distribution of π electron pairs around any planar ring is a function only of the number of such pairs and is independent of the number of atoms making up the ring.

Use the Homework Drawing Board in Node Game and the appropriate column of Table 7.7 to develop our delocalized views of *cyclo*-$C_5H_5^-$.

The molecular motions of *cyclo*-$C_5H_5^-$ can also be obtained from Figures 7.12 and 7.13. These motions are linear combinations of the SMs of the carbons *and the hydrogens*, which these figures represent.

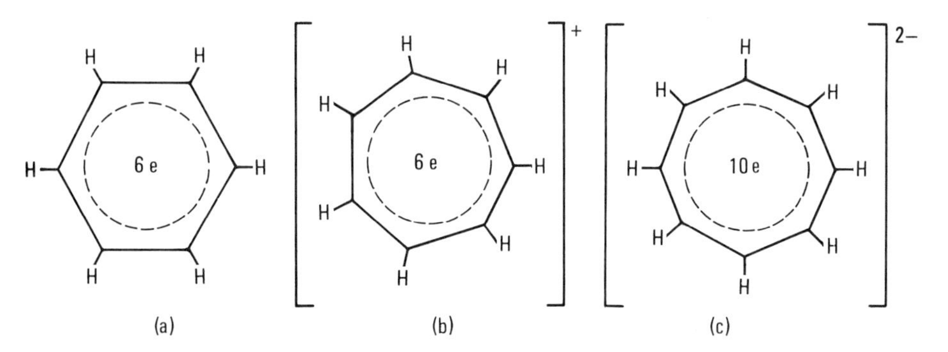

FIGURE 7.17 Average electron dash structures for *cyclo*-C_6H_6 (a), *cyclo*-$C_7H_7^+$ (b), and *cyclo*-$C_8H_8^{2-}$ (c).

7.4. *cyclo*-C_6H_6, *cyclo*-$C_7H_7^+$, *cyclo*-$C_8H_8^{2-}$

The average electron dash structures of benzene, the tropylium ion, and the cyclooctatetraenide ion shown in Figure 7.17a–c, respectively, tell us that all of the title species are aromatic. That is, in the $4n + 2$ rule $n = 1$ for *cyclo*-C_6H_6 and *cyclo*-$C_7H_7^+$ while in *cyclo*-$C_8H_8^{2-}$ $n = 2$. The bonding treatment of these molecules using the axis systems in Figure 7.18 is entirely analogous to that of *cyclo*-$C_5H_5^-$, and so the symmetry orbitals and motions in Figures 7.19 to 7.21 generate the generator. Tables 7.4 to 7.6 are

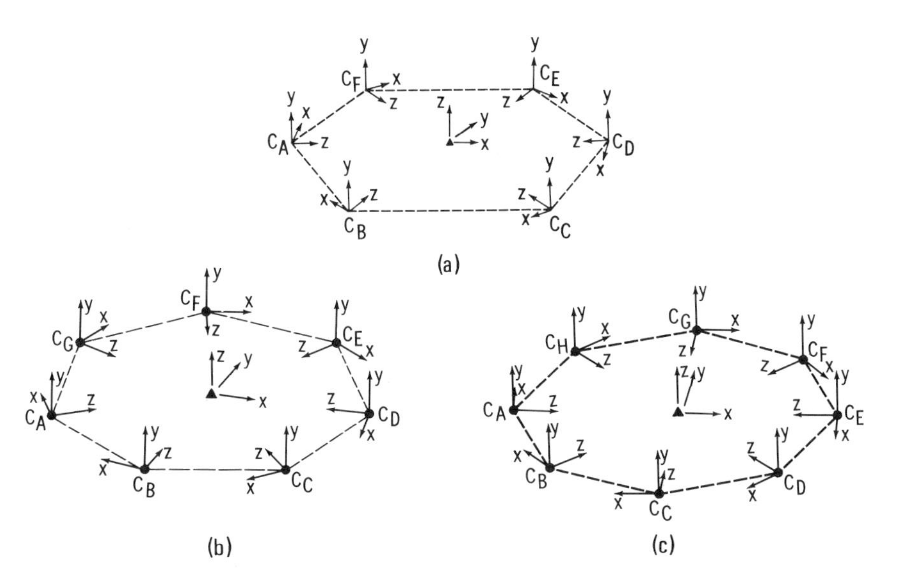

FIGURE 7.18 Axis systems for *cyclo*-C_6H_6 (a), *cyclo*-$C_7H_7^+$ (b), and *cyclo*-$C_8H_8^{2-}$ (c).

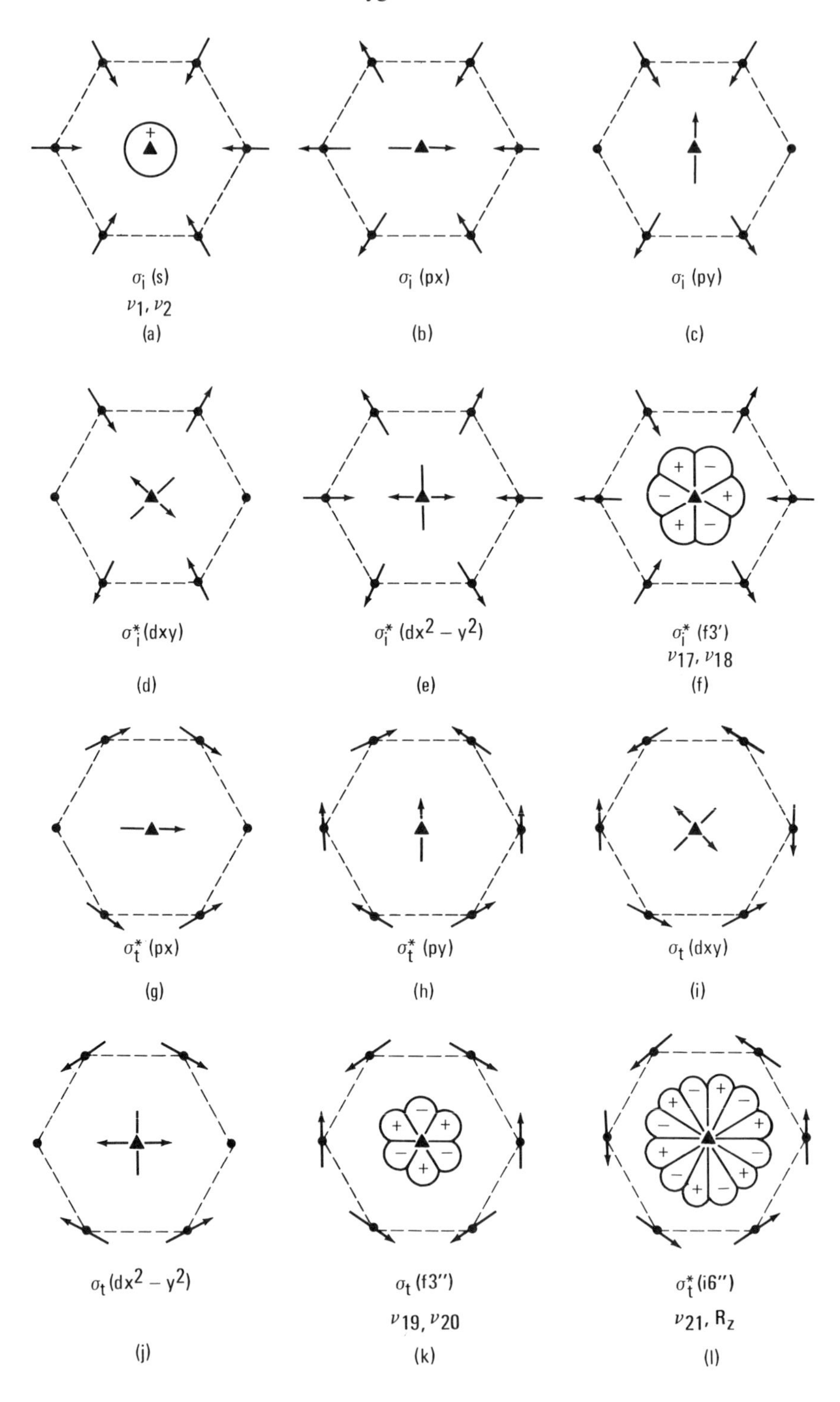

+
σ_i (s)
ν_1, ν_2
(a)
σ_i (px)
(b)
σ_i (py)
(c)
σ^*_i(dxy)
(d)
σ^*_i (dx^2 – y^2)
(e)
+ – – + + –
σ^*_i (f3′)
ν_{17}, ν_{18}
(f)
σ^*_t (px)
(g)
σ^*_t (py)
(h)
σ_t (dxy)
(i)
σ_t (dx^2 – y^2)
(j)
– + + – – +
σ_t (f3″)
ν_{19}, ν_{20}
(k)
– + + – – + + – – + + –
σ^*_t (i6″)
ν_{21}, R_z
(l)

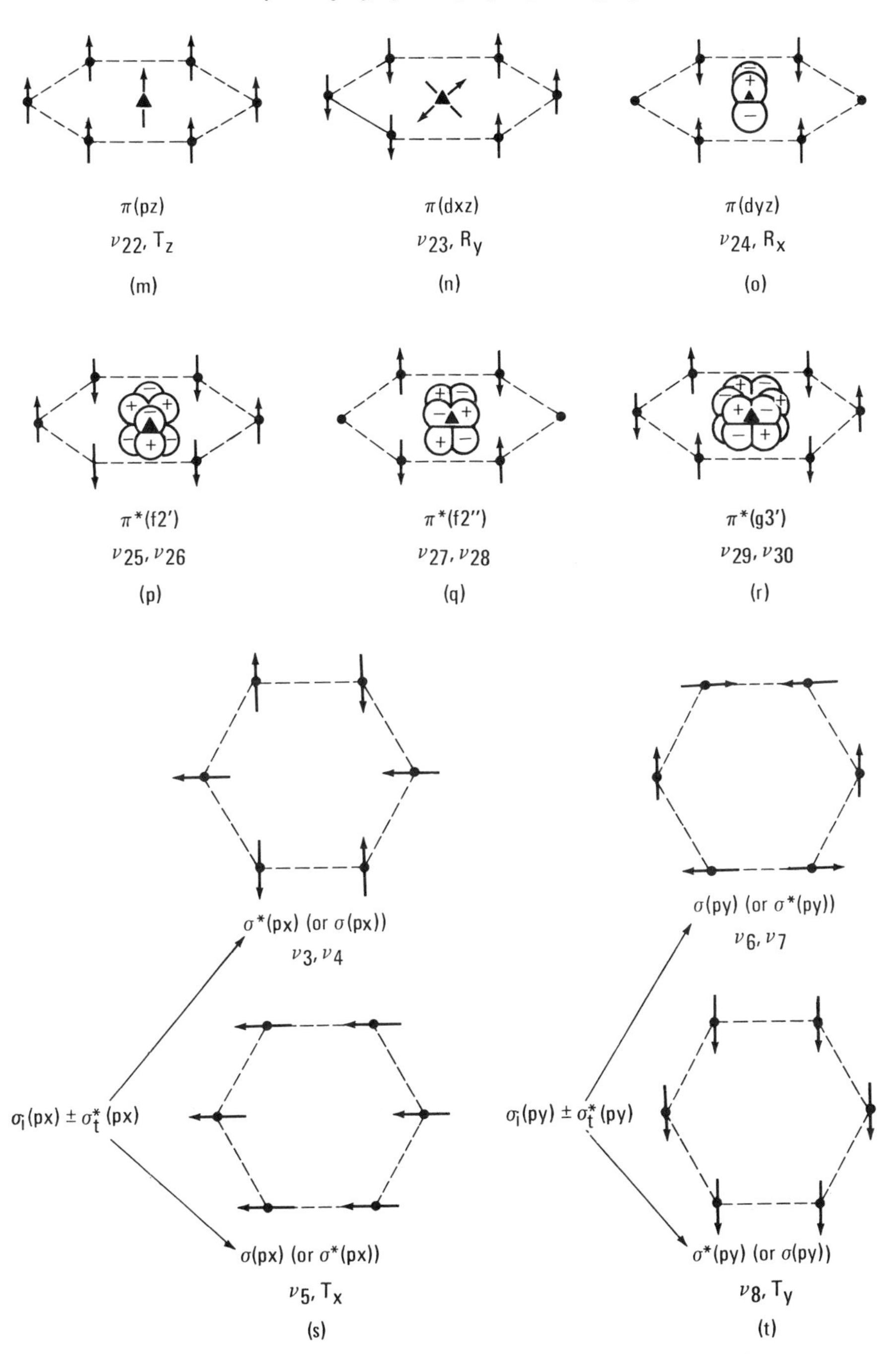

FIGURE 7.19 Drawings of the SOs of *cyclo*-C_6H_6 and their linear combinations (see also Table 7.4). These drawings also represent the SMs of a hexagonal array of atoms. The subscripts *i* and *t* refer to "inward" pointing and "tangential" σ MOs, respectively. Many of the molecular motions are not labeled because they arise from linear combinations of SMs generated by the same GO as is shown for the *px* and *py*-generated SMs in (*s*) and (*t*).

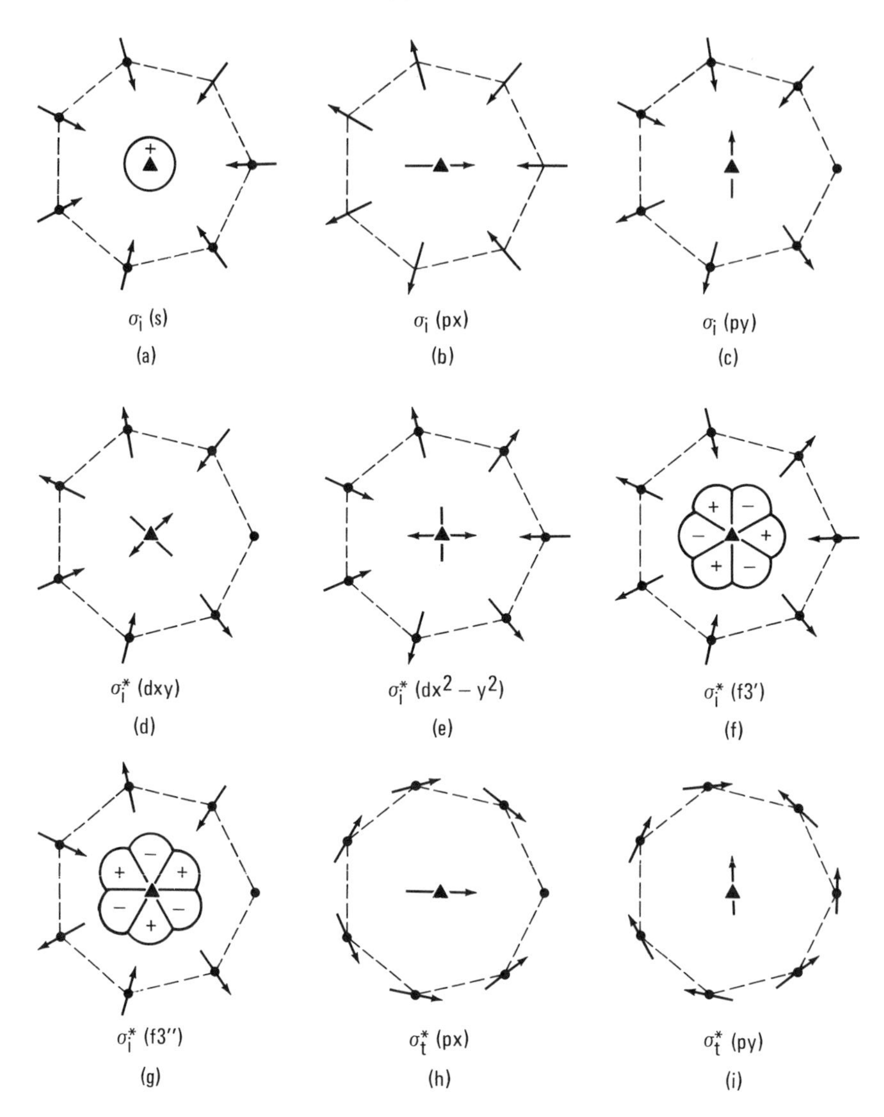

FIGURE 7.20 Drawings of SOs of *cyclo*-$C_7H_7^+$ as given in Table 7.5. These drawings also represent the SMs of a heptagonal array of atoms. The subscripts *i* and *t* refer to "inward" pointing and "tangential" σ MOs.

given without further comment. You should verify the entries in these tables, however. In Figures 7.19 to 7.21 no attempt has been made to show relative sizes of orbitals or to indicate hybrid character in the (C2*spi*) VAOs. When drawing the SOs or SMs generated by the various GOs, be careful to observe the geometric relationships between given AOs or AVs, respectively, and the GO. The drawings of the delocalized MO energy level diagrams for these molecules are left as an exercise.

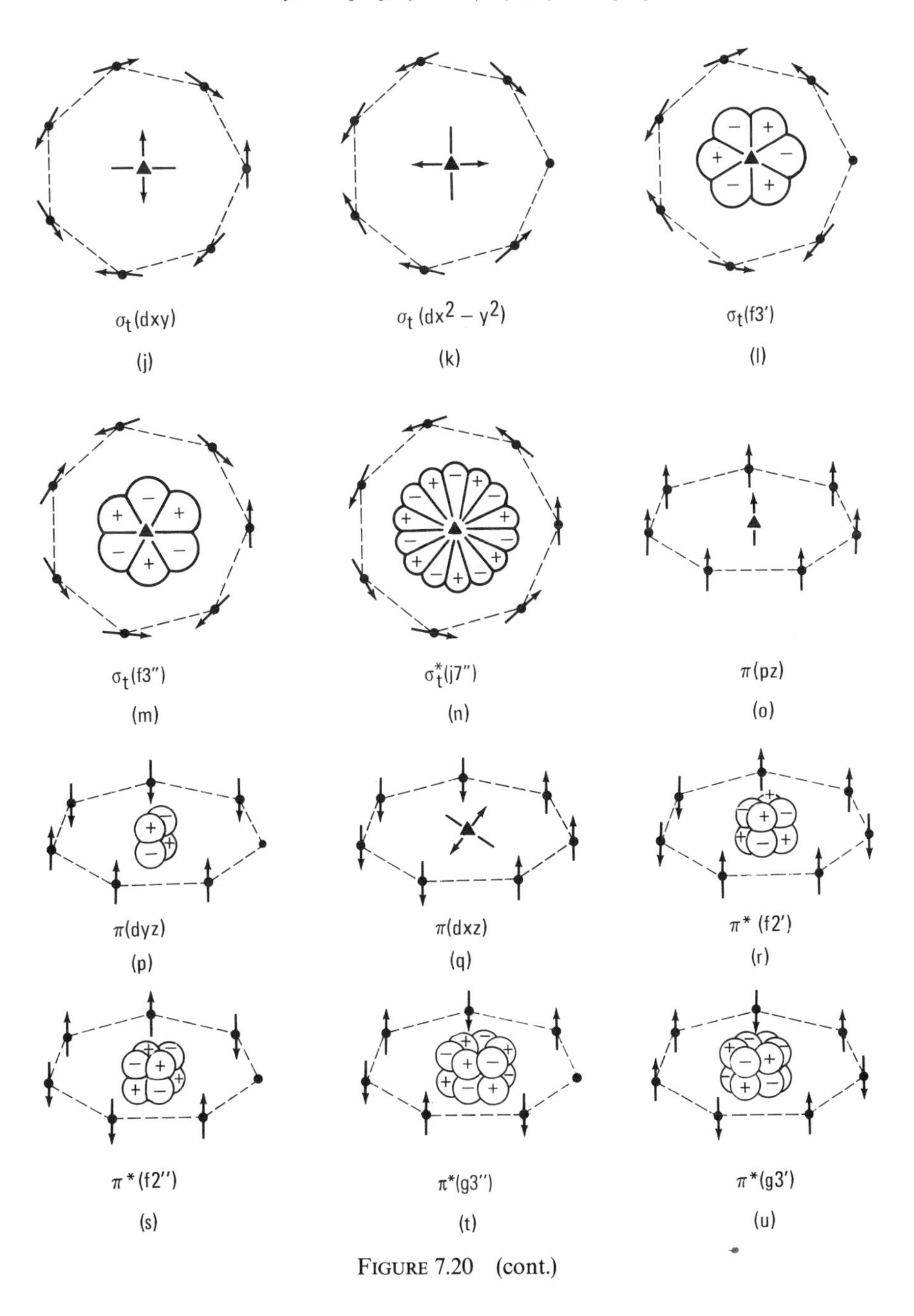

FIGURE 7.20 (cont.)

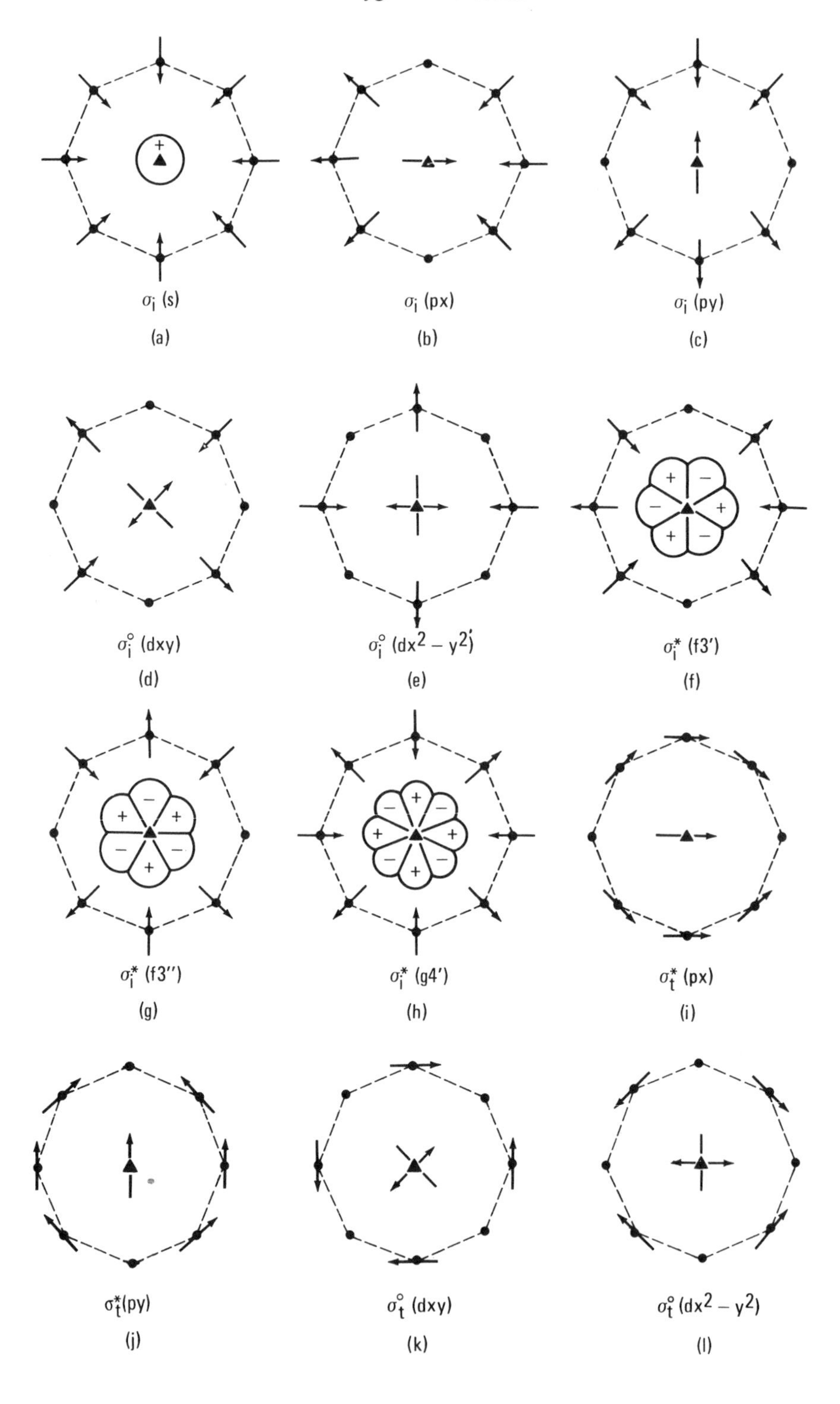

σ_i (s)
(a)

σ_i (px)
(b)

σ_i (py)
(c)

σ_i° (dxy)
(d)

σ_i° ($dx^2 - y^2$)
(e)

σ_i^* (f3′)
(f)

σ_i^* (f3″)
(g)

σ_i^* (g4′)
(h)

σ_t^* (px)
(i)

σ_t^*(py)
(j)

σ_t° (dxy)
(k)

σ_t° ($dx^2 - y^2$)
(l)

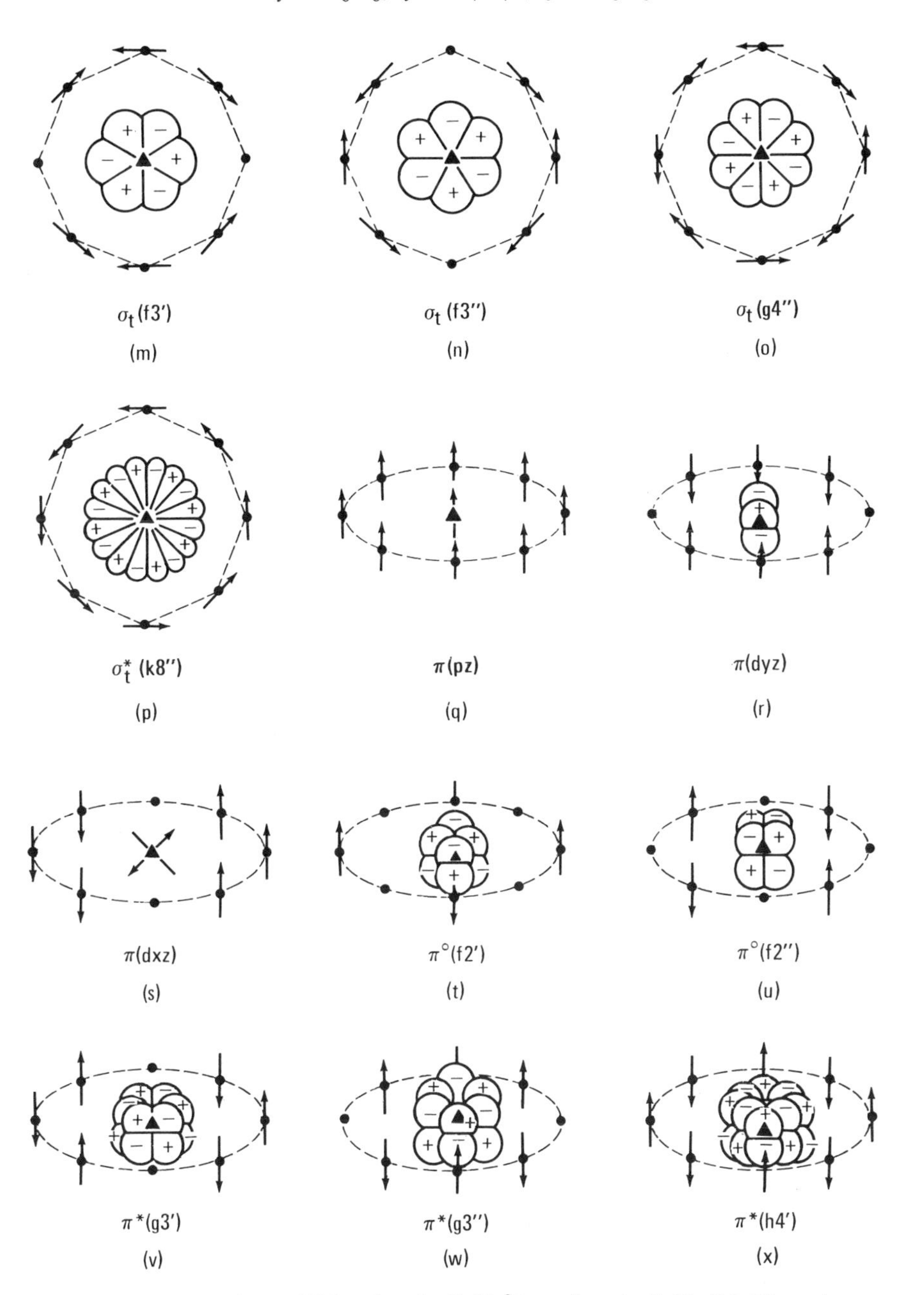

FIGURE 7.21 Drawings of SOs of *cyclo*-$C_8H_8^{2-}$ as given in Table 7.6. These drawings also represent the SMs of an octagonal array of atoms. The subscripts *i* and *t* refer to "inward" pointing and "tangential" σ MOs.

TABLE 7.4 Generator Table for C_6 Framework in *cyclo*-C_6H_6 ($18\ e^-$)

VAO Equivalence Sets	GOs													
	s	px	py	pz	dxy	dx^2-y^2	dxz	dyz	$f2'$	$f2''$	$f3'$	$f3''$	$g3'$	$i6''$
C*spi*	b	b	b		a	a					a			
C*px*		a	a		b	b						b		a
C*pπ*				b			b	b	a	a			a	

TABLE 7.5 Generator Table for C_7 Framework in *cyclo*-$C_7H_7^+$ ($20\ e^-$)

VAO Equivalence Sets	GOs														
	s	px	py	pz	dxy	dx^2-y^2	dxz	dyz	$f2'$	$f2''$	$f3'$	$f3''$	$g3'$	$g3''$	$j7''$
C*spi*	b	b	b		a	a					a	a			
C*px*		a	a		b	b					b	b			a
C*pπ*				b			b	b	a	a			a	a	

TABLE 7.6 Generator Table for C_8 Framework in *cyclo*-$C_8H_8^{2-}$ ($26\ e^-$)

VAO Equivalence Sets	GOs																	
	s	px	py	pz	dxy	dx^2-y^2	dxz	dyz	$f2'$	$f2''$	$f3'$	$f3''$	$g3'$	$g3''$	$g4'$	$g4''$	$h4'$	$k8''$
C*spi*	b	b	b		n	n					a	a			a			
C*px*		a	a		n	n					b	b				b		a
C*pπ*				b			b	b	n	n			a	a			a	

TABLE 7.7 A Useful Summary of Procedures for Generating Pictorial Representations of Delocalized and Localized Bonding Views of a Molecule and for Visualizing its Vibrational Modes

Delocalized View	Localized View	Normal Modes
1. Decide on best electron dash structure using lowest possible formal charges and drawing resonance forms. In such forms, some lone pairs may be involved in a π system.	1. Decide on which set(s) of doubly occupied delocalized MOs to localize.	1. Identify the motional atomic vector (AV) equivalence sets and list them in the generator table.
2. Establish the molecular geometry from VSEPR theory if possible.	2. Hybridize the GOs which give rise to the sets(s) in step 1.	2. Using GOs, draw the symmetry motions (SMs).
3. After preassigning lone pairs *and localized bond pairs* to (s) and suitably oriented (p) VAOs, identify all the VAO equivalence sets (hybrids may also be used) and list them in the generator table.	3. Using the hybridized GOs, draw the SOs.	3. Fill in the rest of the generator table (i.e., GOs and m designations).
4. Using GOs located at a molecular center or centroid, draw all the SOs, recalling that #SOs = #VAOs.	4. Draw the MOs, being aware of the consequences of a central atom on which the full set of SOs (i.e., hybridized VAOs) is not generated by the hybridized GO set. If necessary, rehybridize peripheral atom VAOs *and rotate hybrid* GOs to locate bp's between atoms.	4. Take appropriate linear combinations of SMs generated by the same GO and draw all the molecular motions.
5. Fill in rest of generator table (i.e., the necessary GOs and the n, a, or b designations). Establish partner GO and SO sets.	5. Draw the resulting MO energy level diagram and occupy the appropriate levels with electrons.	5. Identify the normal vibrational modes.
6. Take linear combinations of SOs generated by the same GO and draw all the MOs.	6. Verify that the bond order per link is the same as in the delocalized view.	
7. Draw the resulting MO energy level diagram and occupy the appropriate levels with electrons.		
8. Deduce the bond order per link.		

Use the Homework Drawing Board in Node Game and the appropriate column of Table 7.7 to develop our delocalized views of *cyclo*-C_6H_6, *cyclo*-$C_7H_7^+$ and *cyclo*-$C_8H_8^{2-}$.

Summary

In this chapter we gained an appreciation for delocalized and localized bonding systems in planar ring-like molecules such as Te_4^{2+}, *cyclo*-C_4H_4, *cyclo*-$C_5H_5^-$, *cyclo*-C_6H_6, *cyclo*-$C_7H_7^+$, and *cyclo*-$C_8H_8^{2-}$. All of these examples are aromatic (i.e., they obey the $4n + 2$ rule) except *cyclo*-C_4H_4 which is anti-aromatic. In studying the latter molecule, the importance of Jahn–Teller distortions in removing orbital degeneracy was brought out.

In Table 7.7 is contained our latest amendment to the recipe for using the GO approach.

Exercises

1. Draw Lewis structures with resonance forms for the following cyclic species and indicate which are aromatic, which are anti-aromatic, and which are neither. (a) $(CH_2)_3$, (b) $(CH)_3^+$, (c) $(CH)_3^-$, and (d) $(CH)_8^-$ which is made by shining light on $K_2(CH)_8$ at low temperature.

2. Draw the localized MO energy level diagrams for the cyclic species (a) Te_4^{2+}, (b) rectangular $(CH)_4$, and (c) $(CH)_5^-$.

3. The unknown molecule $(HSiN)_3$ has been calculated to have some aromatic character. Develop the delocalized view of the π bonding system in the alternating Si—N bonds making up the six-membered ring. (Hint: Treat the system as a combination of an Si_3 part and an N_3 moiety.)

4. Sketch the molecular vibrations of square planar cyclobutadiene and give the corresponding generator table.

5. Sketch the delocalized MO energy level diagrams for the cyclic species (a) C_6H_6, (b) $C_7H_7^+$, and (c) $C_8H_8^{2-}$.

6. Give the localized views of the species in Problem 5.

7. The cyclic species $C_6Cl_6^{2+}$ is made in the reaction $2C_6Cl_6 + Cl_2 + 2SbF_5 \rightarrow 2C_6Cl_6^- + 2SbF_5Cl^- \xrightarrow{h\nu} 2C_6Cl_6^{2+} + 2SbF_5Cl^- + 4e^-$. It is paramagnetic and has a triplet ground state. Draw its delocalized MO energy diagram assuming that no Jahn–Teller distortion occurs. Make a sketch showing how this molecule could distort to remove orbital degeneracy and show how the MO energy level diagram change as a consequence.

8. Develop the delocalized and localized views of Te_4^{2+} using an axis system in which the x and y axes point at the atoms. Compare the results with those obtained in the text.

9. The cyclic species $C_5H_5^+$, which is made in the reaction *cyclo*-$C_5H_5I + SbF_5 +$ *cyclo*-$C_5H_5^+ + SbF_5Cl^-$, has a triplet ground state. Draw its delocalized MO energy level diagram assuming no Jahn–Teller distortion occurs.

10. The cyclic molecule N_2S_2 has been shown to have a planar diamond-shaped structure. (a) On the basis of electron dash structures, would you expect this molecule to have an N—N and an S—S link, or alternating S—N links? (b) Develop the delocalized view of this molecule.

11. One of the singlet excited state structures of cyclobutadiene is square planar. Develop the delocalized view of this molecule. Comment on the relationship of the electrons in the degenerate nonbonding levels.

12. Draw the permitted C≡O vibrations of the molecule below. (For more information see Problem 14 in Chapter 4.)

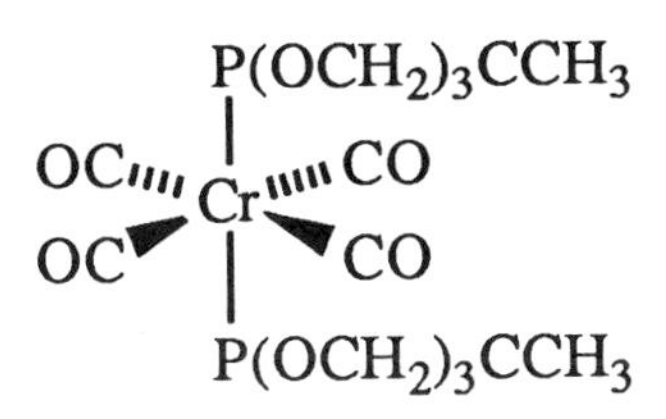

13. Azides (i.e., N_3^- compounds) can be decomposed under controlled conditions allowing N_4^- ions to be trapped. This cyclic species has been calculated to be rectangular. Develop a delocalized view for this ion and comment on whether it is dia- or paramagnetic.

14. Evidence for *cyclo*-N_6 was obtained when the azide complex below was irradiated. Develop a delocalized bonding view of this molecule. Is it aromatic?

$$cis\text{-}[Pt(N_3)_2(PPh_3)_2] \rightarrow N_6 + [Pt(PPh_3)_2]$$

CHAPTER 8

Octahedral and Related Molecules

Octahedral species include not only examples such as PF_6^-, CoF_6^{3-} and $P[AuP(C_6H_5)_3]_6^{3+}$, which contain central atoms, but also those which do not, such as the cluster systems $B_6H_6^{2-}$ and $Mo_6Cl_8^{4+}$. We will also find it of interest to consider BrF_5, ICl_4^-, and B_5H_9, whose structures, depicted in Figure 8.1, can be considered as missing one or two octahedral vertex atoms.

To accelerate your understanding of the application of the GO approach to the bonding views we develop in this chapter, you should repeat the steps we take by using the appropriate column in Table 8.10 and the Homework Drawing Board in Node Game. Notice that the delocalized view of PF_6^- in this chapter is also the subject of a Practice Exercise in Node Game.

8.1. ICl_4^-

We will find it worthwhile to begin with the square planar ICl_4^- ion. The shape of this ion is easily predicted by examining its electron dash structure in Figure 8.2. The two lone pairs on iodine repel one another more strongly than lone pairs repel bond pairs, and so the lone pairs are 180° apart while the bond pairs are forced to be separated by only 90°. Why do we place these lone pairs on iodine rather than distribute them among the chlorines? Iodine is a larger atom, and it is thus better able to accommodate electron-pair repulsions than chlorine.

To obtain the delocalized view of ICl_4^-, we first assign the chlorine lone pairs to (Cl3*s*), (Cl3*px*), and (Cl3*pz*) VAOs and the iodine lone pairs to (I5*s*) and (I5*pz*) VAOs using the axis system in Figure 7.2. The remaining chlorine equivalence set Cl*pr*, by analogy to that in Te_4^{2+}, leads to the four SOs listed in Generator Table 8.1. The eight electrons are easily seen to reside in a pair of degenerate five-center two-electron BMOs generated by *px* and *py* GOs, and in a pair of nondegenerate NBMOs generated by *s* and *dxy* GOs. Note that the MO generated by the *s* GO is nonbonding because the (I5*s*) was reserved for a lone pair. The MO generated by the *dxy* GO is nonbonding because we choose not to use 5*d* VAOs on the iodine. As with Te_4^{2+}, our axis

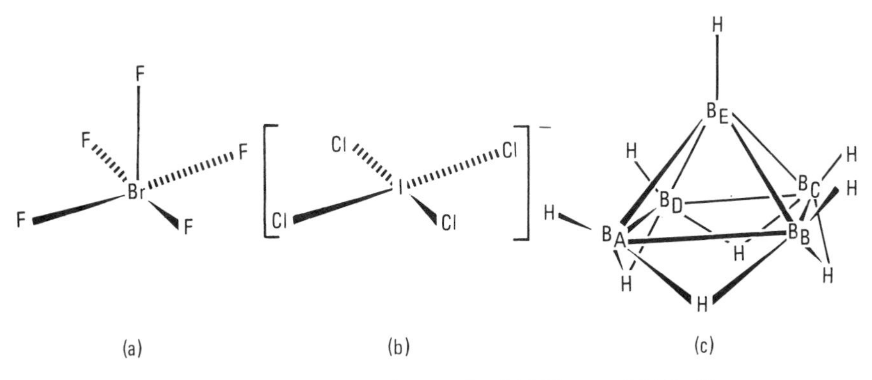

FIGURE 8.1 Structures of (a) BrF_5, (b) ICl_4^-, and (c) B_5H_9.

system for ICl_4^- points the x and y GO axes *between* the peripheral atoms to maximize delocalization. It is left as an exercise to show that directing these axes toward two of the chlorines gives rise to a three-center rather than a five-center bond system. Since we prefer the most extensively delocalized bonding pattern we can construct, we will opt for the five-center bond system.

Localization of the four occupied delocalized MOs requires an sp^2d GO set. A computer-generated contour diagram of a member of such a set is shown in Figure 8.3. The major lobe of each member of such a set points to the corner of a square. This implies that we can orient the set so that the

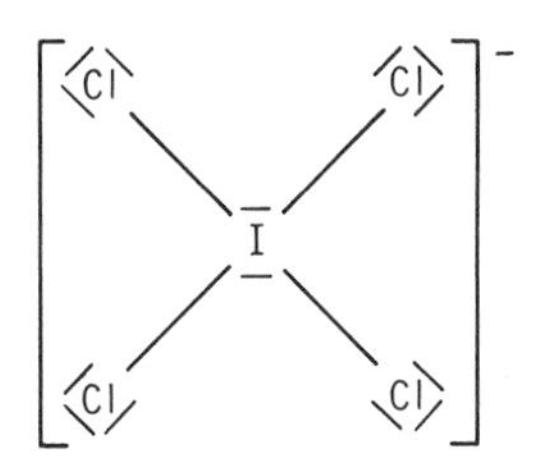

FIGURE 8.2 Electron dash structure for ICl_4^-.

TABLE 8.1 Generator Table for ICl_4^- without Iodine (d) VAOs (8 e^-)

VAO Equivalence Sets	GOs			
	s	*px*	*py*	*dxy*
I*pr*		n	n	
Cl*pr*	n	n	n	n

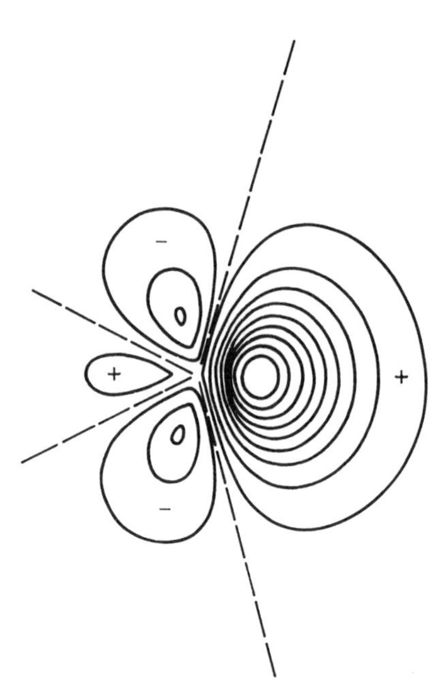

FIGURE 8.3 Computer-generated plot of an sp^2d orbital.

major lobes point to the chlorines. A member of this GO set is schematically represented in Figure 8.4a as is the SO it generates in the Cl*pr* equivalence set. Since the average bond order in the delocalized BMOs is 0.5, an antibonding contribution to each of the four localized MOs is present. It is seen in Figure 8.4b that this antibonding contribution comes from the (Cl3*py*) VAO diagonally opposite to the (Cl3*py*) VAO making the major contribu-

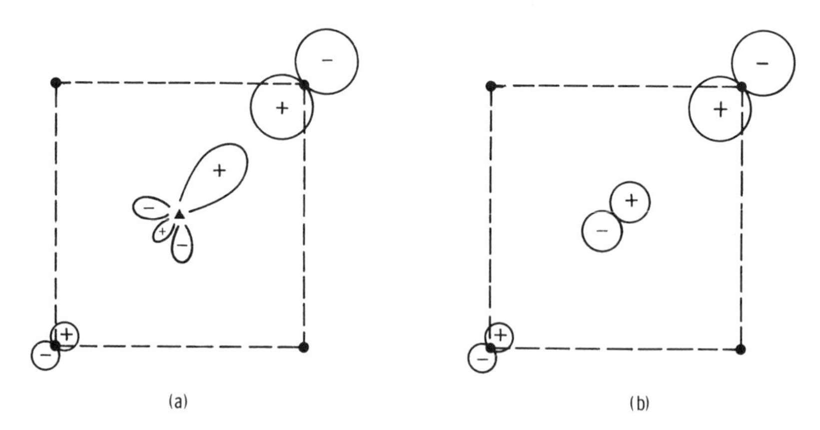

FIGURE 8.4 Chlorine (3*py*) VAOs called in by an sp^2d GO (a) and the (5*p*) VAO such a GO calls in on the iodine (b).

tion to the localized bond. As a consequence of pointing the GOs toward the chlorine atoms, the ($5p$) VAOs of the iodine have merely been rotated by 45° from the x and y axes. Because of the location of the node in the GO in Figure 8.4a, the remaining two (Cl3py) VAOs adjacent to the major (Cl3py) VAO are not called into the localized MO. A pair of digonal ($5spz$) VAO hybrids can be used to localize the iodine lone pairs.

Many molecules exist in which an atom is surrounded by more than the usual octet of electrons in sigma-type MOs plus any lone pairs that may be present. Such atoms are said to be *hypervalent* or *hypercoordinate*.

For hypervalent atoms, it is, of course, always possible to invoke the use of (d) VAOs for the purpose of eventually making localized two-center two-electron bonds. In the present example containing hypervalent iodine, all five of the (I5d) VAOs are generated by the full set of five d GOs (recall XeF_2 in Chapter 4). However, only two of these GOs (dz^2 and dxy) interact with our Cl SOs, and that means that (Xe5dz^2) and (Xe5dxy) can form MOs with these Cl SOs. (In the dz^2 orbital, the "doughnut" lies in the plane of the chlorine (py) VAOs.) We see immediately that dz^2 generates an SO in Clpr identical to that generated by s. Since four SOs have already been generated in the Clpr set (see Table 8.1) we know that the SO $\tilde{\sigma}(dz^2|\text{Cl}pr)$ cannot be independent. As we have seen in the case of XeF_2, it is appropriate to label the corresponding column of the GO table by the most inclusive GO (in this case dz^2). From Table 8.2 we see that a BMO and an ABMO can be obtained by linearly combining the SOs generated by the same GO. The four delocalized BMOs that result are occupied by the eight electrons available for bonding.

For the corresponding localized view we note that the hybrid GO set associated with the occupied delocalized BMOs is p^2d^2. (Notice that we do not include the s GO in our hybrid GO set because the delocalized MO it generated is nonbonding.) As might be expected, the contours of a member of such a set look qualitatively very much like those of the sp^2d hybrid orbital shown in Figure 8.3. This GO hybrid set generates the corresponding (p^2d^2) iodine VAO set, whose main lobes point toward the chlorines. (Although we are more accustomed to mixing ($dx^2 - y^2$) with (s), (px) and (py) to obtain square oriented (sp^2d) hybrids, we can use (dxy) instead of ($dx^2 - y^2$) as long as we rotate our (px), (py) set by 45°. In this way the main lobes of our hybrid

TABLE 8.2 Generator Table for ICl_4^- including Iodine (d) VAOs (8 e^-)

	GOs			
VAO Equivalence Sets	dz^2	px	py	dxy
Ipr = (I5px), (I5py)		n	n	
Idz^2 = (I5dz^2)	n			
Idxy = (I5dxy)				n
Clpr = (Cl_A3py), (Cl_B3py), (Cl_C3py), (Cl_D3py)	n	n	n	n

set point toward the chlorines.) Four localized BMOs can be formed from these iodine (p^2d^2) hybrids and the (py) VAOs of the chlorines.

The molecular motions of square planar ZY_4 species are obtained by considering the SMs of the three AVs on each member of the Y_4 moiety (see Te_4^{2+} in Chapter 7) and the three AVs on Z. The visualization of these motions is left as an exercise at the end of the chapter.

8.2. BrF_5

From the electron dash structure in Figure 8.5, we conclude via the VSEPR rules that this molecule is tetragonal pyramidal. In contrast to ICl_4^-, in which two atoms are missing from octahedral vertices, BrF_5 has only one such atom missing. Other examples of such structures are XF_5 (X = Cl, I), ChF_5^- (Ch = Chalcogen = S, Se, Te and $SbCl_5^{2-}$) in which the unique atom, like the bromine in BrF_5, is also hypervalent. In the molecular structure of BrF_5, the basal fluorines are bent out of the plane toward the apical fluorine, and their angle with the apical fluorine is 84.4°. This distortion can be rationalized as stemming from lp—bp repulsions which exceed bp—bp repulsions. Another interesting feature of this molecule is the difference in length between the axial bond (1.744 Å) and the equatorial bonds (1.689 Å). It is apparent from the structure of BrF_5 that the apical fluorine experiences a molecular potential different from that experienced by the four equivalent basal fluorines. We must therefore treat the two types of fluorines separately in generating our SOs. Because of fluorine's greater electronegativity, there is some ionic character in the Br—F bonds.

We begin developing our delocalized view by using the same central axis system for the GO center (and bromine) as we did for ICl_4^-, but using inward-pointing z axes for all the peripheral F atoms. We also assign all the fluorine lone pairs to (2s), (2px), and (2py) VAOs and the bromine lone pair to the (4s) VAO. For simplicity we will assume that the Br and basal fluorine atoms are coplanar. The GO center lies somewhere along the Br—F (apical) internuclear axis. After drawing the SOs in the usual way, we construct Generator Table 8.3 wherein the labels "a" and "b" in the VAO equivalence sets denote the apical and basal positions. In this table we see that there are three n entries in the s GO column. *Whenever three n entries are seen in a*

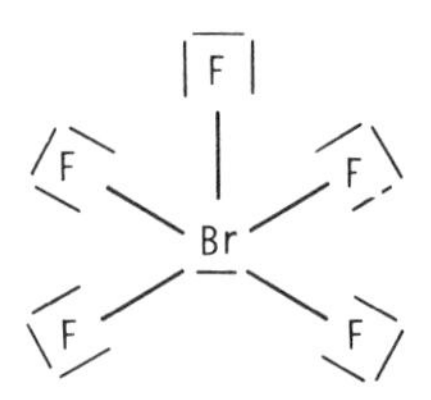

FIGURE 8.5 Electron dash structure of BrF_5.

TABLE 8.3 Generator Table for BrF_5 (10 e^-)

	GOs			
VAO Equivalence Sets	*s*	*px*	*py*	*dxy*
Br*p*a = (Br4*pz*)	n			
Br*p*b = (Br4*px*), (Br4*py*)		n	n	
F*p*a = (F_E2*pz*)	n			
F*p*b = (F_A2*pz*), (F_B2*pz*), (F_C2*pz*), (F_D2*pz*)	n	n	n	n

column in such a table, we will expect one of the three linear combinations to provide a nonbonding MO, while the two others give an antibonding and bonding MO. We can visualize this process in BrF_5 by observing that the main contribution to the NBMO stems from the $\tilde{\sigma}(s|Fpb)$ SO. That this is so can be realized by noting that the two remaining SOs strongly interact (owing to the colinear configuration of these orbitals along the central *z* axis) and they linearly combine to give a BMO and an ABMO. On the other hand, a linear combination of two orbitals that includes $\tilde{\sigma}(s|Fpb)$ as one of the contributors leads to relatively little interaction owing to very poor overlap with either $\tilde{\sigma}(s|Brpa)$ or $\tilde{\sigma}(s|Fpa)$. The remaining two BMOs come from linearly combining the SOs in the *px* and *py* GO columns.

The ten electrons available for our delocalized MOs fill the three BMOs and two NBMOs, giving an average bond order of 1/2 in each link in the basal plane and a bond order of 1.0 in the apical link. This conclusion is at variance with the rule (usually valid) that strong bonds are associated with short bond distances, because the apical bond distance is slightly longer than the equatorial bond distance. This anomaly presumably originates in nonbonding repulsions arising from the proximity of the basal fluorines to the apical fluorine. This in turn results from the fact that the basal fluorines are actually bent out of the plane toward the apical fluorine due to the predominance of lp—lp bond pair repulsions over bp—bp repulsions. In order for the bromine lone pair to exert this effect, it must possess some (4*pz*) character which gives it directionality away from the apical fluorine. Correspondingly, the VAO in the VAO equivalence set, Br*pa*, has some (4*s*) character.

Since $\sigma(s)$ is already a localized occupied two-center MO binding the bromine to the apical fluorine, we confine our attention to the remaining occupied MOs which bond in the basal plane. These are associated with the GO set *s*, *px*, *py*, and *dxy*, which when hybridized (remembering to rotate the (*px*), (*py*) set by 45°) produce four equivalent hybrid GOs in a square planar arrangement with the main lobes pointed toward the fluorine. Since the Br*p*b VAO equivalence set contains only two orbitals, the four hybrid GOs cannot call in orthogonal contributions from the bromine atom, and consequently the localized MOs have substantial antibonding character, reducing the bond order in each basal Br—F link to 1/2.

A discussion of the molecular motions of square pyramidal ZY_5 molecules such as BrF_5 is deferred until the end of the next section.

8.3. B_5H_9

Most neutral boron hydrides such as B_5H_9 can be obtained by heating the simplest member, gaseous B_2H_6, and eliminating hydrogen gas. In the present discussion our goal is to obtain an understanding of the bonding in B_5H_9. The structure of this molecule in Figure 8.1c shows that two kinds of hydrogens are present, namely, those which are bound to one boron and those which are linked to two boron atoms. The hydrogens of the former type, of which there are five, are called *terminal hydrogens* and those of the latter variety, of which there are four, are referred to as *bridging hydrogens.* Because all the hydrogens are not of the terminal type, as in *cyclo*-$C_5H_5^-$, we must first appreciate how the bonding in a bridging B—H—B bond differs from that in a terminal B—H bond before determining which boron VAOs to consider and how many electrons are available to bind the square pyramidal cluster of borons together. As we will see later in this section, there is an insufficient number of electrons to bind each adjacent boron atom with two electrons. That is why a normal Lewis structure cannot be drawn for this molecule. Models for predicting boron hydride structures have been suggested but they are still in a developmental stage.

Bridging hydrogen geometries can be linear, as in FHF^-, or bent, as in many boron and transition metal hydrides. Two bent bridging hydrogens occur in the simplest boron hydride B_2H_6, a gas under ordinary conditions, which can be made from BCl_3 and $LiAlH_4$. The bonding in B_2H_6 is useful to examine before we study B_5H_9. The structure of B_2H_6, depicted in Figure 8.6, consists of two boron atoms surrounded by four hydrogens in an approximately tetrahedral fashion. The nearly tetrahedral stereochemistry around the borons makes it convenient to utilize two (Bsp^{α}) VAOs on each boron which are directed at the (1*s*) VAO of each of the four terminal hydrogens. The four two-center BMOs that can be formed from each pair of neighboring

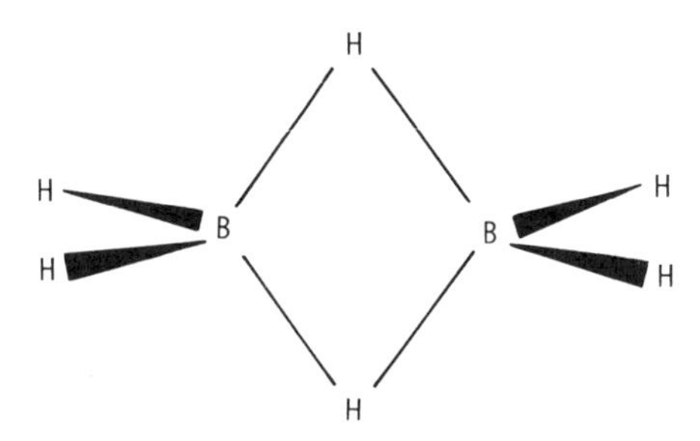

FIGURE 8.6 Structure of B_2H_6. The molecule has four terminal and two bridging hydrogens.

(H1*s*) and (Bsp^{α}) VAOs will be reserved for four pairs of bonding electrons holding the terminal hydrogens to the boron. The remaining four valence electrons of the total of twelve available in this molecule are used to bind the borons together via the two bridging hydrogens. The two remaining VAOs (Bsp^{β}) on each boron are completely determined by orthogonality if we choose them to be equivalent. From Figure 8.7 we see that they are directed approximately toward the bridging hydrogens. Using Figure 8.7 we can draw all the SOs in our usual way and then construct Generator Table 8.4. From this table, we see that linear combinations of the SOs generated by *s* and *py* each give a bonding and an antibonding MO. The corresponding MO energy diagram in Figure 8.8 tells us that both four-center BMOs are filled with the four valence electrons we have available, giving an average bond order of 0.5 per B—H link in the B_2H_2 ring system. It should be noted that the delocalized bonding picture of these bridging hydrogens differs from that in FHF^-, wherein the NBMO is also occupied. In B_2H_6, the NBMO associated with each bridging hydrogen is empty.

To obtain a localized picture of the bonding in the B_2H_2 ring of B_2H_6, we proceed in the usual manner by hybridizing the GOs associated with the occupied MOs in Figure 8.8. The resultant digonal *spy* hybrid GOs have their main lobes directed at H_A and H_B, which tells us that one bonding pair of electrons is localized near H_A and the others near H_B. These bond pairs must each link both borons and one bridging hydrogen in a three-center

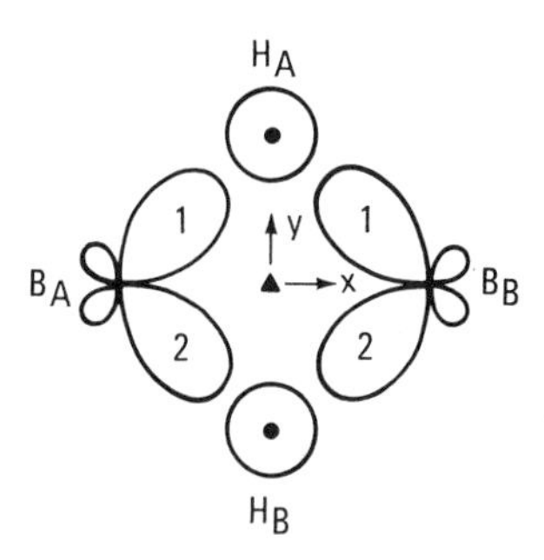

FIGURE 8.7 Boron and bridging hydrogen VAOs available for the MO system of the B_2H_2 ring of B_2H_6.

TABLE 8.4 Generator Table for B_2H_2 Ring in B_2H_6 (4 e^-)

	GOs			
VAO Equivalence Sets	*s*	*px*	*py*	*dxy*
B*sp*-in = ($B_Asp^{\beta}1$), ($B_Asp^{\beta}2$), ($B_Bsp^{\beta}1$), ($B_Bsp^{\beta}2$)	n	n	n	n
H1*s* = (H_A1s), (H_B1s)	n		n	

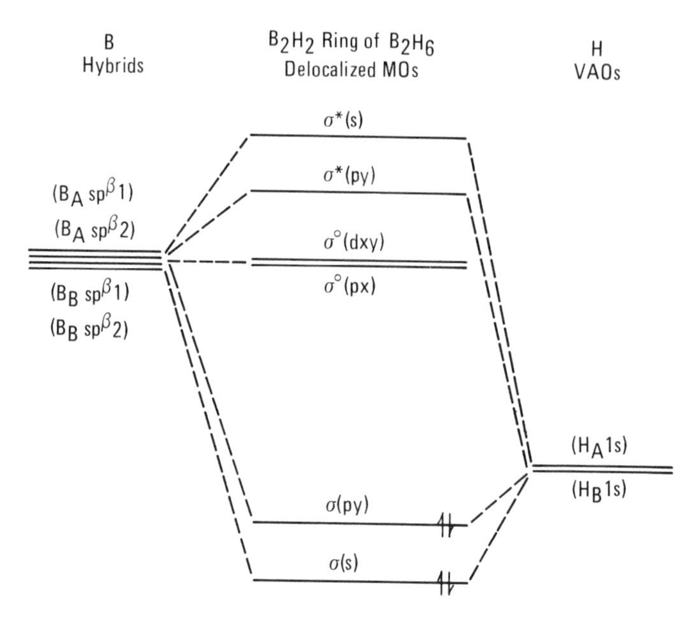

FIGURE 8.8 Delocalized MO energy level diagram for the B_2H_2 ring of B_2H_6.

two-electron bond formed by the overlap of two boron (sp^β) hybrid VAOs and a hydrogen (1*s*) valence orbital. Note that we did not choose here to point the lobes of the *spy* GO set between diagonally located B—H links. Such a choice is, of course available and this is left as an exercise. As expected, the average bond order is 0.5 per B—H link. Again, notice the difference in the localized bond pictures of FHF^- and B_2H_6. In FHF^-, *two* localized pairs of electrons give rise to an overall bond order of 0.5 per H—F link. This is because of the partial antibonding contribution in each localized three-center bond in FHF^-.

One of the reasons boron hydrides are typically reactive to Lewis bases is that the NBMOs in the B—H—B linkages are empty. These "frontier" orbitals are vulnerable to attack by electron pair donors. Thus B_2H_6 is cleaved by two $(CH_3)_3N$ molecules to give two molecules of $(CH_3)_3NBH_3$ in which no NBMOs are present as can be seen by drawing a Lewis structure.

Three-center bonds containing a bridging hydrogen between two borons always contain a pair of electrons in a delocalized BMO of the type just described. With this generality it becomes possible to assign a pair of electrons to each such three-center system we encounter in a new molecule. We will now do this in our treatment of B_5H_9. When B—H—B linkages occur in boron hydride systems, a complete description of their bonding in the localized view always leads to three-center two-electron MOs of the type just discussed. Anticipating this event we can reserve the hydrogen (1*s*) VAOs and the appropriate (sp^β) hybrid VAO on the borons to form such bonds when they are present. Thus for B_5H_9 we reserve two (Bsp^{β}) VAOs on each basal boron for the formation of the four localized B—H—B bonds. These B—H—B bonds will require a total of eight of the twenty-four available valence electrons.

For the apical atom, B_E, if we reserve the upwardly directed ($sp^{\alpha'}$) hybrid AO for bonding to the apical hydrogen, then the inwardly directed, orthogonal hybrid ($sp^{\beta'}$) is available for bonding the boron cluster and it is the only member of the VAO equivalence set B*sp*ai. The AOs (B_E2px) and (B_E2py) form the VAO equivalence set B*pt*. For each basal boron atom we reserve an (sp^{α}) hybrid orbital for bonding to the terminal hydrogen atom and two (sp^{β}) hybrid orbitals for contributions to two three-centered B—H—B bonds. There remains a linear combination of the (2*s*) and (2*p*) boron orbitals which is orthogonal to the reserved orbitals and is available for bonding the boron cluster. These inwardly directed hybrid orbitals, which we call sp^{γ}, belong to the VAO equivalence set B*sp*bi. Using the axis system in Figure 8.9 in which as expected, the GO center lies above the basal boron plane, Generator Table 8.5 is easily constructed after drawing the SOs in our usual way. The SO $\tilde{\sigma}(dxy|Bspbi)$ is the antibonding MO $\sigma^*(dxy)$ and by linear combination of the remaining SOs we form three BMO–ABMO pairs. Two of these pairs are degenerate since they arise from *px* and *py* GO partners. The five localized BH MOs associated with the terminal hydrogens, and the four BHB MOs associated with the bridging hydrogens together require eighteen electrons. The remaining six valance electrons fill the three five-center delocalized

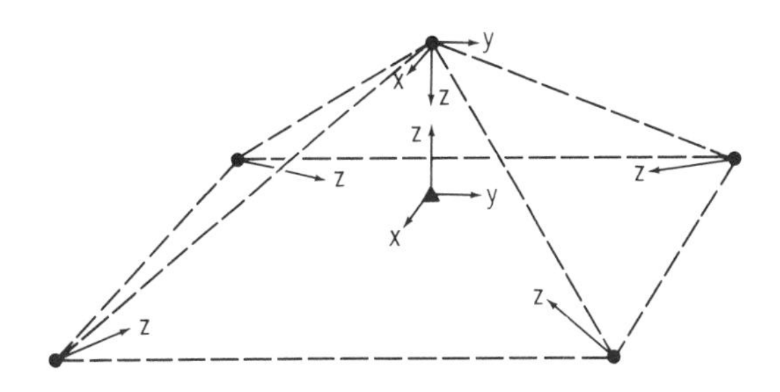

FIGURE 8.9 Axis system for the square pyramidal B_5 cluster of B_5H_9.

TABLE 8.5 Generator Table for B_5 Cluster in B_5H_9 (6 e^-)

	GOs			
VAO Equivalence Sets	*s*	*px*	*py*	*dxy*
B*sp*ai = ($B_Esp^{\beta'}$)	n			
B*pt* = (B_E2py)(B_E2px)		n	n	
B*sp*bi = (B_Asp^{γ}), (B_Bsp^{γ}), (B_Csp^{γ}), (B_Dsp^{γ})	b	n	n	a

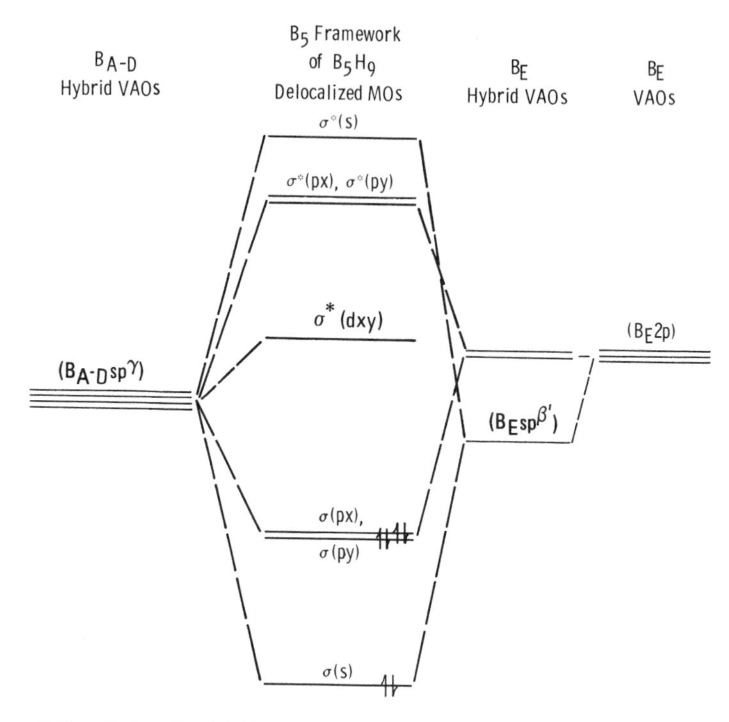

FIGURE 8.10 Delocalized MO energy level diagram for the B_5 framework of B_5H_9.

BMOs, as shown in the delocalized energy level diagram for B_5 cluster in Figure 8.10. The overall bond order is 3/8 since there are eight B—B links in the square pyramidal boron cluster.

The three bonding pairs in Figure 8.10 are localized by means of an sp^2 GO hybrid set. The resulting localized MOs are multicentered because of the unusually low bond order of the boron cluster. This molecule serves as yet another example of our observation that as the localized orbitals become increasingly multicentered, they generally become more difficult to visualize. This "delocalization" renders the localized view less useful as a complement to the delocalized view, and so we will not try to develop a localized view in such cases.

The molecular motion of square and tetragonal pyramidal molecules will now be briefly discussed. The motions of ZY_5, of which BrF_5 is an example, are obtained by taking linear combinations of the SMs of the basal Y atoms, the apical Y atom, and the Z atom that are generated by the same GO. The motions of a square pyramidal Z_5 cluster are similarly obtained. Since the motions of the component moieties of such molecules have already been examined, the generation of their LCAVs will be left as an exercise.

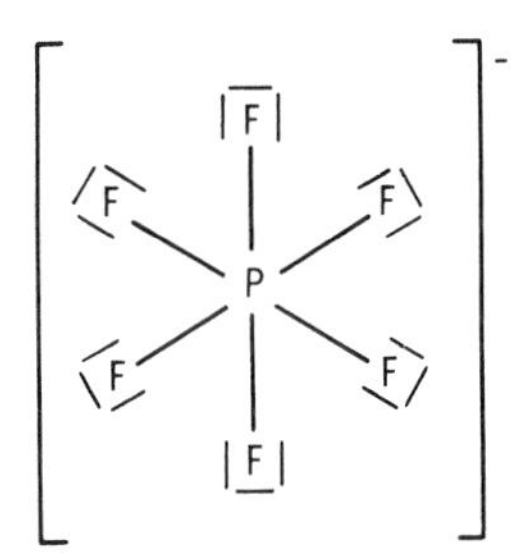

FIGURE 8.11 Electron dash structure of PF_6^-.

8.4. PF_6^-

The electron dash structure of PF_6^- in Figure 8.11 tells us that there are no lp's on phosphorus and that this hypervalent atom has twelve electrons surrounding it. VSEPR theory predicts that these six equivalent electron pairs will direct themselves toward the apices of a regular octahedron. A planar hexagonal array of fluorines around a phosphorus would be much less favorable because, such a geometry requires 60° between electron pairs.

We first assign the fluorine lone pairs to their (2*s*), (2*px*), and (2*py*) VAOs. Using the axis system in Figure 8.12 we can draw the SOs in our usual manner and construct Generator Table 8.6. From that table we recognize that four of the six electron pairs available for bonding are housed in four BMOs arising from linearly combining four pairs of SOs generated by the same GO. The remaining two electron pairs are housed in the nonbonding SOs (i.e., NBMOs) $\tilde{\sigma}(dx^2 - y^2)|\text{F}pr)$ and $\tilde{\sigma}(dz^2|\text{F}pr)$. The bond order in each

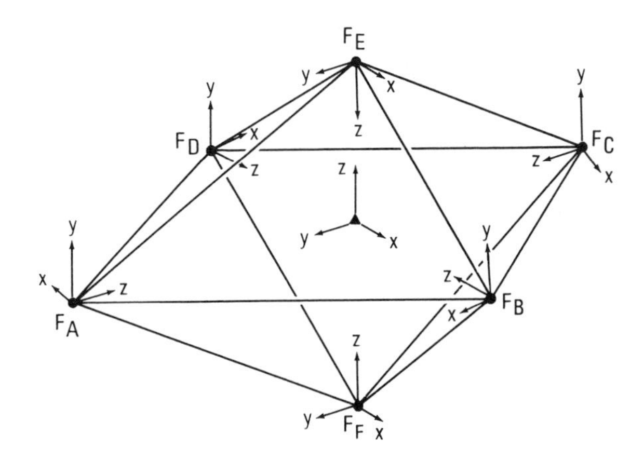

FIGURE 8.12 Axis system for PF_6^-.

TABLE 8.6 Generator Table for PF_6^- (12 e^-)

VAO Equivalence Sets	GOs					
	s	*px*	*py*	*pz*	$dx^2 - y^2$	dz^2
*P*s = (P3*s*)	n					
P*pr* = (P3*px*), (3*py*), (P3*pz*)		n	n	n		
F*pr* = (F_A2pz), (F_B2pz), (F_C2pz), (F_D2pz), (F_E2pz), (F_F2pz)	n	n	n	n	n	n

link is 2/3. Owing to the high electronegativity of fluorine, however, there is considerable ionic character in the P—F bonds.

Contrary to our usual procedure, all of the GO axes in Figure 8.12 are pointed toward (rather than between) atoms. Although we are geometrically unable to direct all three Cartesian axes between atoms, we could have pointed two of them in such directions (e.g., *x* and *y*). Three reasons why the axis system shown is more convenient for the octahedron are: (1) the equivalence of the GO Cartesian axes is more apparent, (2) the SOs among tangential orbitals (e.g., those among the F*pt* set in CoF_6^{3-} to be discussed shortly) are easier to visualize, and (3) the two partner sets within the (d) VAO set on a central atom are more readily visualized (see later).

Take this opportunity to review the above delocalized view of PF_6^- by doing the corresponding Practice Exercise in Node Game.

Inspection of Table 8.6 reveals that the hybrid GO set for localizing the occupied delocalized MOs is sp^3d^2. The main lobes of the members of such a set point to the apices of an octahedron. This enables us to construct a localized MO in each P—F link. The only SOs on phosphorus which can be called in by the sp^3d^2 GO hybrid set are in the P*s* and P*pr* equivalence sets. Thus, each main lobe of an sp^3d^2 hybrid GO can call in one member of a phosphorus digonal *sp* hybrid directed toward a fluorine. Each of these digonal hybrids is constructed from the (P3*s*) and one of the (P3*p*) VAOs, leading to the linear combinations $a(\text{P}3s) \pm b(\text{P}3px)$, $a(\text{P}3s) \pm b(\text{P}3py)$, and $a(\text{P}3s) \pm b(\text{P}3pz)$ ($b/a \approx 3^{1/2}$). Because full participation of the VAOs on phosphorus is not possible, these linear combinations are not orthogonal. To assure orthogonality of the localized MOs, each MO is bonding in one link and partially antibonding in the opposite link resulting in an average bond order of 2/3 per P—F link.

Species with forty-eight valence electrons such as PF_6^- are expected to be octahedral on the basis of VSEPR theory. Currently there is no simple way to predict the structure of fifty-electron hypervalent systems such as $SbBr_6^{3-}$, $SeBr_6^{2-}$, and XeF_6. The electron dash structure for XeF_6 in Figure 8.13 features seven electron pairs around the xenon atom. Experimentally it is found in several fifty-electron systems, such as $SeCl_6^{2-}$, $TeCl_6^{2-}$ and $SbCl_6^{3-}$, that the structures are essentially octahedral with little, if any,

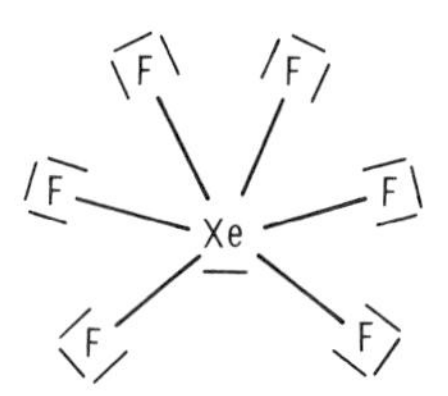

FIGURE 8.13 Electron dash structure for XeF_6.

significant distortion. By contrast, the fifty-electron system XeF_6 in the gas phase displays structural and spectroscopic properties that are consistent with some distortion. Because such a distorted octahedral system possesses very low symmetry, we will not consider it further.

The bonding in fifty-electron molecules that are truly octahedral can be rationalized in the delocalized view by allowing the two additional electrons to occupy an ABMO. This occupied ABMO must be *s*-generated since only the *s* orbital of all the GOs contains the fully symmetry of the octahedron. If these two electrons were placed in a triply degenerate *p*-generated ABMO set, only two of the MOs could be occupied by one electron each, and the full symmetry of the octahedron could not be contained in two half-occupied MOs because of the lack of occupation of the third member of the partner set. Not only would the molecule be paramagnetic (in contradiction to experiment) but it would also not be strictly octahedral because of Jahn–Teller distortion. Alternatively, we can localize the extra pair of electrons in a valence *s* orbital of the central atom. The spherical geometry of such an orbital distributes the lone pair density evenly around the central atom so that no distortion of the molecule can occur.

It would seem logical for XeF_6 to have the pentagonal pyramidal structure depicted in Figure 8.14a because such a structure is consistent with the experimentally observed structure of IF_7, in which the central atom is surrounded by seven bond pairs as shown in Figure 8.14b. However, the best

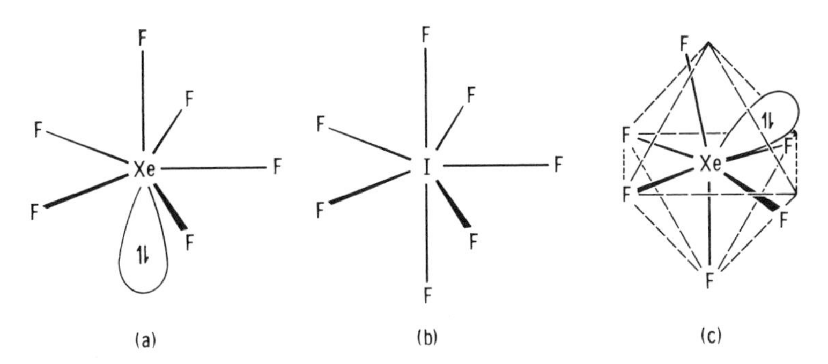

FIGURE 8.14 Possible structure for XeF_6 (a). Experimentally observed structures for IF_7 (b) and XeF_6 (c).

information we presently have indicates that the structure of XeF_6 in the gas phase contains distorted octahedral isomers of which one is shown in Figure 8.14c. Here the lone pair apparently protrudes through a triangular face of the octahedron. There is evidence that such distorted structures represent vibrationally excited states and that the ground state of XeF_6 is actually an undistorted octahedron.

The molecular motions of octahedral ZY_6 species will be discussed at the end of this chapter.

8.5. Octahedral Transition Metal Complexes—Some Special Considerations

Transition metal complexes are capable of using their incompletely filled (d) valence orbitals in the $n - 1$ shell for interaction with *ligand* orbitals. A *ligand* is an atom or molecule that binds to the metals. Recall that n refers to the principal quantum number of the valence shell. For example, in addition to the ($4s$) and ($4p$) valence orbitals of the first-row transition elements scandium through copper, we also include the ($3d$) orbitals as valence orbitals because they are of similar energy. The same relationship holds for the (d) orbitals in the penultimate shell and the AOs in the outermost shell in the two succeeding transition metal series of the periodic chart. Presumably this is also true for the next transition series, whose known members include elements 104 to 111. Little is known about the electronic structures of these elements, however, because of their radioactive instability. The lanthanides (cerium through lutetium) and the actinides (thorium through lawrencium) of the inner transition series are characterized by the filling of the ($4f$) and ($5f$) shells, respectively. The extent of (f) orbital participation in bonding, however, is presently not as well understood as that of (d) VAOs in the transition elements.

Lewis and others initiated the idea that metal atoms could accept lone pairs from ligands such as Cl^- and H_2O to form a *coordinate covalent link*. The terms "*coordination compounds*" or "*coordination complexes*" are used to describe such species. A localized bonding approach that stemmed from these ideas was popularized by L. Pauling during the 1930s and 1940s. However, the delocalized view of many coordination compounds is best visualized through the *ligand field theory* (LFT) which was developed by several theoretical physicists and chemists between 1930 and 1950. LFT is a refinement of an electrostatic theory called "*crystal field theory*" (CFT) which treats the ligands as negative point charges or point dipoles surrounding a positively charged metal ion. We note at this point that the crystal field model considers the metal–ligand interactions as strictly ionic. It is, of course, certainly more realistic to include, at least to some degree, covalent interaction which involves electron sharing via orbital overlaps and MO formation. This is the refinement that characterizes LFT which we will discuss after CFT. In further defense of LFT it should be noted that coordinated atoms are not points, as is assumed in CFT, since their sizes are quite comparable to the metal atom to which they are attached. It should also be noted that VSEPR considerations are not very useful in discussing structures of transition metal complexes.

To appreciate CFT let us briefly consider an octahedral complex such as $TiF_6{}^{3-}$. Anions of this type are generally formed by treating the neutral metal halide with an excess of halide ions. In $TiF_6{}^{3-}$ the trivalent metal is surrounded by six fluoride ions in an octahedral array. In the "naked" gaseous Ti^{3+} ion (i.e., no ligands are present) the single valence electron can occupy any of the five (3*d*) VAOs with equal probability (Figure 8.15a). The same would be true if six negative ligand charges were "smeared out" in a sphere around the metal, except that all the (*d*) VAOs would be equally raised in energy owing to repulsions between the ligand charges and the (*d*) electron on the metal (Figure 8.15b). If these six negative ligand charges arrange themselves into a well-defined octahedral array, however, they will tend to repel the electron density only in those (*d*) orbitals whose lobes are directed toward the charges [e.g., the $(dx^2 - y^2)$ and (dz^2) VAOs in the axis system shown in Figure 8.12. Less repulsion will occur with orbital lobes directed between the charged ligands [i.e., the (*dxy*), (*dxz*), and (*dyz*) VAOs]. It can be intuitively appreciated that merely rearranging the charges from spherical symmetry to octahedral symmetry, while keeping the metal–ligand distance constant, should not change the overall energy of the (*d*) VAOs. Consequently, two of the (*d*) orbitals will rise in energy by $(3/5)\Delta$ each while three descend by $(2/5)\Delta$ each (Figure 8.15c). In other words, if the overall energy splitting caused by the octahedral charge field is Δ, the energy of the spherical charge field is preserved if two orbitals go up in energy by $3/5\,\Delta$ (for a positive total of $6/5\,\Delta$) while three orbitals drop in energy by $2/5\,\Delta$ (for a balancing negative total of $-6/5\,\Delta$). The (3*dxy*), (3*dxz*), and (3*dyz*) VAOs are partners in an octahedral field because they are all in equivalent potentials in an octahedral field, and none of them by itself possesses the full symmetry of the octahedron. Since neither the $(3dz^2)$ nor the $(3dx^2 - y^2)$ orbitals have the full symmetry of the octahedron, they must from a second degenerate partner set.

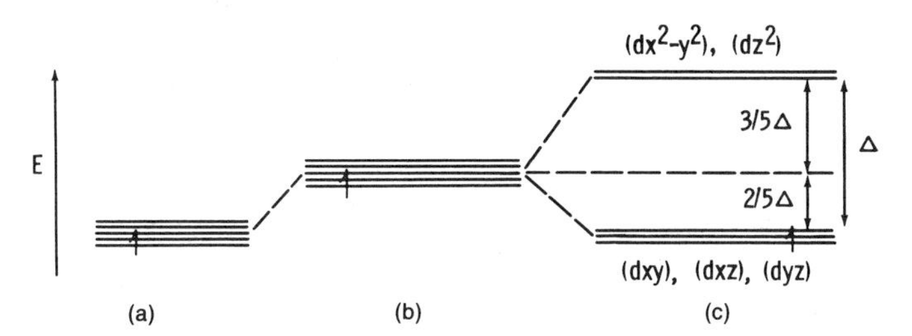

FIGURE 8.15 The relative energies of the (*d*) VAOs of a gaseous Ti^{3+} ion that is (a) free, (b) surrounded by six ligand charges in a spherical array, and (c) surrounded by six ligand charges in an octahedral geometry.

The titanium (*d*) VAO electron resides in the three degenerate orbitals since they lie at lower energy than the doubly degenerate set. Excitation of this electron by ultraviolet light of energy Δ promotes the electron to the higher-energy degenerate pair of orbitals wherein the electron encounters more severe repulsions with the ligand charges because of the closer proximity of the ligands to these orbitals.

An interesting consequence of the splitting of the (*d*) VAOs is that an octahedral complex with four (*d*) VAO electrons, for instance, may contain either four or two unpaired (*d*) VAO electrons, depending on whether or not Δ is smaller or larger than the electron pairing energy. Coordination complexes in which spin pairing occurs in the lower (*d*) orbitals are called *strong field* (i.e., Δ is large relative to the spin pairing energy) while those in which electrons remain unpaired by occupying the upper orbital set are referred to as *weak field* complexes.

The need for covalency in coordination compounds can be appreciated by realizing that it is unrealistic to consider the existence of an isolated Mo^{5+} ion in $MoCl_5$, for example, because the complete transfer of five electrons from a molybdenum atom to the chlorine ligands would result in an intolerably large positive charge on the metal. Substantial electron sharing between the chlorine atoms and the metal is thus expected in order to reduce this cationic charge.

There are many results from a variety of experiments which strongly demonstrate that electrons associated with the metal atom or ion in the uncomplexed state are partially delocalized onto ligand atoms in the complex and that ligand electron density is simultaneously shifted to the metal. In LFT, this sharing of electron density is accounted for by MO formation between ligand and metal atom VAOs. We have now returned to familiar ground, because this procedure is just the one that we have been following for a variety of geometries containing a central atom, including the octahedron. The major difference is that we now *must include penultimate* (*i.e., next to the last*) (*d*) *orbitals in MO formations* because in transition metals these AOs are energetically compatible with the valence (*s*) and (*p*) VAO set and are therefore also included in their VAO sets.

8.6. $CoF_6{}^{3-}$ and $Co[P(OCH_3)_3]_6{}^{3+}$

We will discuss these two transition metal complexes in some detail because they display rather contrasting electronic properties despite the fact that both have an octahedral arrangement of the ligating atoms around the cobalt and also contain the same oxidation state (3+) of the metal. Formally the $CoF_6{}^{3-}$ ion can be thought of as being made up of a Co^{3+} ion and six fluoride ions. Alternatively, it can be conceived of as possessing a cobalt atom, six covalently bound fluorine atoms, and three negative charges. The truth lies somewhere between these extremes. A Lewis structure can be

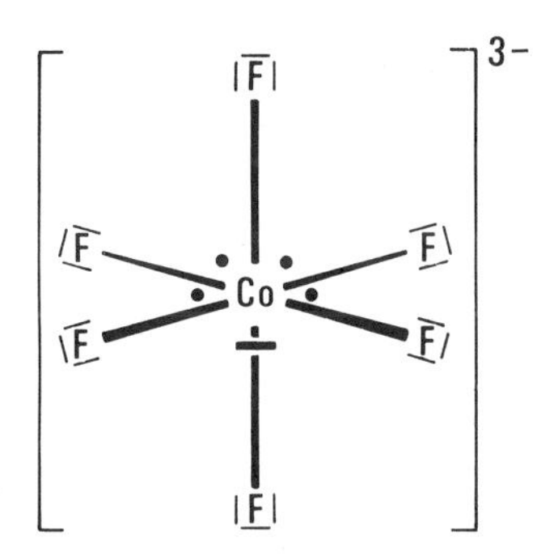

FIGURE 8.16 Lewis structure for CoF_6^{3-}.

drawn, as shown in Figure 8.16, which emphasizes the covalent character of this ion. This structure tells us that VSEPR theory is not helpful in predicting the structure of CoF_6^{3-} because the electrons we are forced to place around the cobalt would suggest a geometrical distortion from the experimentally observed octahedral one, and strict adherence to the normal Lewis rules would force us to pair up all the lone electrons on the cobalt to give a diamagnetic complex, contrary to the experimental observation of paramagnetism with four unpaired electrons. In the crystal field approach to this ion, all of the bond pairs shown in Figure 8.16 would become fluoride lone pairs, and the cobalt lone pair and lone singles would reside in (3*d*) VAOs on the Co^{3+} ion. The CFT picture stresses the contribution of the ionic interactions between the cobalt metal atom and the electronegative fluorines. The covalent electron dash structure in Figure 8.16 tells us that the formal charge on each fluorine is zero. The formal charge on cobalt, which formally possesses twelve electrons (six nonbonding and six bonding electrons) is -3, however, since a neutral cobalt atom has only nine electrons. The -3 formal charge accounts for the overall charge on the ion.

As we will soon see, the inclusion of (3*d*) orbitals in the cobalt VAO set permits π interactions to occur between some of these VAOs and the tangential (2*p*) VAOs of the fluorines. Therefore, we reserve only the (F2*s*) VAOs for a lone pair on each of the ligands. Although we would normally be inclined to reserve the (4*s*) VAO on cobalt for a lone pair, we will not do so because the (4*s*) VAO is not the lowest-energy VAO on the metal. This is because lower-energy (3*d*) VAOs are available and we will return to this possibility later.

In the axis system shown in Figure 8.12, the GO axes are directed toward the ligand atom centers. This arrangement is advantageous because it maintains the axis system used in the CFT model in which the ligands lie along the axes of the metal center. In Figures 8.17a–f are shown the SOs generated in the radial 2*pz* AO set and in Figures 8.18g–r are depicted the 12 SOs generated in the tangentially oriented (2*px*) and (2*py*) VAO set.

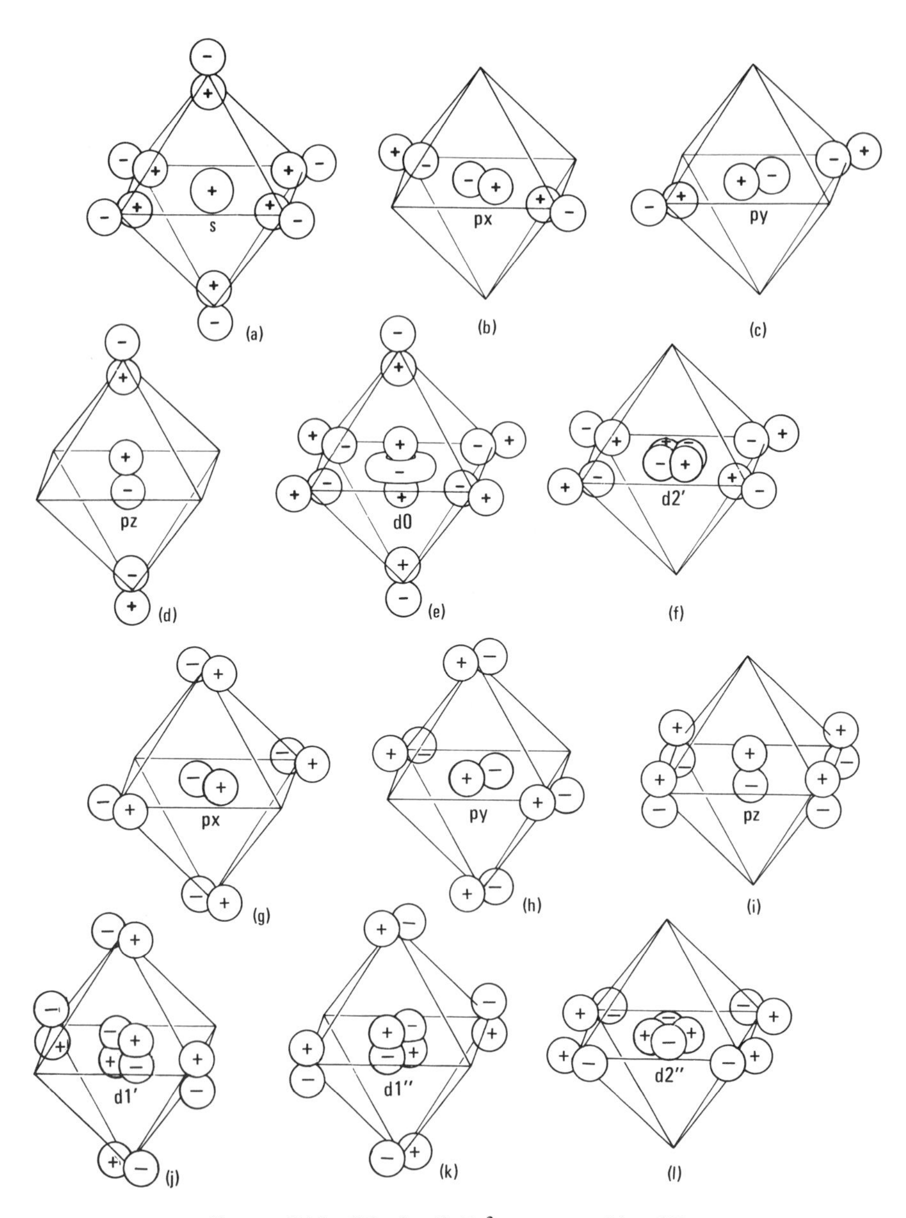

FIGURE 8.17 SOs for CoF_6^{3-} generated by GOs.

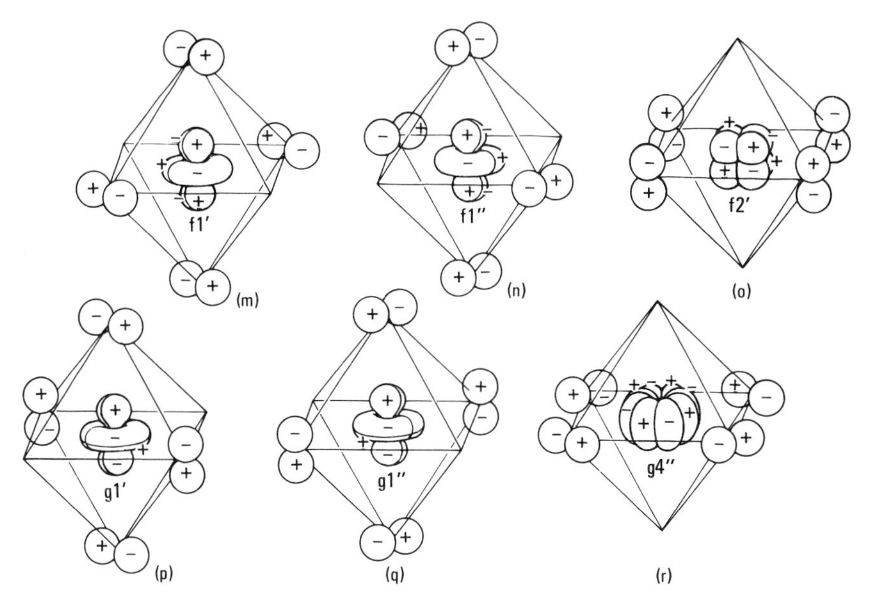

FIGURE 8.17 (cont.)

The pictorially obtained information in Figure 8.17 is summarized in generator Table 8.7. As we saw in the case of Generator Table 8.3 for BrF_5, the NBMO stemming from the three-entry s GO column was chosen to consist mainly of the $\sigma(\mathrm{s}|\mathrm{F}p\mathrm{b})$ SO, because that allowed (mainly) the remaining two SOs, which experienced good overlap with one another, to linearly combine to give a BMO–ABMO pair. Let us consider first the px GO column in Table 8.7. The pair of SOs having the best overlap between the cobalt and fluorines is $\tilde{\sigma}$ $(px|\mathrm{F}pr)$ and $\tilde{\sigma}(px|\mathrm{Co}pr)$. This pair when linearly combined (with very little contribution from $\tilde{\sigma}(px|\mathrm{F}pt)$) clearly gives an ABMO and a BMO. The NBMO must therefore consist mainly of the $\sigma(px|pt)$ SO (Figure 8.17g), which experiences very little overlap owing to the distance between F atoms in the complex. These same considerations apply to the py- and pz-generated columns in Table 8.7.

Let us now see how we arrive at the ordering of the energy levels in the MO diagram in Figure 8.18. From the relative energies in this figure of the Co (3d) VAOs and the F (2p) VAOs with which they interact, we see that the MOs with mainly cobalt (3d) character are $\sigma^*(dx^2 - y^2)$, $\sigma^*(dz^2)$, $\pi^*(dxy)$, $\pi^*(dxz)$ and $\pi^*(dyz)$ while the corresponding BMOs have mainly fluorine (4p) orbital character. Notice that the cobalt d VAOs are divided into two equivalence sets because they break up into two partner sets in the octahedral potential field. These partner sets are $dx^2 - y^2$, dz^2 and dxy, dxz, dyz. The former set engages in σ interactions with radially directed fluorine VAOs (Figures 8.17e and f) while the latter set can π bond with tangential fluorine

TABLE 8.7 Generator Table for CoF_6^{3-} (42 e^-)

VAO Equivalence Sets	GOs														
	s	px	py	pz	dx^2-y^2	dz^2	dxy	dxz	dyz	$f1'$	$f1''$	$f2'$	$g1'$	$g1''$	$g4''$
Codr = (Co3dx^2-y^2), (Co3dz^2)					n	n									
Coda = (Co3dxy), (Co3dxz), (Co3dyz)							n	n	n						
Cos = (Co4s)	n														
Copr = (Co4px), (Co4py), (Co4pz)		n	n	n											
Fpr = (F_A2pz), (F_B2pz), (F_C2pz), (F_D2pz), (F_E2pz), (F_F2pz)	n	n	n	n	n	n									
Fpt = (F_A2px), (F_A2py), (F_B2px), (F_B2py), (F_C2px), (F_C2py), (F_D2px), (F_D2py), (F_E2px), (F_E2py), (F_F2px), (F_F2py)		n	n	n			n	n	n	n	n	n	n	n	n

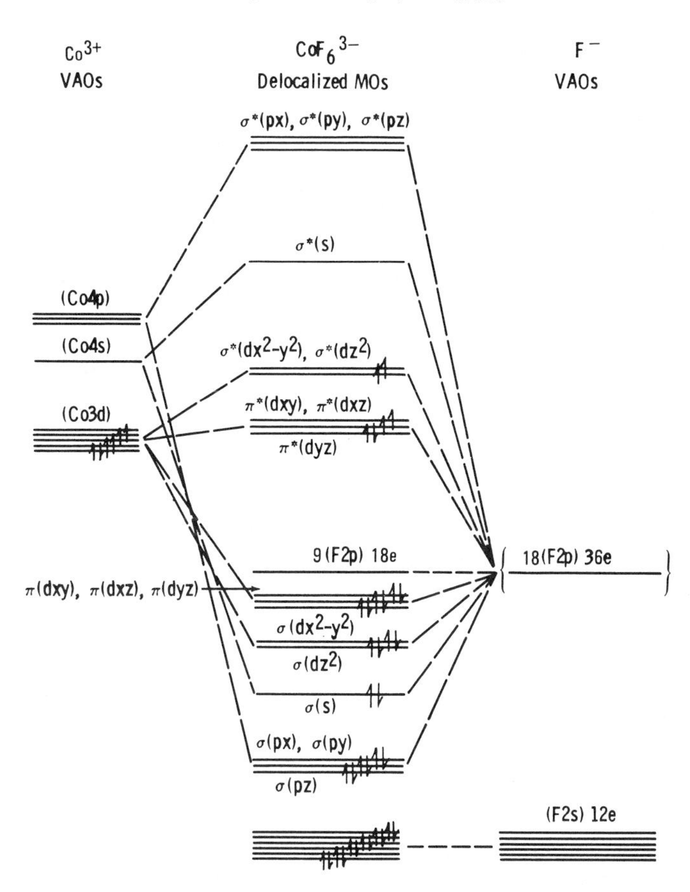

FIGURE 8.18 Delocalized MO energy level diagram for CoF_6^{3-}.

VAOs (Figures 8.17j–l). It is reasonable to place the four ABMOs arising from the linear combination of the *s*, *px*-, *py*-, and *pz*-generated SOs at higher energies than the ABMOs stemming from the linear combination of any of the *d*-generated SOs, since the cobalt (4*s*) and (4*p*) VAOs lie at higher energies than the Co (3*d*) VAOs. Furthermore, splittings of π MOs tend to be small compared with σ interactions, as we saw in Chapter 3 (Figure 3.13). After preassigning one lone pair on each fluorine to an F (2*s*) orbital, the remaining 42 electrons are seen to occupy nine BMOs, nine NBMOs and the lowest five ABMOs. We see that the lowest five ABMOs must be occupied as shown because we know from experiment that this complex has four unpaired electrons. It may be noted that six of the nine NBMOs in Figure 8.18 are generated by *f* and *g* GOs (Figure 8.17m–r) and three are generated by the *p* GOs as just discussed.

As we saw earlier, the CoF_6^{3-} ion is a weak-field complex. In that case, we would place each of the first five electrons of this d^6 ion in separate cobalt (d) VAOs in the crystal field orbital diagram in Figure 8.15, and the sixth electron would pair with an electron in the (dxy), (dxz), (dyz) set. Notice first of all, that this occupation pattern is very much like that in the LFT MO energy diagram in Figure 8.18, and secondly, that in the CFT approach these electrons are localized on the cobalt ion, whereas in the LFT view these electrons are in *antibonding* MOs. The energy difference Δ as calculated from ligand field theory is, of course, different from that obtained from crystal field theory. The more ionic the interaction is, however, the more the two values will tend to agree. The doubly and triply degenerate (d) orbitals in the CFT model for octahedral systems arise from ionic interactions, whereas in the LFT approach the corresponding triply degenerate MOs are largely nonbonding (partially antibonding) metal VAOs, and the corresponding doubly degenerate MOs possess substantial antibonding character. Covalency is also evident in the BMOs because the ligand electrons do attain some metal character owing to orbital overlap in the MOs. Finally, we note that the bond order is difficult to determine because of the partial occupation of σ and π ABMOs. If we make the crude approximation that the six antibonding electrons cancel six bonding electrons, the bond order per Co-F link is 1.0.

Before proceeding to the localized bond picture for CoF_6^{3-}, it should be realized that ligands such as PH_3 or AsF_3 have no lone pairs on the ligating Group V atom for π interactions with the metal (d) VAOs. However, there are empty (d) VAOs on phosphorus and arsenic in such ligands, respectively, and there is evidence that such interactions may occur between these orbitals and the metal (d) VAOs. The example with which we choose to illustrate this concept is the octahedral, colorless diamagnetic trivalent cobalt complex $Co[P(OCH_3)_3]_6^{3+}$ whose electron dash structure appears in Figure 8.19. Inspection of the set of five d VAOs on any of the six phosphorus atoms (taking into account the axis system in Figure 8.12) tells us that only the ($P3dxz$) and ($P3dyz$) VAOs have the correct symmetry for π interactions with the metal. We also see that the two lobes of these VAOs which are directed into the octahedron for such interactions have the same symmetry with respect to the octahedron as do the ($P3px$) and ($P3py$) VAOs. Thus the SOs in Figures 8.17g–r drawn for two-lobed (p) VAOs are seen to be analogous to

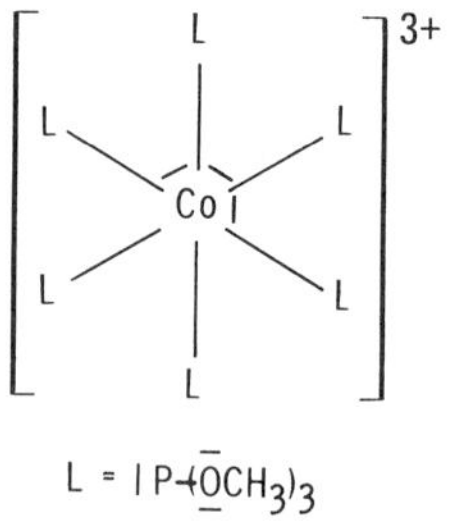

$L = |P(\overline{O}CH_3)_3$

FIGURE 8.19 Electron dash structure for $Co[P(OCH_3)_3]_6^{3+}$.

those expected for a set of four-lobed (dxz) and (dyz) ligand orbitals. The SOs for the (P3dxy) AOs are depicted in Figure 8.20. The interactions of these

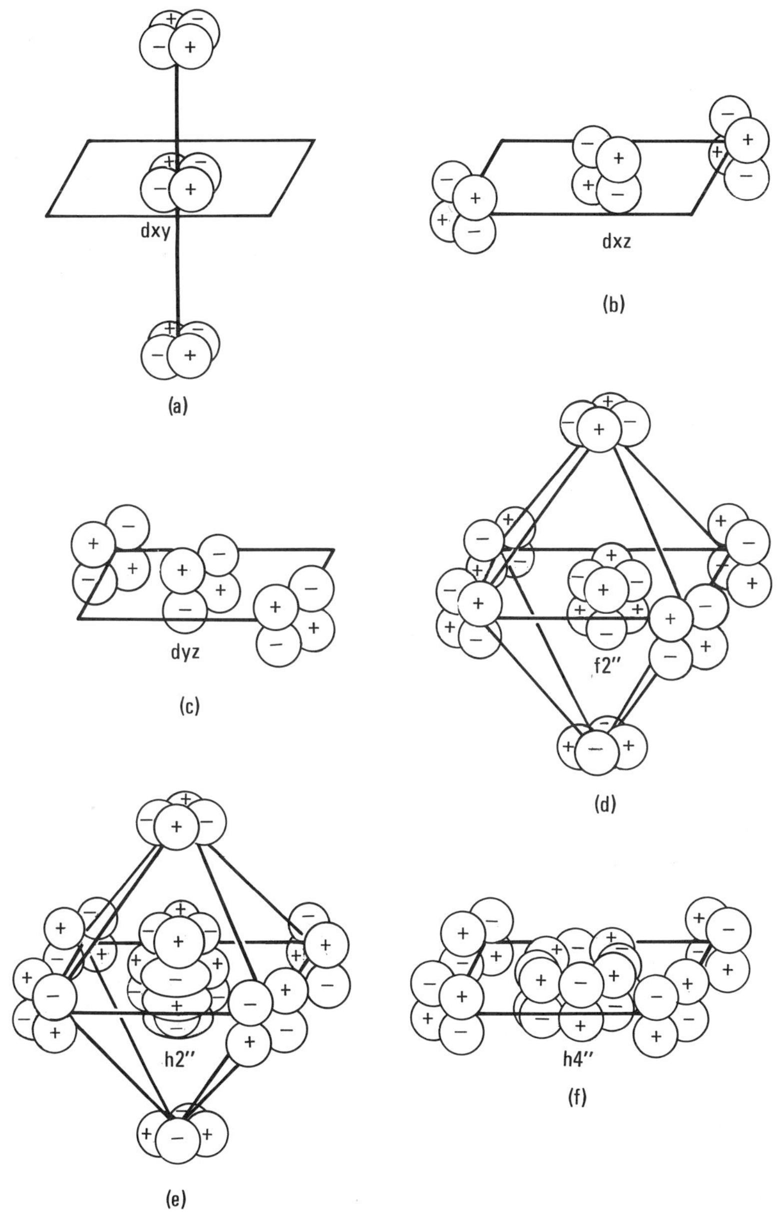

FIGURE 8.20 SOs for the phosphorus (3dxy) AOs in $Co[P(OCH_3)_3]_6^{3+}$.

AOs with the (dxz), (dyz), and (dxy) VAOs of cobalt are δ, rather than π. A δ interaction is seen to be weaker than π, and we will view it as negligible in the present example. From Figures 8.17g–r and 8.20, we conclude that, except for Figures 8.17j–l, all the ligand SOs form NBMOs to a first approximation. Thus the SOs in Figures 8.17g–i form largely NBMOs because the metal $(4p)$ VAOs are mainly involved in σ bonding. The other SOs which become nonbonding (Figures 8.17m–r) derive this property from the fact that the GOs that generated them possess symmetries that are incompatible with any metal AOs in the valence set.

In general, valence (d) *VAOs on ligands lie above the metal* (d) *VAOs.* Thus the splitting between the resulting triply degenerate BMO and doubly degenerate ABMO levels (see Figures 8.17j–l and 8.17e, f, respectively) tends to *increase* Δ. To understand this point, consider first a ligand such as NH_3, which does no π bonding at all. In that case the $\pi^*(dxy)$, $\pi^*(dxz)$, $\pi^*(dyz)$ MOs would lie at the same energy as the (Co3d) set, becoming $\pi^0(dxy)$, $\pi^0(dxz)$, $\pi^0(dyz)$ NBMOs. In addition, the $\pi(dxy)$, $\pi(dxz)$, $\pi(dyz)$ BMOs would disappear [as well as the nine (F2p) lone pair orbitals] because NH_3 has no π-type lone pairs on the nitrogen. If we now "turn on" π interactions between the $\pi^0(dxy)$, $\pi^0(dxz)$, $\pi^0(dyz)$ orbitals and a set of *higher-lying* (dxy), (dxz), (dyz) orbitals on the phosphorus atoms in $Co[P(OCH_3)_3]_6{}^{3+}$, the splitting that results causes the $\pi^0(dxy)$, $\pi^0(dxz)$, $\pi^0(dyz)$ orbitals to descend in energy, while a corresponding set of three antibonding MOs appears above the phosphorus $(3d)$ VAO set. As a consequence, Δ (the energy difference between the $\sigma^*(dx^2 - y^2)$, $\sigma^*(dz^2)$ MOs and the $\pi(dxy)$, $\pi(dxz)$, $\pi(dyz)$ sets is larger than it was when no π interactions were present. In contrast, turning on π interactions with *lower-lying* (p) orbitals on the ligand decreases Δ (Figure 8.18). In all three cases, the higher-energy doubly-degenerate MOs $\sigma^*(dx^2 - y^2$, $\sigma^*(dz^2)$ marking the upper energy boundary for Δ are *always* σ antibonding MOs, while the lower-lying triply degenerate orbitals are somewhat π antibonding in the case of ligand (p) VAO interactions (Figure 8.18), nonbonding when no π interactions occur, and somewhat bonding when there are ligand (d) VAO interactions. Because of the usual Lewis octet rules, ligands with electron pairs in their (p) VAOs (such as F^-) can act as π electron pair "donors" to the metal for making π BMOs [see $\pi(dxy)$, $\pi(dxz)$, $\pi(dyz)$ in Figure 8.18] while ligands containing empty (d) VAOs [e.g., phosphorus in $P(OCH_3)_3$] act as π electron "acceptors" from the metal for constructing π BMOs. As a result, ligands that accept π electron density from the metal strengthen the ligand field (i.e., increase covalent bonding), while π-donating ligands (such as F^-) weaken the ligand field (i.e., favor ionicity).

In formulating a localized approach for $CoF_6{}^{3-}$, we note in Figure 8.18 that an s, three p, and five d GOs are associated with the occupied BMOs in the delocalized MO scheme for $CoF_6{}^{3-}$. However, it is always simpler to localize the σ and the π MOs separately. The six delocalized σ BMOs are generated by the s, the three p, the dz^2, and the $dx^2 - y^2$ GOs. First, we consider six delocalized sigma BMOs, which are generated by the s, the three

p, the dz^2, and $dx^2 - y^2$ GOs. These six GOs can be hybridized so that the main lobes form an octahedrally oriented d^2sp^3 GO set. The hybrid GOs generated an identical set of cobalt VAOs and they also call in a ($2pz$) VAO on each fluorine. Six localized MOs can then be formed from the six pairs of cobalt–fluorine VAOs. The Lewis structure in Figure 8.16 is in accord with this logic because this structure suggests that there are six σ localized BMOs.

The doubly occupied ABMO in Figure 8.18 is composed predominantly of one of the (d) VAOs in the triply degenerate set, or some linear combination of these VAOs. For this reason we are unable to further localize this electron pair. The same is true for the remaining two unpaired electrons in this orbital set and the two in the doubly degenerate set since they reside in MOs that arise from linear combinations of SOs generated by GOs that lie in the same set. Because the doubly occupied π BMOs also arise from SOs generated by dxy, dxz, dyz, their electron pairs are not localized. The electron pairs in the nine NBMOs are already largely localized in fluorine orbitals and so we do not attempt further localization. In Figure 8.21 is an energy-level

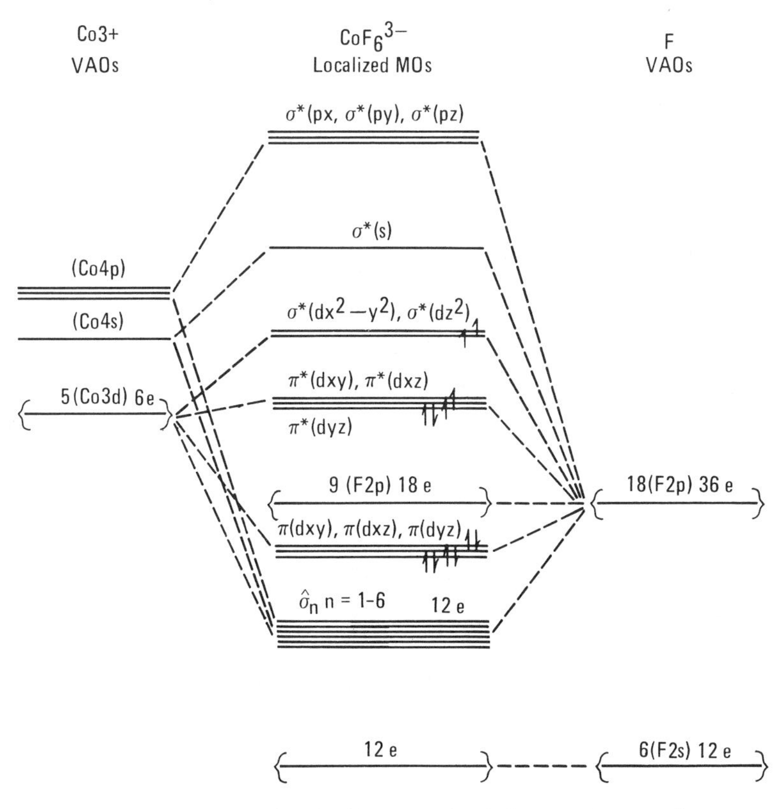

FIGURE 8.21 Localized MO energy level diagram for CoF_6^{3-}.

diagram reflecting our localization scheme. Notice that the bond order in Figure 8.21 (1.0) is the same as the one we deduced earlier from Figure 8.18 for the delocalized view.

8.7. $B_6H_6^{2-}$

Like B_5H_9 and other electron-deficient boron hydrides, the structure of the $B_6H_6^{2-}$ ion is not predicted by drawing Lewis structures. We therefore must be satisfied with the experimental finding that $B_6H_6^{2-}$ consists of an octahedral cluster of boron atoms, each possessing a terminal hydrogen as shown in Figure 8.22.

We first assign the B—H bond pairs in $B_6H_6^{-2}$ to localized MOs, each of which is composed of an (H1*s*) and a (B*sp*$^{\alpha}$) VAO. The remaining VAOs on each boron are made up of a digonal (B*sp*$^{\beta}$) hybrid pointed into the center of the octahedron and a (B2*px*) and a (B2*py*) orbital oriented tangentially around the octahedron along the axes shown in Figure 8.12. The SOs in the inward-pointing and tangential sets are visualized as depicted in Figure 8.17 and we then summarize them along with their corresponding GOs in Generator Table 8.8. It may be noted that the equivalence sets in this table are analogous to the F*pr* and F*pt* equivalence sets in Table 8.7. Although the GOs are also the same, the b and a entries under them in Table 8.8 reflect the fact that the boron atom VAOs interact with one another because they are close enough to each other. Examination of the SOs in Figure 8.17b–d and g–i shows us that the *p*-generated BMOs are linear combinations dominated by the SOs of the B*pt* set (Figures 8.17g–i) because the VAOs they contain are on adjacent atoms. By contrast, the VAOs in the SOs shown in Figures 18.17b–d are on opposite borons and so these nonbonding SOs dominate the antibonding linear combination. The three BMOs from the linear combinations are π in appearance (Figure 8.17g–i) and the corresponding three ABMOs are σ^* in appearance (Figure 8.17b–d). In the energy level diagram

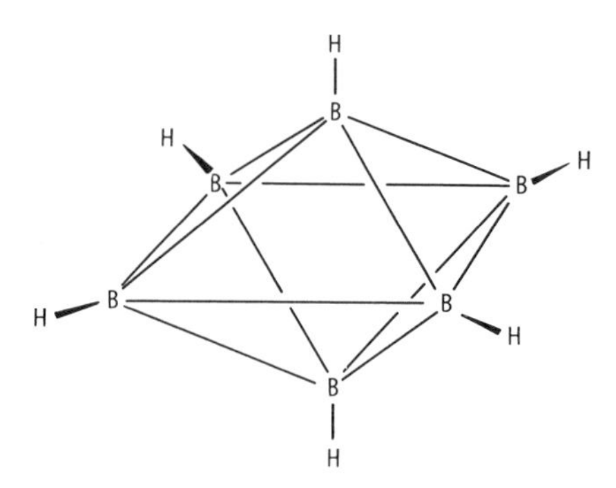

FIGURE 8.22 Structure of $B_6H_6^{2-}$.

TABLE 8.8 Generator Table for B_6 Cluster in $B_6H_6^{2-}$ (14 e^-)

	GOs														
VAO Equivalence Sets	s	px	py	pz	dxy	$dx^2 - y^2$	dxz	dyz	dz^2	$f1'$	$f1''$	$f2'$	$g1'$	$g1''$	$g4''$
Bsp-in = ($B_A sp^\beta$), ($B_B sp^\beta$), ($B_C sp^\beta$), ($B_D sp^\beta$), ($B_E sp^\beta$), ($B_F sp^\beta$)	b	n	n	n		a			a						
Bpt = ($B_A 2px$), ($B_A 2py$), ($B_B 2px$), ($B_B 2py$), ($B_C 2px$), ($B_C 2py$), ($B_D 2px$), ($B_D 2py$), ($B_E 2px$), ($B_E 2py$), ($B_F 2px$), ($B_F 2py$)		b	b	b	b		b	b		a	a	a	a	a	a

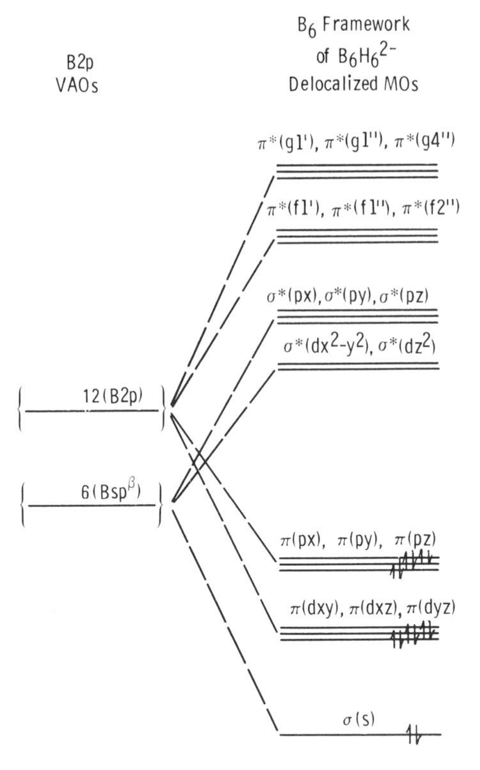

FIGURE 8.23 Delocalized MO energy level diagram for the B_6 framework of $B_6H_6^{2-}$.

in Figure 8.23, the electron occupation leads us to conclude that the average bond order in the B_6 cluster is 7/12 per B—B link.

From Figure 8.23 we see that seven GOs are associated with the filled BMOs. If we treat the inwardly directed σ and the tangential π MOs separately, we find that, although no localization of the single bond pair in $\sigma(s)$ stemming from the inwardly directed MO is possible, it is possible to localize the six tangentially oriented bonding MO pairs. The relevant GO hybrid set is p^3d^3, whose members point to the vertices of a trigonal antiprism. Such an antiprism can be visualized in the octahedral array of boron atoms by picking any pair of oppositely oriented triangular faces in the octahedron and imagining the main lobes of the p^3d^3 hybrid set to point almost toward the midpoints of the edges of the triangles as indicated in Figure 8.24. Six bond pairs are localized in the regions toward which these six lobes point. It must be realized that there are antibonding contributions from the other boron ($2px$) and ($2py$) VAOs, which confer an overall B—B bond order of 7/12. The natures of these antibonding interactions are complicated by the rather complex geometry of the octahedron and we will not discuss

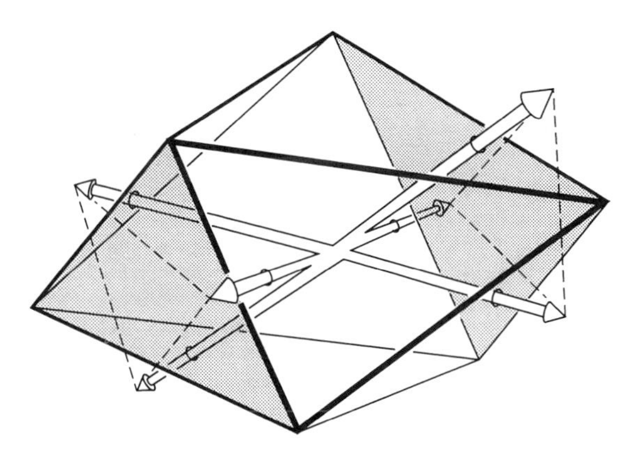

FIGURE 8.24 Sketch of p^3d^3 hybrid GO set inscribed in an octahedron.

them further. The energy level diagram that corresponds to this localized view is shown in Figure 8.25. By localizing the six bond pairs near the edges of the opposite triangular faces as shown in Figure 8.24, we see that these opposite faces do not appear to be bound to one another. Because the manner in which we orient the trigonal antiprismatic array of bond pairs is arbitrary, however, three additional resonance forms can be drawn which involve the remaining three pairs of opposite triangular faces. Recall that a similar set of σ-bonded resonance structures was encountered for the electron-deficient ion H_3^+. In contrast to $B_6H_6^{2-}$, in which some localization can be achieved, no localization was realized for the two bonding electrons in H_3^+.

8.8. $Mo_6Cl_8^{4+}$

Reaction of $MoCl_4$ with molybdenum metal at 600°C gives Mo_6Cl_{12} from which four chloride ions are easily lost to give the $Mo_6Cl_8^{4+}$ cation. Although a Lewis structure that is not electron deficient can be drawn, the prediction of the experimentally determined structure by any set of simple VSEPR rules is not feasible. Many octahedral metal cluster compounds are known, but other symmetrical geometries, including triangular, tetrahedral, square pyramidal, trigonal bipyramidal, trigonal prismatic, and icosahedral have also been observed. In the structure in Figure 8.26, each line represents an electron pair so that the Lewis structure is also represented. Each Cl—Mo bond can be viewed as a localized two-center two-electron bond composed of a (p) VAO on the chlorine and one member of a square planar set of molybdenum ($spxpydxy$) hybrid VAOs. [The chlorine lone pair is assigned to a (Cl3s) VAO.] The VAOs remaining on each molybdenum for binding the

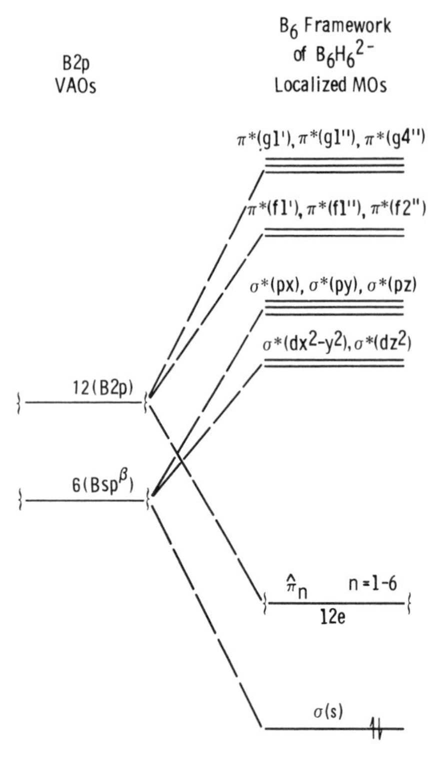

FIGURE 8.25 Localized MO energy level diagram for the B_6 framework of $B_6H_6^{2-}$.

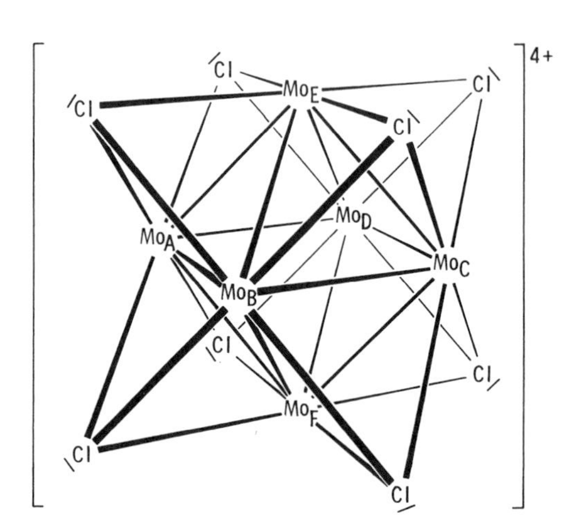

FIGURE 8.26 Structure of $Mo_6Cl_8^{4+}$.

metal cluster are ($5pz$), ($4dz^2$), ($4dx^2 - y^2$), ($4dxz$) and ($4dyz$). From the first two orbitals, which are oriented along the (radial) z axis, we can construct an Mo*pr* equivalence set composed of inward pointing Mo$pzdz^{2\beta}$ and outward pointing Mo$pzdz^{2\alpha}$ hybrids. Only the members of the inwardly directed hybrid set will be close enough to interact with one another and they will pictorially generate SOs analogous to those in Figure 8.17a–f. Similarly, the (Mo4dxz) and (Mo4dyz) VAOs generate SOs analogous to those in Figures 8.17g–r. The SOs for the (Mo4$dx^2 - y^2$) VAOs are depicted in Figure 8.27. Notice in this figure how the *nodal* properties of the GO carry over into SOs (MOs) formed by the molybdenum ($4dx^2 - y^2$) VAOs. This is particularly obvious in Figure 8.27e. In Figure 8.27b, notice that each adjacent pair of lobes along an edge of the horizontal plane have opposite signs as expected when crossing a nodal plane of the GO. At first glance, it may seem that there should be no (Mo4$dx^2 - y^2$) VAOs on the horizontal plane in Figure 8.27b. Remember, however, that the lobes of the $dx^2 - y^2$ GO are not really spheres but are shaped like kidney beans tucked "stem first" into the corners of the nodal planes as shown in Figure 2.6. Therefore, the signs of the $Mo_6Cl_8^{4+}$ molybdenum ($4dx^2 - y^2$) VAO lobes that lie above and below the horizontal plane in Figure 8.27b are determined by the GO lobe facing those lobes. In Figure 8.27a the two cones of the dz^2 GO call in VAO lobes above and below the plane of the same sign and lobes of the opposite sign all around the edge of this plane. Thus the upper positive lobes are separated from the negative lobes by a cone (whose surface undulates in a regular manner around the z axis). The lower positive lobes have a similar conal node associated with them. The same is true in Figures 8.27c and d except that each GO and SO also contains a vertical planar node. In Figure 8.27f, there are two additional cones as well as another plane. The question may be asked why the additional two cones are necessary. The answer is that if they were absent, the SO generated in Figure 8.27b would be duplicated. Having generated all the SOs available for binding the Mo_6 cluster, we generate Table 8.9. This table looks very much like the one for $B_6H_6^{2-}$ (Table 8.8) in our GO requirements except that an $i2'$ GO is needed to accommodate the use of a ($4dx^2 - y^2$) VAO on each molybdenum. The 24 electrons available to bind the molybdenum atom cluster are housed in 12 bonding orbitals, one from each of the first 12 columns of SOs in Table 8.9. As expected from the Lewis structure, the bond order for $Mo_6Cl_8^{4+}$ indicated by the energy level diagram in Figure 8.28 is 1.0. Without doing calculations, it is not possible to determine the order of the levels in this energy level diagram. In fact, when such calculations are carried out, the degenerate $\psi(dx^2 - y^2)$, $\psi(dz^2)$ levels turn out to be the highest-energy occupied MOs. Because chlorine is more electronegative than molybdenum, there is considerable ionicity in the Mo—Cl bonds.

The SOs associated with the occupied delocalized MOs can be mixed to form an $sp^3d^5f^3$ hybrid GO set whose members direct their main lobes to the edges of an octahedron. This GO hybrid set calls in contributions from the inward-pointing Mo$pzdz^{2\beta}$ hybrid VAOs and the (Mo5s), (Mo4dxz),

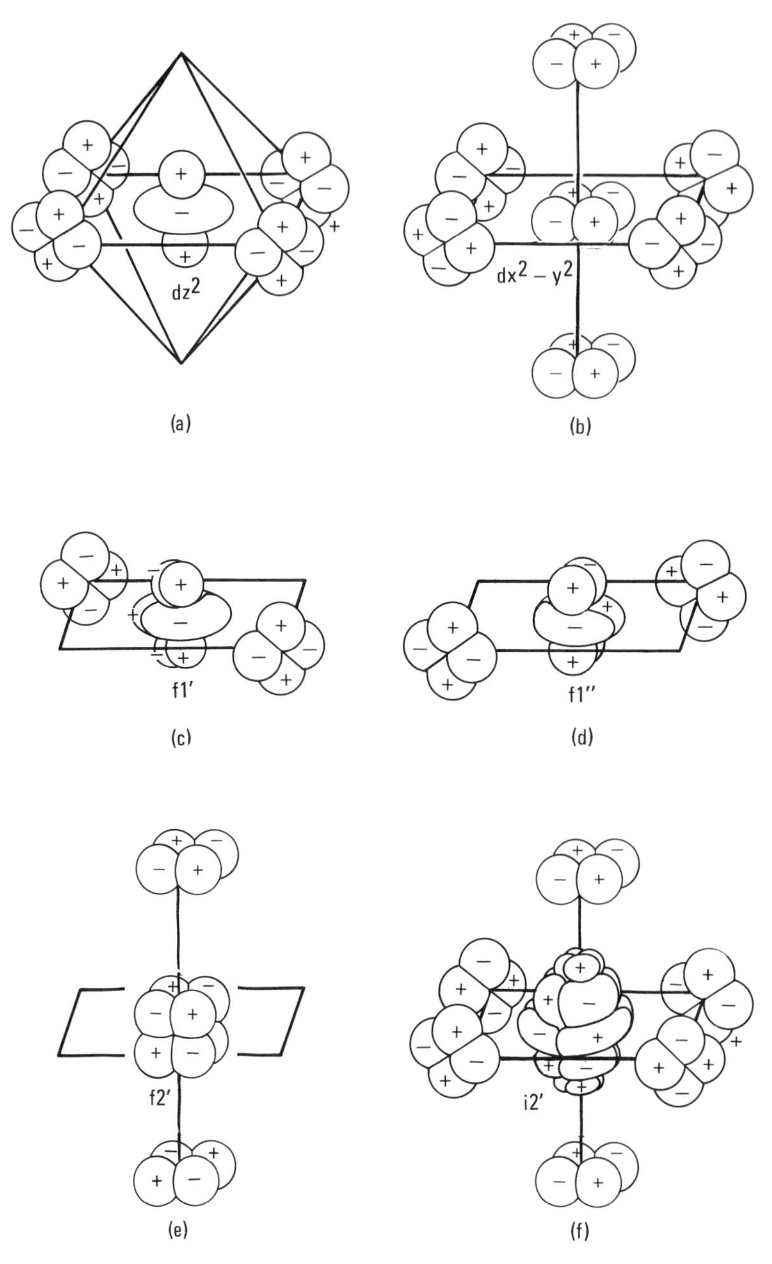

FIGURE 8.27 SOs for the molybdenum ($4dx^2 - y^2$) VAOs in $Mo_6Cl_8{}^{4+}$.

TABLE 8.9 Generator Table for Mo_6 Cluster in $Mo_6Cl_8^{4+}$ (24 e^-)

VAO Equivalence Sets	GOs															
	s	px	py	pz	dxy	dx^2-y^2	dxz	dyz	dz^2	$f1'$	$f1''$	$f2'$	$g1'$	$g1''$	$i2'$	$g4''$
Mppr = ($Mo_A pzdz^{2\beta}$), ($Mo_B pzdz^{2\beta}$), ($Mo_C pzdz^{2\beta}$), ($Mo_D pzdz^{2\beta}$), ($Mo_E pzdz^{2\beta}$), ($Mo_F pzdz^{2\beta}$)	b	n	n	n		a			a							
Moda = ($Mo_A 4dx^2 - y^2$), ($Mo_B 4dx^2 - y^2$), ($Mo_C 4dx^2 - y^2$), ($Mo_D 4dx^2 - y^2$), ($Mo_E 4dx^2 - y^2$), ($Mo_F 4dx^2 - y^2$)						b			b	n	n	n			a	
Modb = ($Mo_A 4dxz$), ($Mo_A 4dyz$), ($Mo_B 4dxz$), ($Mo_B 4dyz$), ($Mo_C 4dyz$), ($Mo_C 4dxz$), ($Mo_D 4dxz$), ($Mo_D 4dyz$), ($Mo_E 4dxz$), ($Mo_E 4dyz$), ($Mo_F 4dyz$), ($Mo_F 4dxz$)		b	b	b	b		b	b		a	a	a	a	a		a

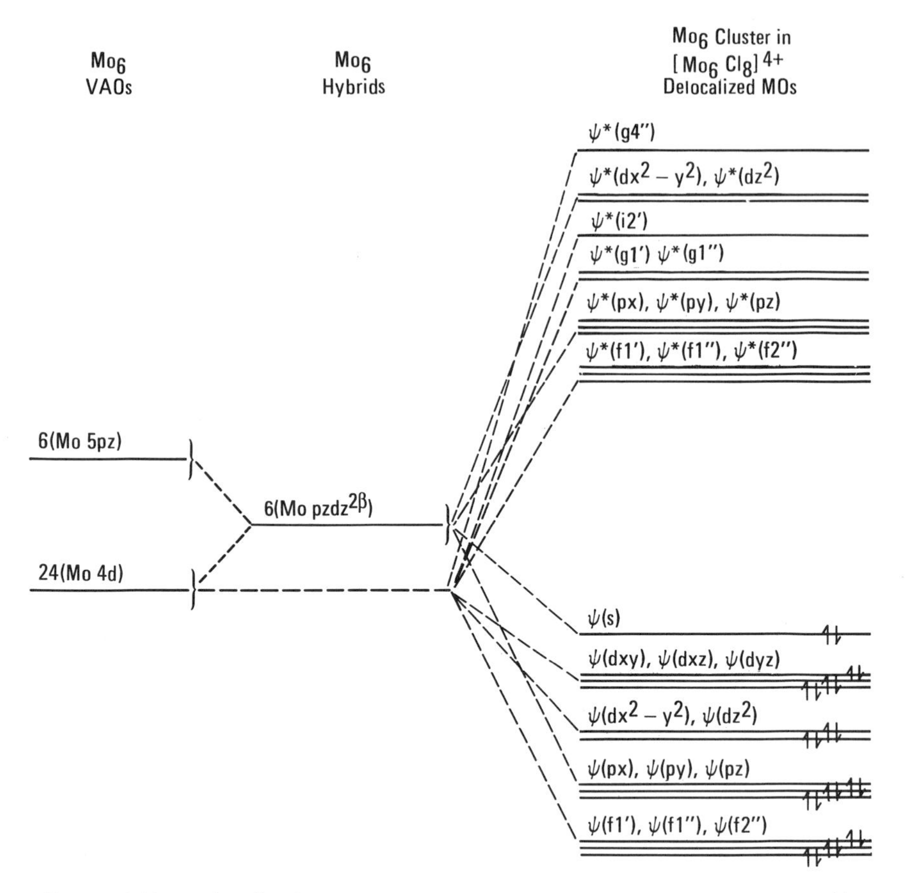

FIGURE 8.28 Delocalized MO energy level diagram for Mo_6 cluster of $Mo_6Cl_8{}^{4+}$.

(Mo4*dyz*), and (Mo4$dx^2 - y^2$) VAOs. [Recall that the (Mo4*dxy*) VAO was used to bind the chlorines]. Although the nature of these contributions is complicated, the result is that a bond pair is localized to a large extend at the midpoint of each of the 12 edges of the octahedral metal cluster. This localized picture conforms nicely with the Lewis structure.

The molecular motions of Z_6 and YZ_6 species are obtained in the usual way. The same is true for a cluster system such as $[MoCl_8]^{4+}$, although it would be a somewhat complicated process. The molecular motions of octahedral Z_6 molecules are obtained by taking appropriate linear combinations of the symmetry motions for Z_6 represented in Figure 8.17. For ZY_6, the SMs of the central atom AVs must also be taken into account. Drawing these motions is left as an exercise.

TABLE 8.10 A Useful Summary of Procedures for Generating Pictorial Representations of Delocalized and Localized Bonding Views of a Molecule and for Visualizing its Vibrational Modes

Delocalized View	Localized View	Normal Modes
1. Decide on best electron dash structure using lowest possible formal charges and drawing resonance forms. In such forms, some lone pairs may be involved in a π system.	1. Decide on which set(s) of doubly occupied delocalized MOs to localize.	1. Identify the motional atomic vector (AV) equivalence sets and list them in the generator table.
2. Establish the molecular geometry from VSEPR theory if possible. *For transition metal complexes and boron hydrides, other considerations apply.*	2. Hybridize the GOs which give rise to the set(s) in step 1.	2. Using GOs, draw the symmetry motions (SMs).
3. After preassigning lone pairs to (*s*) and suitably oriented (*p*) VAOs, identify all the VAO equivalence sets (hybrids may also be used) and list them in the generator table.	3. Using the hybridized GOs, draw the SOs.	3. Fill in the rest of the generator table (i.e., GOs and m designations).
4. Using GOs located at a molecular center or centroid, draw all the SOs, recalling that #SOs = #VAOs.	4. Draw the MOs, being aware of the consequences of a central atom on which the full set of SOs (i.e., hybridized VAOs) is not generated by the hybridized GO set. If necessary, rehybridize peripheral atom VAOs and rotate hybrid GOs to locate bond pairs between atoms.	4. Take appropriate linear combinations of SMs generated by the same GO and draw all the molecular motions.
5. Fill in rest of generator table (i.e., the necessary GOs and the n, a, or b designations). Establish partner GO and SO sets.	5. Draw the resulting MO energy level diagram and occupy the appropriate levels with electrons.	5. Identify the normal vibrational modes.
6. Take linear combinations of SOs generated by the same GO and draw all the MOs.	6. Verify that the bond order per link is the same as in the delocalized view.	
7. Draw the resulting MO energy level diagram and occupy the appropriate levels with electrons.		
8. Deduce the bond order per link.		

Summary

In this chapter square planar ICl_4^-, tetragonal pyramidal BrF_5, and octahedral PF_6^- were found to be examples of diamagnetic main-group molecules whose central atoms are hypervalent. The transition metal complex CoF_6^{3-}, however, is paramagnetic with four unpaired electrons, owing to the participation of (*d*) VAOs in weak-field bonding. In studying this ion, the importance of crystal and ligand field concepts was presented. Metal-to-ligand π bonding is very likely to occur in the related diamagnetic complex $Co[P(OCH_3)_3]_6^{3+}$. Of the three cluster systems investigated, namely, B_5H_9, $B_6H_6^{2-}$, and $Mo_6Cl_8^{4+}$, the first two are electron deficient while the last one is not. In the discussion of B_5H_9, the differences between bonds to bridging and terminal hydrogens were elucidated (using B_2H_6 as an example) to simplify the bonding treatment of molecules in which such hydrogens occur.

In Table 8.10 is a slightly amended version of our procedures using the GO approach. The change reflects the fact that structures of transition metal complexes and metallic and nonmetallic cluster compounds are not governed by VSEPR theory.

Exercises

1. Draw the SOs for the delocalized view of ICl_4^-. Do the same with the axis system of the GOs rotated by 45° in the *xy* plane.
2. Develop the delocalized and localized views of the stable compound XeF_4.
3. Draw the molecular motions of BrF_4^- and give the corresponding GO table. In your drawings, also show the SMs as combinations of *resultant* AVs on the fluorine atoms.
4. Sketch the delocalized and localized MO energy level diagrams for (a) ICl_4^-, (b) BrF_5, (c) PF_6^-.
5. Draw the SMs for BrF_5.
6. Develop the localized view of BrF_5 including (*d*) orbital participation.
7. The structure of $InCl_5^{2-}$ is tetragonal pyramidal with the indium atom slightly above the plane of the square array of chlorines. Give a reason for this observation based on VSEPR theory. Is $InCl_5^{2-}$ isoelectronic with $SbCl_5^{2-}$?
8. Account for the fact that the bond lengths decrease on going from IF_5 to XeF_5^- (equatorial, 1.869 to 1.845 Å; axial, 1.844 to 1.793 Å) in spite of the fact that Xe is larger than I.
9. The $[(C_6H_5)_3PAu]_4As^+$ cation contains a square pyramidal Au_4As cluster with the As atom at the apex. Develop the delocalized view of the Au_4As^+ cluster assuming all the bonds external to the cluster are localized. The Au atoms are close enough to one another to interact.

10. Consider the localized view of the B_2H_2 ring of B_2H_6. Show that by pointing the main lobes of the *sp* GO hybrids toward diagonally oriented B—H links, one of two resonance forms is generated, in each of which the ring system is broken.

11. Draw the SOs of B_5H_9 using the axis system in Figure 8.9.

12. There is some evidence that the structure below can exist as a transient species. In this cation the apical C—H is presumably bonded via carbon to the π orbital system of the cyclobutadiene. Develop the delocalized view of this cation by considering the interaction of the CH fragment with the π system of the C_4 fragment.

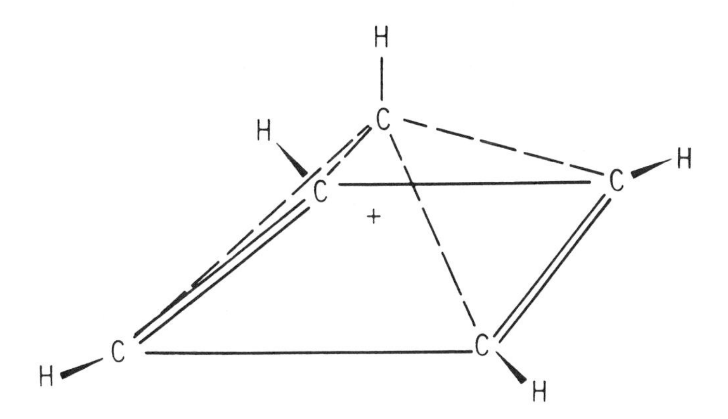

13. Using CFT briefly account for the relative magnitudes of the metal–ligand distance in the following pairs of ions: (a) FeF_6^{3-} (1.90 Å) and FeF_6^{4-} (2.08 Å); (b) NiF_6^{3-} (1.85 Å) and NiF_6^{4-} (2.00 Å).

14. Develop the delocalized and localized views of the strictly octahedral form of XeF_6. Also draw its molecular motions.

15. Using CFT show how the (*d*) orbital energy levels change when the ligands on the *z* axis are compressed in an octahedral complex. Show how such a distortion could destroy the degeneracy in (a) a weak-field d^4 complex; (b) a strong-field d^9 complex.

16. By means of an energy level diagram, show that the presence of empty (*d*) VAOs on the phosphorus in diamagnetic $Co[P(OCH_3)_3]_6^{3+}$ can increase Δ over the value expected in the absence of such π bonding. What is the bond order?

17. By heating LiC≡CLi to about 1000K, the species Li_6C was detected mass spectrometrically in 1992. The remarkable stability of this species was actually predicted computationally ten years before its detection. With an octahedral structure for Li_6C, develop the delocalized and localized views of this molecule assuming (a) the lithiums do not interact with one another; (b) they do. Which option do you prefer and why?

18. The Cr in $Cr(CO)_6$ is believed to back-donate electron density from its (dxy), (dxz), (dyz) VAOs to the π^* MOs of the CO ligands. (a) Generate the SOs for these interactions and sketch an MO energy level diagram for the CrC_6 moiety. (b) Indicate from this diagram how the CrC multiple bonding stabilizes the system and why a higher CO stretching frequency is expected in the absence of the π back-bonding. (c) Draw the stretching SMs of the C—O bonds. Assume that these motions are dominated by movement of the oxygens.

19. (a) Draw the C—O stretching SMs for the C—O bonds in *trans*-$[(C_6H_5)_3P]_2Mo(CO)_4$. Here the phosphorus ligands are at opposite vertices of the octahedron. (b) Do the same for *cis*-$(CO)_2Mo[P(C_6H_5)_3]_4$ in which the COs are on adjacent vertices. (c) Do the same for *mer*-$(CO)_3Mo[P(C_6H_5)_3]_3$ in which two of the COs are on opposite vertices and one is located on the 90° corner between the other two COs such that the three COs lie along a meridian of the octahedron. Since two of the COs are different from the third, treat the two sets separately.

20. Sketch the localized energy level diagram for $Mo_6Cl_8{}^{4+}$.

21. Reaction of bromine with $[W_6Br_8]Br_4$ gives $[W_6Br_8]Br_6$, which apparently retains the same cation structure as the starting material. In contrast, reaction of chlorine with $[W_6Cl_8]Cl_4$ gives $[W_6Cl_{12}]Cl_6$, in which the cation has the structure shown below. Sketch the delocalized MO diagrams for the cations in both tungsten products. Consider the metals to be in the planes of the chlorines to which they are bound.

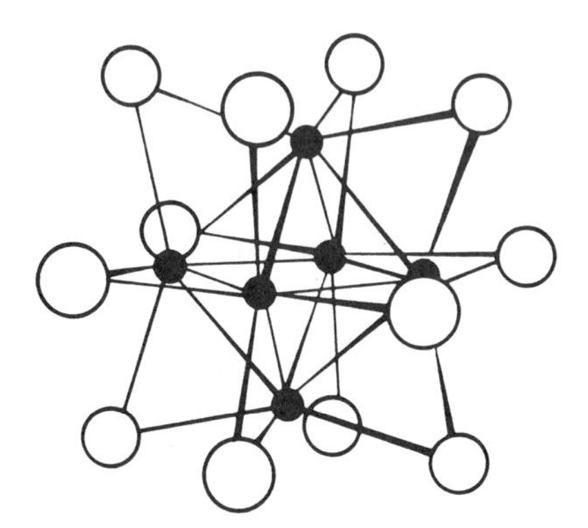

22. Develop the delocalized view of the neutral diamagnetic octahedral nickel cluster of $(\eta^5\text{-}C_5H_5)_6Ni_6$, in which each $\eta^5\text{-}C_5H_5{}^-$ ion can be considered to donate three pairs of π electrons to three (sd^3) metal hybrids obtained by hybridizing $(4s)$ with $(3dxy)$, $(3dxz)$, and $(3dyz)$. Such a hybrid set possesses main lobes that point to the corners of a tetrahedron. Why is it better to pick an sd^3 rather than an sp^3 hybrid set?

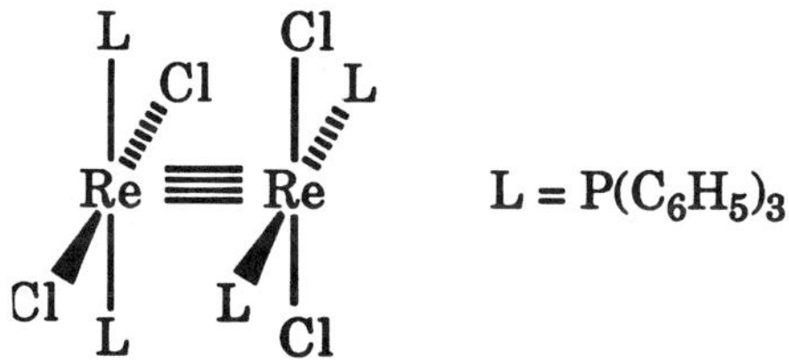

$L = P(C_6H_5)_3$

23. The molecule below has been shown to possess a quadruple Re≣Re bond. Consider the Cl and $P(C_6H_5)_3$ ligands as linked to the Re atoms by localized bonds using an $sp^2dx^2 - y^2$ square planar hybrid orbital set on each of the metal atoms. The Re-Re sigma bonding system is composed of pdz^2 hybrids. (a) Which VAOs on each Re contribute to the two π bonding systems? (b) Which VAO on each Re gives rise to the fourth bond, the so-called δ bonding system? (c) Which matching VAOs on each Re provide better overlap, the π oriented ones or the δ oriented? (d) Using GOs develop the delocalized view of the Re_2 fragment in the above molecule.

24. Develop the delocalized view for the octahedral cluster in $[(C_6H_5)_3PAu]_6C^{2+}$. Assume localized bonds in the $(C_6H_5)_3P$ ligand and in the Au-P bonds.

25. Alkali metals can be dissolved in liquid ammonia giving blue solutions:

$$M \xrightarrow{NH_{3(l)}} \underset{\text{colorless}}{[M(NH_3)_x]^{1+}} + \underset{\text{blue}}{[e(H_3N)_y]^{1-}}$$

The exact number of solvent molecules forming the "cage" around the electron is not known with certainty and it undoubtedly is variable. For simplicity, assume that in some of these anions it can be as low as two and that the hydrogens surround the electron octahedrally. (a) Develop the delocalized view of $[e(H_3N)_2]^{1-}$, assuming that since the sigma N-H BMOs are occupied, the N-H sigma ABMOs are used. (b) At higher concentrations of M in NH_3(l) the solution loses some of its paramagnetism due to $[e_2(NH_3)_y]^{2-}$ formation. Rationalize this result in terms of your energy level diagram. (c) Why do we write the formula for the ammonia molecule NH_3 in the cation and H_3N in the anion? (d) Develop a delocalized view for the octahedral $Na(NH_3)_6{}^+$ ion.

26. Draw the permitted CO stretching modes for $Mo(CO)_6$.

27. Develop a delocalized view for the B_4C_2 cluster in $B_4C_2H_6$ in which one carbon lies above and one lies below the square plane of borons.

28. Show that by pointing the GO axis system to the atom positions in $CoF_6{}^{3-}$ that the d-generated SOs involving the tangential (p) VAOs on the fluorines do not readily reveal the partner relationships among the cobalt (d) VAOs.

29. Draw the molecular motions of $CoF_6{}^{3-}$.

30. All of the following species are known: $SF_6{}^-$, SF_6, $SF_5{}^-$, SF_5, $SF_5{}^+$, $SF_4{}^-$, SF_4, $SF_4{}^+$, $SF_3{}^-$, SF_3, $SF_3{}^+$, $SF_2{}^-$, SF_2, $SF_2{}^+$, SF^-, SF, SF^+. Using sketches, account

for the following observations: (a) SF_6^- (like SF_6) is octahedral; (b) Square pyramidal SF_5^- has a smaller FSF basal bond angle than SF_5 and the S-F_{ax} bond length in SF_5^- is longer; (c) SF_3^+ is not trigonal planar; (d) The bond angle decreases in the order: SF_2^- (160.6°), SF_2^+ (101.9°), SF_2 (99.3°), (e) The bond length decreases in the order SF^- (1.744 Å), SF (1.627 Å), SF^+ (1.537 Å).

CHAPTER 9

Tetrahedral and Related Molecules

In this chapter we examine the P_4 tetrahedral cluster. Other examples in this class are the isoelectronic species Tl_4^{8-}, Si_4^{4-}, Ge_4^{4-}, Pb_4^{4-}, As_4, and Te_4^{4+}. We will also study PO_4^{3-} and CH_4 in which a central atom is present within a tetrahedral array of peripheral atoms. Additional examples in this broad class of molecules are transition metal complexes such as CrO_4^{3-}, $CoCl_4^{2-}$, VCl_4^-, $FeCl_4^-$, and $TiCl_4$. Since transition metal complexes possess interesting bonding features, we will briefly study VCl_4^-. We begin, however, with PnH_3 (Pn = pnicogen = N, P, As, Sb, Bi) whose structure is trigonal pyramidal; a tetrahedron with one vertex atom missing.

For each bonding situation discussed in this chapter, you should use the appropriate column in Table 8.10 and the Homework Drawing Board in Node Game to repeat the steps presented in the text.

9.1. PnH_3

The first three members of the pnicogen hydrides are thermally stable gases, while SbH_3 and BiH_3 easily decompose to their constituent elements. From the normal electron dash structure of PnH_3 in Figure 9.1 and from VSEPR considerations we conclude that these molecules are pyramidal. The experimentally determined bond angles tend to be near 90°, except for NH_3, in which this angle is 107°. The larger angle in NH_3 can be rationalized from the fact that nitrogen is the smallest pnicogen and therefore the bp–bp repulsions assume greater importance. In addition, the energy separation between (n*s*) and (n*p*) VAOs increases with *n* (owing to greater nuclear shielding effects) so that *sp* mixing is less likely in congeners heavier than N. Mixing of (*s*) and (*p*) VAOs is typical in elements of the second row and we have already seen an example of this behavior in H_2O, which possesses a wider bond angle than its analogues H_2S or H_2Se.

After reserving an *ns* VAO on the pnicogen for the lone pair, the delocalized MOs are obtained pictorially in the usual manner using GOs. Figure 9.2 shows a convenient way to show the geometry and an axis system. Table 9.1

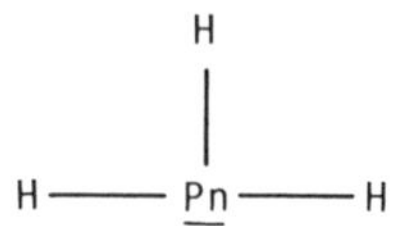

FIGURE 9.1 Electron dash structure for PnH_3, where Pn = N, P, As, Sb, and Bi.

lists the VAO equivalence sets and the SO entries. Two of the BMO–ABMO pairs are shown as degenerate in the energy level diagram in Figure 9.3. This degeneracy, which arises from the fact that the SOs are generated by the *px* and *py* GOs, is analogous to that discussed in detail for the case of H_3^+. As expected from the electron dash structure, the bond order is 1.0.

Since the occupied BMOs in Figure 9.3 are associated with the GO set *s*, *px*, and *py*, the hybrid GO set sp^2 is required for the localized view. This hybrid GO set can be oriented on the GO center such that the main lobe of each member points to a line connecting a hydrogen and the pnicogen as shown in Figure 9.4 for one of these hybrids. The three bonding pairs resulting from this localization form two-center bonds composed of an (H1*s*) VAO and a (Pn*p*) VAO. These (*p*) VAOs are directed toward the (H1*s*) lobes, as shown for one of them in Figure 9.4, by simply rehybridizing them from the orientation they possess in our delocalized view (Figure 9.2). In the localized energy level diagram in Figure 9.5 we see that the average bond order (1.0) is the same as that determined in our delocalized view.

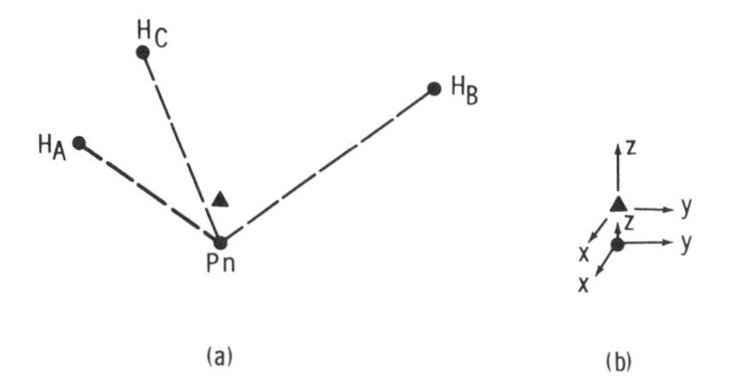

FIGURE 9.2 Geometry of PnH_3 (a) and axis system for GO center and Pn orbitals (b).

TABLE 9.1 Generator Table for PnH_3 (6 e^-)

	GOs		
VAO Equivalence Sets	*s*	*px*	*py*
Pn pt = (Pn *npx*), (Pn *npy*)		n	n
Pn *pr* = (Pn *npz*)	n		
H*s* = (H_A1*s*), (H_B1*s*), (H_C1*s*)	n	n	n

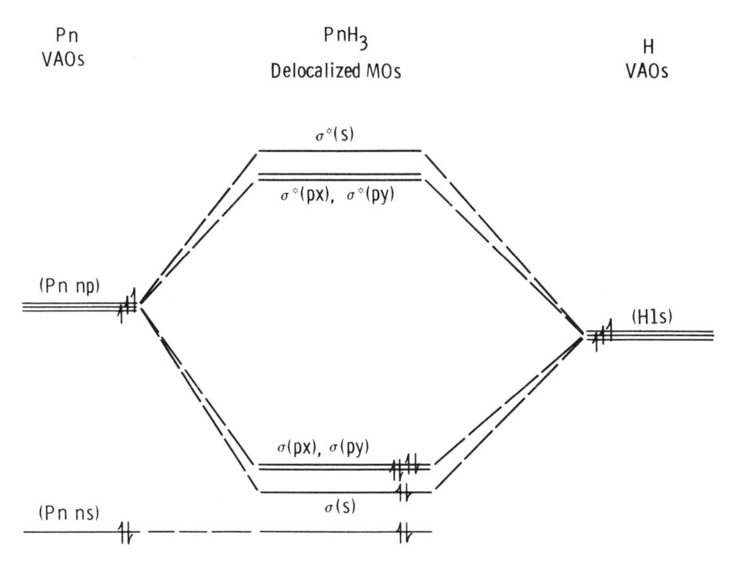

FIGURE 9.3 Delocalized MO energy level diagram for PnH_3.

Experimentally, NH_3 is found to have a bond angle of 107°, which is nearly the tetrahedral angle. In the localized view of this molecule, more favorable overlap in the two-center N—H bonds can be achieved by mixing the (N2*s*) and (N2*p*) VAOs so that the hybrids pointing toward the (H1*s*) lobes are separated by 107° instead of 90°. Such rehybridization, which results in a nearly (sp^3) VAO set on nitrogen, places almost 25% (*p*) character in the fourth hybrid to preserve approximate orthogonality of the nitrogen contributions to the localized MOs. This fourth hybrid, which points downward along the negative *z* axis of the nitrogen in the coordinate system of Figure 9.2b, houses the lone pair. A point to be appreciated from this discussion is that the reservation of a pure (N2*s*) VAO for the lp is only an approximation and that in the present case the lone pair actually possesses a somewhat higher energy than that expected for a pure (N2*s*) VAO since there is considerable (2*p*) character mixed with it. By not initially assigning the lone pair to

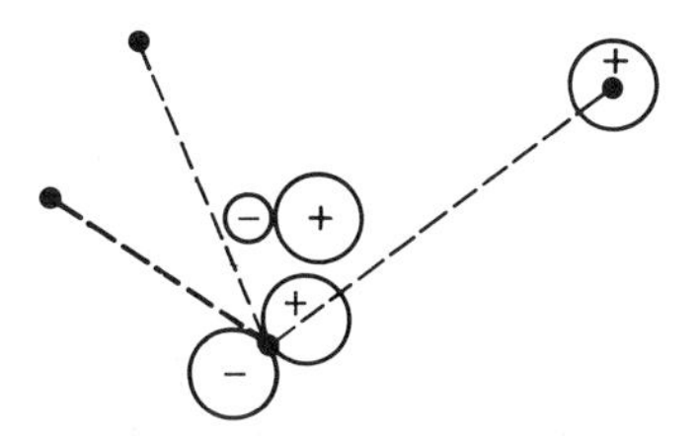

FIGURE 9.4 The orientation of one sp^2 GO hybrid and the VAOs it calls in on one hydrogen and on the Pn atom in PnH_3.

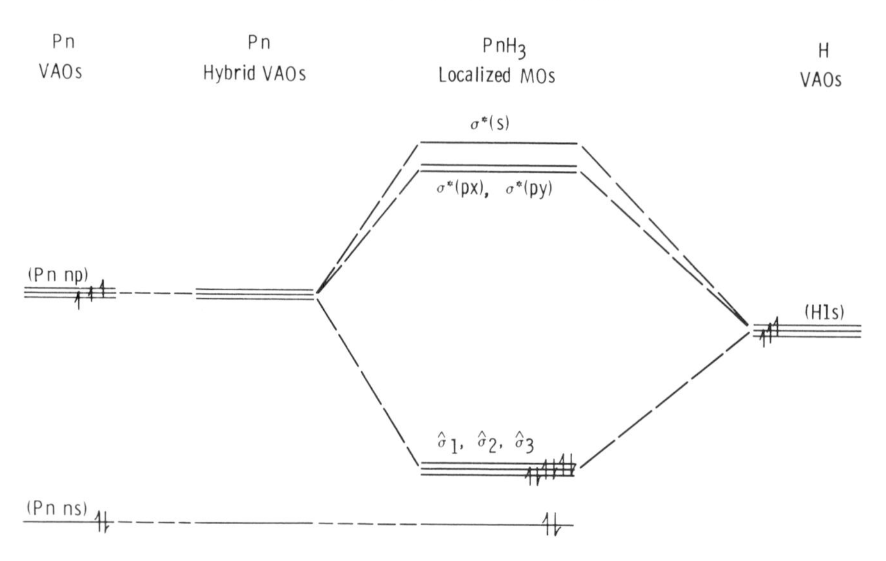

FIGURE 9.5 Localized MO energy level diagram for PnH_3.

an (N2*s*) VAO, this orbital is introduced into Table 9.1 as an equivalence set. Such a procedure is easily seen in Table 9.1 to give three SOs generated by an *s* GO, which linearly combine to form an NBMO and a BMO–ABMO pair. Although the (N2*s*) VAO is expected to contribute substantially to the NBMO, a considerable contribution is also expected from (N2*pz*) in view of the (*sp*) mixing, which was concluded to occur in the localized view. Thus the lone pair in the delocalized view of NH_3 is expected to be higher in energy than an (N2*s*) VAO because of a gain in (*p*) character. In that case we could have reserved an approximately (sp^3) hybrid orbital for the lone pair on the nitrogen in NH_3 at the outset. The increase in lone-pair energy due to hybridization with the (*p*) VAO is offset by a decrease in lp–bp repulsions since *p* character gives the lone pair hybrid lobe a smaller cone angle than a spherical *s* VAO.

The molecular motions of PnH_3 are obtained from the GO-generated SMs of the AVs and this is left as an exercise.

9.2. P_4

Elemental phosphorus exists in more than ten allotropic forms. The structures of most of these are still undetermined. For P_4, however, three electron dash structures that obey the octet rule can be drawn as shown in Figure 9.6. In contrast to second-row elements, those in the third row and below do not readily form π bonds among themselves. This could arise because these atoms are larger and their (*p*) orbitals, which are in higher valence shells, are more diffuse and consequently are less able to overlap effectively. There-

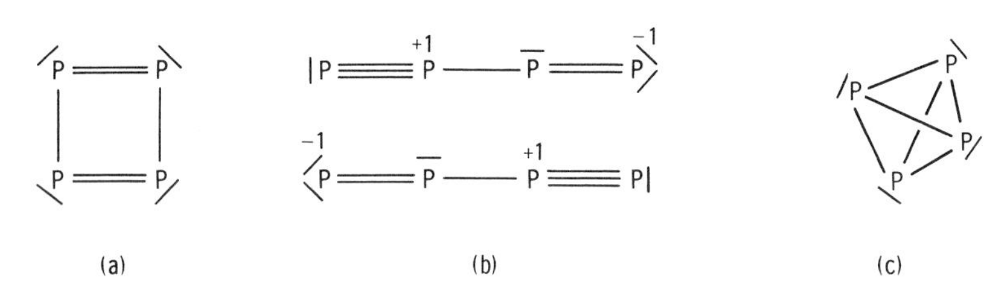

FIGURE 9.6 Electron dash structures for cyclic (a), chain (b), and cage (c) forms of P_4.

fore, the tetrahedral structure becomes the one of choice on this basis (as well as on formal charge arguments in the case of the linear forms by P_4 shown in Figure 9.6b) and indeed it matches the experimentally determined configuration.

The spatial orientation we adopt for the P_4 molecule is shown in Figure 9.7a and the axis system shown in Figure 9.7b is convenient to use. For a better appreciation of the perspective with which we are viewing this molecule, it has been inscribed inside a cube in this figure. As is usual in our delocalized views, one axis of each peripheral atom points to the center of the molecular species, regardless of whether there is a central atom or not. In the present case, the molecular center is also the GO center, and this is true in all cases where the GO center is unique.

Although following our general practice of assigning lone pairs to (*s*) VAOs is, of course, permissible, it is not mandatory. A more precise picture results from not making such a designation, but the resulting analysis is somewhat more challenging. Here we will elect not to assign lone pairs to (P3*s*) VAOs for reasons that will become clear later in the discussion. Among

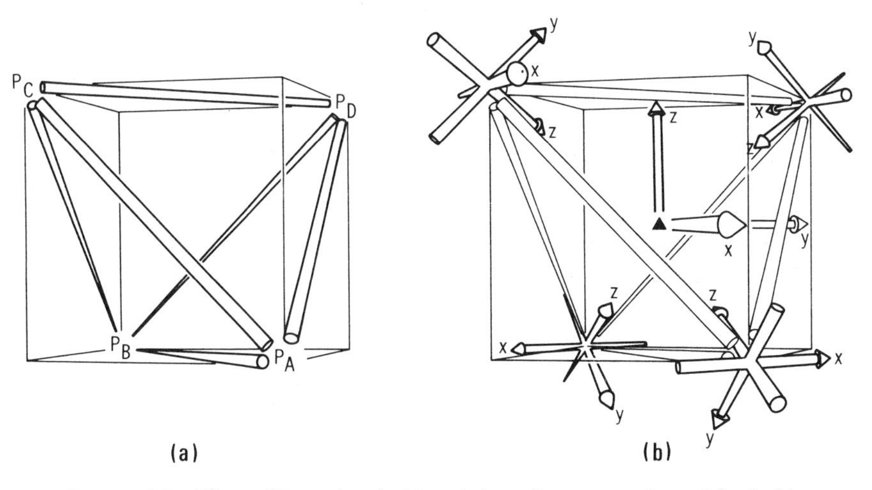

FIGURE 9.7 View of P_4 molecule (a) and the axis system adopted for it (b).

TABLE 9.2 Generator Table for P_4 (20 e^-)

	GOs								
VAO Equivalence Sets	s	px	py	pz	dxy	dz^2	$f3'$	$f3''$	$f2''$
Ps = (P_A3s), (P_B3s), (P_C3s), (P_D3s)	b	a	a	a					
Ppr = (P_A3pz), (P_B3pz), (P_C3pz), (P_D3pz)	b	a	a	a					
Ppt = (P_A3px), (P_A3py), (P_B3px), (P_B3py), (P_C3px), (P_C3py), (P_D3px), (P_C3py)		b	b	b	b	b	a	a	a

the VAO equivalence sets shown in generator Table 9.2, it should be noted that *all eight* tangential (p) VAOs are in *one* equivalence set. That this must be so is realized, for example, by looking down an axis that goes through a corner of the tetrahedron and the center of its opposite face. It is apparent that the two tangential (p) VAOs on that corner bear the same relationship to each other and to this three-fold axis as the (B2px) and (B2py) VAOs have with respect to the three-fold central axis perpendicular to the molecular plane of BF_3.

The SOs generated among the (pz) VAOs are depicted in Figures 9.8a–d. It is evident from Figure 9.8a that the SO generated by the s GO is bonding. None of the three p-generated SOs contains the full symmetry of the molecule and therefore the three of them are degenerate partner MOs. Recall that if we were to sum the squares of the wave functions of three mutually orthogonal (p) orbitals, a spherical density would be produced which, of course, contains the full symmetry of any polyhedron, including the tetrahedron. The two members of the p-generated SO set shown in Figures 9.8b and c are easily seen to be antibonding. That the pz-generated GO in Figure 9.8d is antibonding (and it must be since it is degenerate with the px- and py-generated SOs) can be rationalized by noting that the antibonding interactions between any two adjacent (pz) VAOs outnumber the adjacent bonding interactions by 2 to 1. The SOs in the Ps equivalence set are similarly generated pictorially and you should do this with the Homework Drawing Board in Node Game. The corresponding entries are placed in generator Table 9.2.

The pictorial procedure for generating the eight SOs of the Ppt equivalence set is shown in Figure 9.8e–s. The s GO clearly generates no SO in this set. The three p GOs generate three degenerate partner SOs, because none of these SOs contains the full symmetry of the molecule. Of the d GO set, only dxy and dz^2 can be used since $dx^2 - y^2$, dxz, and dyz duplicate the pz-, px-, and py-generated SOs, respectively. Although it is easy to see how the $dx^2 - y^2$ GO in Figure 9.8j duplicates the SO generated by pz in Figure 9.8g, it is less obvious how dyz (Figure 9.8l) for example, duplicates the py-generated SO (Figure 9.8f). While the GO-induced directionality of the bottom

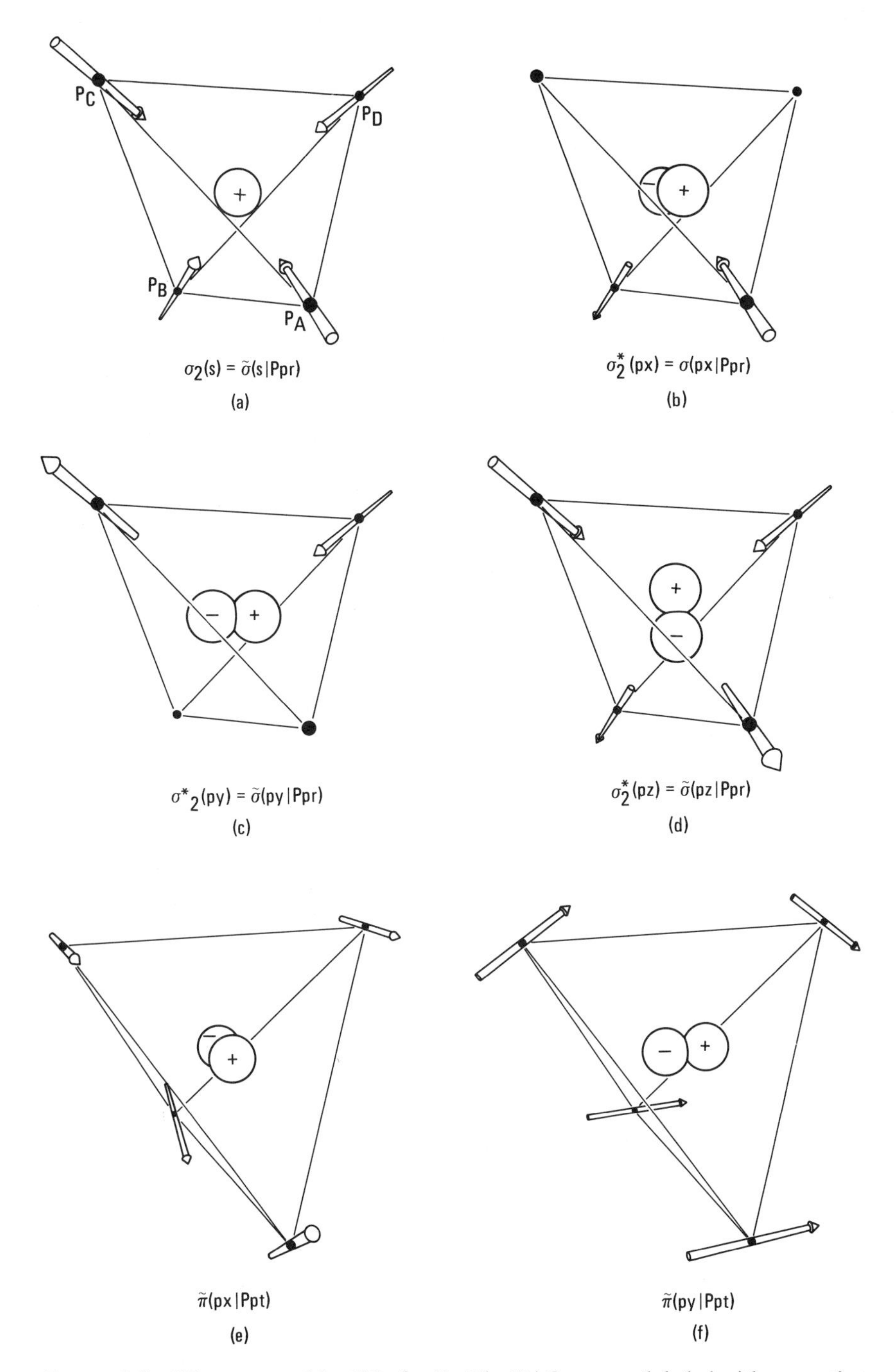

FIGURE 9.8 SOs generated by GOs for P_4. The VAO vectors labeled with a question mark [(n), (o)] are discussed in the text. The arrowhead of ($P_B pt$) in (o) is obscured by the lower lobes of the GO. The sigma MOs are subscripted with a 2 to differentiate them from the sigma MOs generated in the P*s* equivalence set (see text). For clarity, the perspective of the tetrahedron in (a)–(d) is slightly different from that shown in (e)–(s).

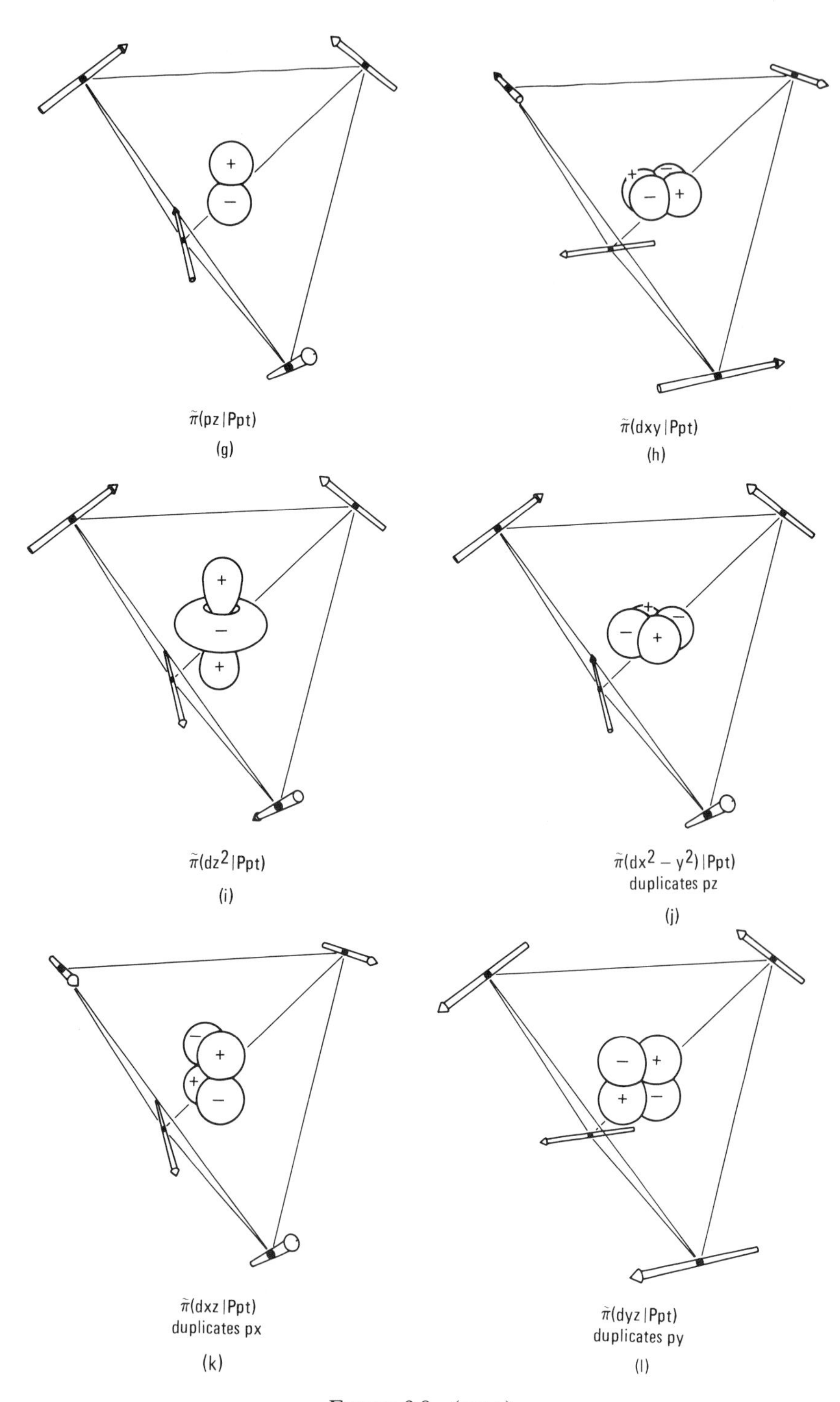

FIGURE 9.8 (cont.)

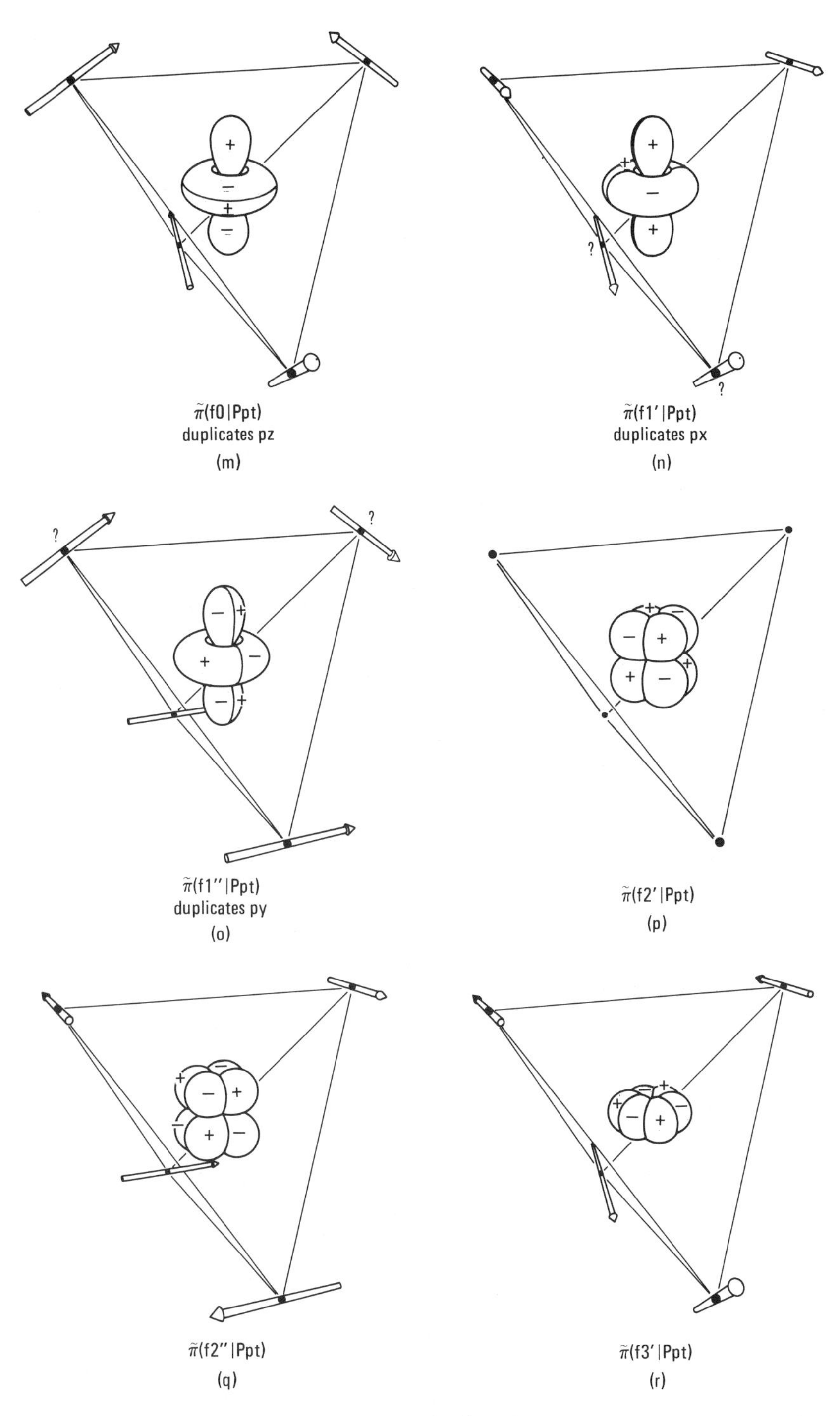

FIGURE 9.8 (cont.)

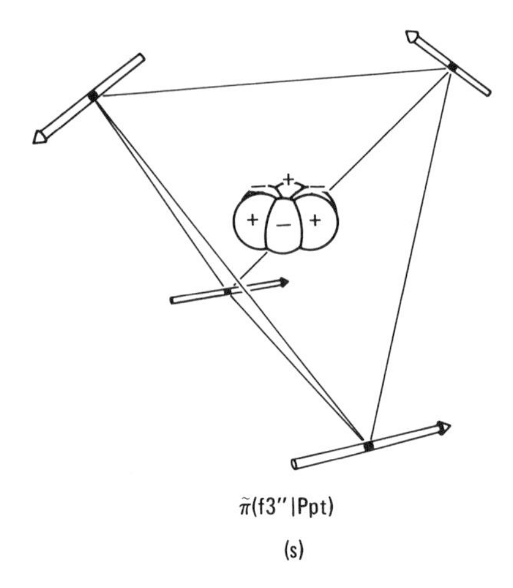

FIGURE 9.8 (cont.)

vectors (p VAOs) in Figure 9.8l is obvious, that of the top vectors is not. The situation does become clear, however, when we consider the fact that the angle between the centers of the upper vectors (p VAOs) and the GO center is 109° 28′, while the upper lobes of the GO are only 90° apart. This means that these upper lobes more strongly call in the corresponding signs on the upper VAO lobes that point toward each other over the top edge of the tetrahedron. Similar arguments apply to Figure 9.8k. The remaining two SOs generated by d GOs stem from dxy and dz^2 (Figures 9.8h and i, respectively). Since neither contains the full symmetry of the molecule, they too form a degenerate partner set.

There remain three Ppt SOs to be found and they are generated by f GOs. To begin with, we note that the lobes of the $f2'$ GO in Figure 9.8p face directly toward the corners of the tetrahedron. This GO therefore cannot generate an SO among the tangential orbitals on phosphorus. Next we observe that the $f0$-generated SO (Figure 9.8m) is the same as that generated by pz (or $dx^2 - y^2$). From this observation we infer that there must exist two other f-generated SOs which complete the partner set to which the $f0$-generated SO belongs. In examining the SOs generated by $f1'$ and $f1''$, the signs with which two of the phosphorus tangential (p) VAOs in each SO are called in are not immediately obvious. Thus if the angle subtended by the conal nodes of the GO is larger than 109°28′, the lower pair of lobes will dominate the manner in which the nearby tangential (p) VAOs will be called in in Figure 9.8n (and similarly for the upper pair of lobes in Figure 9.8o) and the arrowheads would be placed as shown. On the other hand, if the conal angle is less than 109°28′, each of the VAOs would be more strongly influenced by a "half-doughnut" segment in Figure 9.8n and in Figure 9.8o. In this case, the question-marked arrowheads would have to be reversed. We can decide this question in two ways. The most straightforward approach is to calculate the angle θ (see Figure 2.1) from that part of the wave function which governs this angle. From Appendix II.B it is easily shown that $\theta = 63°26'$. Thus the total angle of the cone is 126°52′, which is larger than the tetrahedral angle of 109°28′. We conclude, therefore, that the arrows are correct as shown in Figures 9.8n and o. We also note that the $f1'$ and $f1''$-generated SOs in these figures duplicate the px-and py-generated SOs in Figures 9.8e and f. There is a second way in which we could have reached the same conclusion without calculating θ. It is quite clear that the $f2''$, $f3'$, and $f3''$ SOs are new ones as shown in Figures 9.8q–s. From this observation we can conclude that the $f1'$- and $f1''$-generated SOs must duplicate SOs generated by simpler GOs. The only GOs for which this could possibly be true are the px and py orbitals, respectively, since $f0$ already duplicated pz. These SO duplications require that the phosphorus VAOs labeled with question marks in Figures 9.8n and o must point as shown.

From Figure 9.8 and generator Table 9.2 we see that the SOs generated by the GOs dxy, dz^2, $f3'$, $f3''$, and $f2''$ are also MOs of the system. The $f2''$-generated SO is antibonding as seen from Figure 9.8q and the same is true for its partner SOs generated by $f3'$ and $f3''$ (Figures 9.8r and s, respectively). Similarly, Figure 9.8i shows that the dz^2-generated SO and hence its partner, the dxy-generated SO (Figure 9.8h), are bonding. None of the remaining eleven SOs generated by s, px, py, and pz is an MO by itself and this is realized from the multiple entries in these GO columns in Table 9.2. However, all have bonding or antibonding character because they are constructed from VAOs on neighboring atoms. As was discussed in detail for N_3^+, we anticipate that each MO will have a predominant contribution from one SO only, and we can therefore assume that the character of the MO is determined by that of its corresponding SO. From inspection of symmetry orbitals we drew in Figure 9.8, both s-generated SOs (MOs) are seen to be bonding. The degenerate px-, py-, pz-generated SOs (MOs) are bonding for the Ppt equivalence set but antibonding for the Ps and Ppr sets. Furthermore, the $(3s)$ AO orbital energy in the free P atom lies some 8 eV below the $(3p)$ AO energy and hence we expect that the *antibonding*, as well as the bonding SOs (MOs) of the Ps set lie below all the other MOs because the remaining MOs are all dominated by (p)-type VAOs. Again, the situation is very similar to that in N_3^+.

P_4 has 20 valence electrons and 16 MOs. Thus in the ground state the six MOs of highest energy are empty. These are clearly the six ABMOs of predominantly (p) character, namely, the three f-generated ABMOs and the three p-generated ABMOs (SOs) in the Ppr set. The energy level diagram in Figure 9.9 was obtained by means of a calculation, and it is seen that our qualitative considerations are reasonable.

The occupied MOs in Figure 9.9 arising from the Ps equivalence set are localized by the sp^3 hybrid GO set whose members have their main lobes directed toward the phosphorus atoms. This results in four localized MOs,

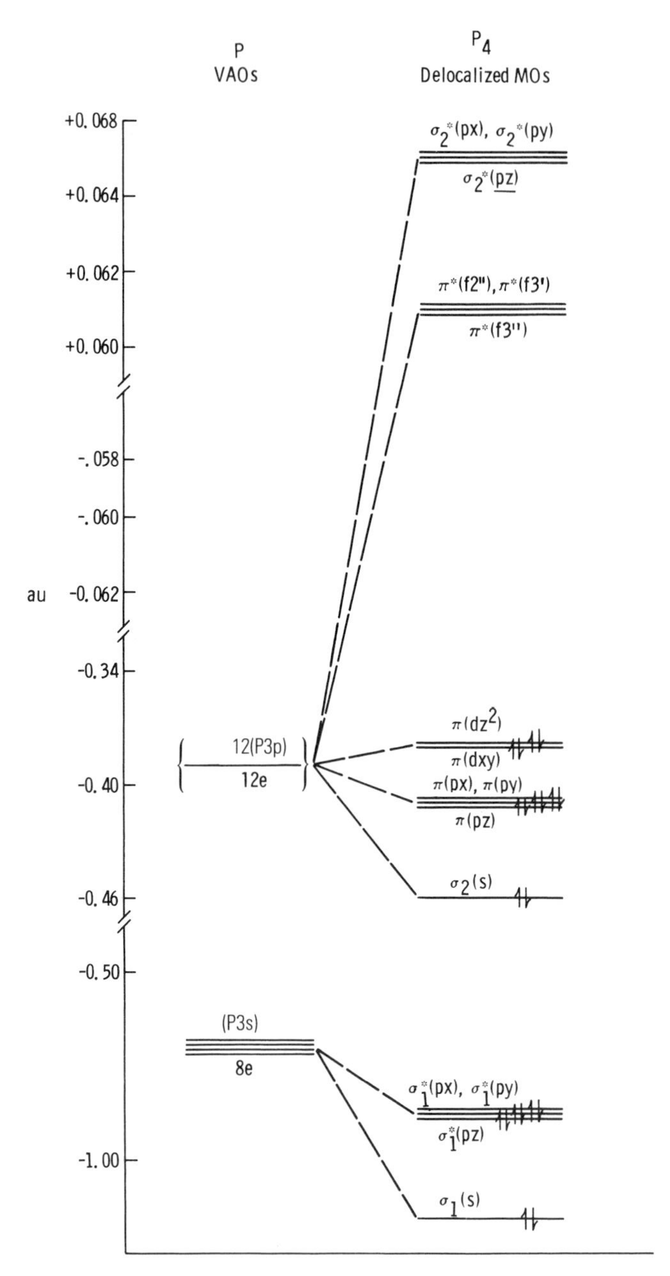

FIGURE 9.9 Delocalized MO energy level diagram for P_4. An au is equal to 27.2116 eV.

each of which consists primarily of a (P3*s*) VAO on a P atom with small antibonding contributions of equal magnitude from each of the three others. These antibonding contributions give rise to nonbonded electronic repulsions among the atoms, which are associated with the requirement of orbital orthogonality and the fact that the (3*s*) VAOs of neighboring atoms have a nonnegligible overlap. If we had reserved the (P3*s*) VAOs for lone pairs at the outset and omitted them from the bonding discussion, we would in fact have been neglecting these nonbonded repulsions. This would not have changed our *qualitative* picture of the molecular orbital scheme but it would have neglected an energy contribution which cannot be omitted for a *quantitative* understanding and which, in fact, might have been important if the bonding of the cluster had been less strong. It should be appreciated that although the (P3*s*) VAOs can be preassigned to house the lone pairs, this is not because these orbitals do not interact, but rather because the bonding and antibonding MOs of the P*s* equivalence set are *all* filled (see Figure 9.9).

The remaining occupied MOs in Figure 9.9 are generated by an *s*, three *p*, and two *d* GOs. These GOs hybridize to form an sp^3d^2 hybrid GO set, whose members point toward the vertices of the octahedron formed by the midpoints of the edges of the P_4 tetrahedron as shown in Figure 9.10a. This yields six localized BMOs, each of which essentially connects two P atoms by a two-center two-electron bond. The effect of the hybrid GO system is to reorient the (*p*) VAOs on each P atom as shown in Figure 9.10b. From this figure we see that the two VAOs that combine to form the bonding lobes of a particular localized BMO do not point directly at each other, and thus form bent bonds. The localized energy level diagram in Figure 9.11 satisfies our expectation from the electron dash structure that the average bond order between phosphorus atoms is 1.0.

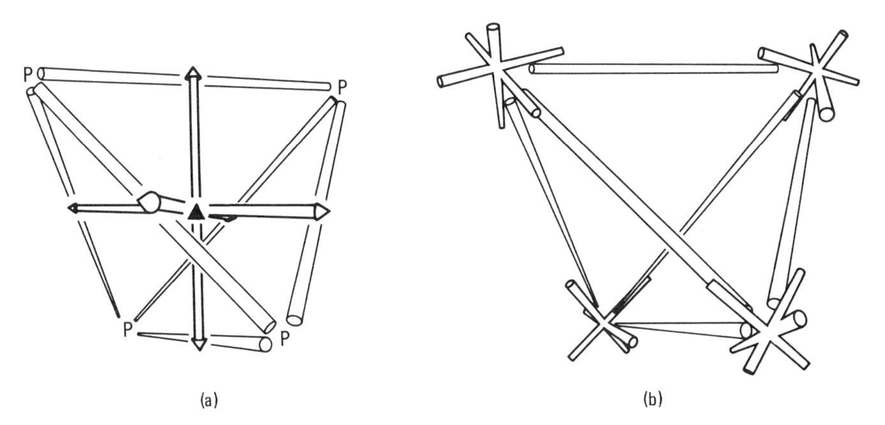

FIGURE 9.10 Octahedral orientation of the main lobes of the sp^3d^2 GO set in P_4 (a). In (b) are shown the phosphorus (*p*) VAO sets reoriented to be most effectively called in by the sp^3d^2 GO set to form two-center bonds over the edges of P_4.

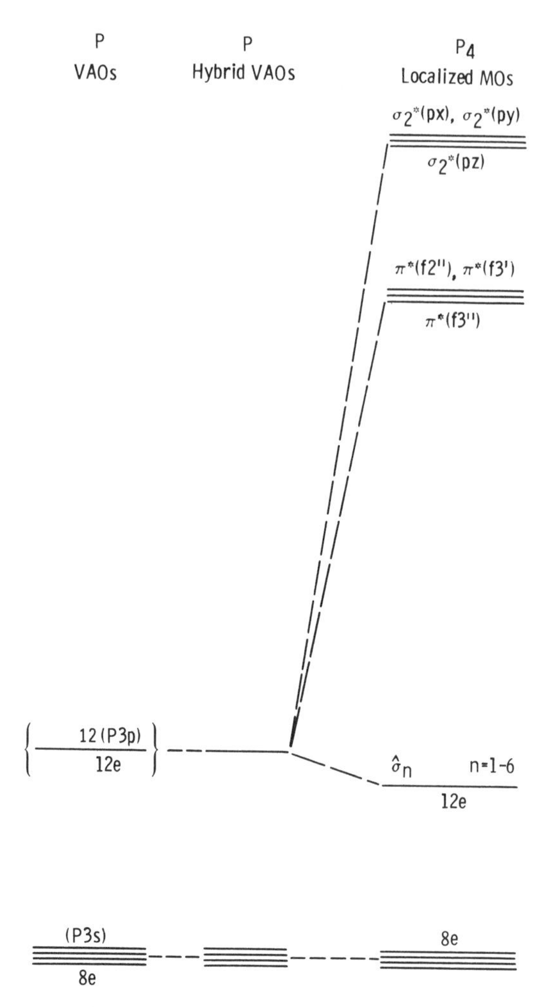

FIGURE 9.11 Localized MO energy level diagram for P_4. Note that the σ and π occupied MOs in Figure 9.9 are given the designation $\hat{\sigma}$ in the present figure. Although the distinction between σ and π among orbitals that overlap at an angle is somewhat arbitrary, the localization of an electron pair between the lobes of two such orbitals has the appearance more like that of σ than π.

The localized view of P_4 that we have developed is a very satisfying one in spite of the fact that we were forced to use atomic orbitals that interact at an angle. The bent bonds in this molecule are often cited as the source of the observation that P_4 is a very highly reactive substance, forming phosphorus compounds in which more normal (less strained) bond angles occur. In fact most of the other forms of elemental phosphorus are much more stable, presumably because they possess bond angles which are substantially larger than 60°.

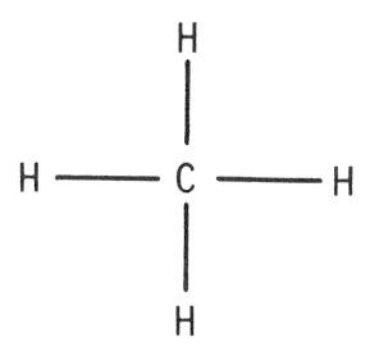

FIGURE 9.12 Lewis structure for CH_4.

TABLE 9.3 Generator Table for CH_4 (8 e^-)

	GOs			
VAO Equivalence Sets	*s*	*px*	*py*	*pz*
C*s* = (C2*s*)	n			
C*p* = (C2*px*), (C2*py*), (C2*pz*)		n	n	n
H*s* = (H_A1*s*), (H_B1*s*), (H_C1*s*), (H_D1*s*)	n	n	n	n

9.3. CH_4

The tetrahedral shape of CH_4 is easily predicted from its Lewis structure (Figure 9.12) and VSEPR considerations. Using the axis system for P_4 in Figure 9.7b to pictorially generate our SOs, we can construct Generator Table 9.3. The H*s* SOs are analogous to the P*s* SOs in P_4. Linear combinations of the pairs of SOs generated by the same GOs give rise to four 5-center delocalized BMO–ABMO pairs whose energy levels are represented in Figure 9.13. The eight valence electrons occupy the four BMOs, telling us that the average bond order is 1.0.

The occupied BMOs in Figure 9.13 are generated from four GOs that, when hybridized, form an sp^3 set. This GO set generates a corresponding tetrahedral sp^3 set of carbon SOs in the carbon VAO set, each main lobe of which is directed toward a (H1*s*) VAOs. The resulting two-center MOs are all equivalent and degenerate. Sketching the energy level diagram is left as an exercise.

9.4. PO_4^{3-}

The electron dash structures for PO_4^{3-} given in Figure 9.14 and the VSEPR rules lead us to predict that the ion should be tetrahedral, and this prediction has been confirmed experimentally. On the basis of formal charges, the structure in Figure 9.14b is favored. This structure has three equivalent partners (which are not shown in Figure 9.14) and together these four resonance structures yield the average resonance form in Figure 9.14c. Provided we

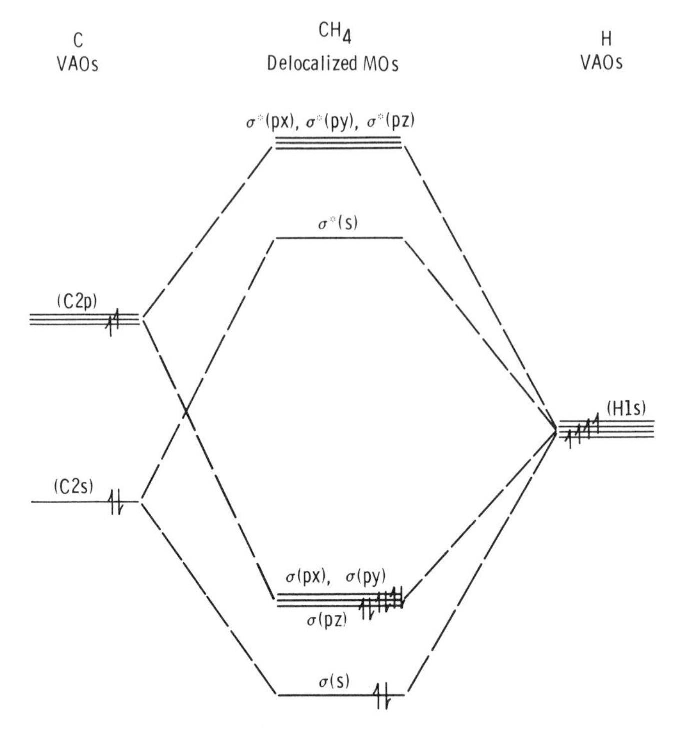

FIGURE 9.13 Delocalized MO energy level diagram for CH_4. (Reproduced with permission from the *Journal of Chemical Education.*)

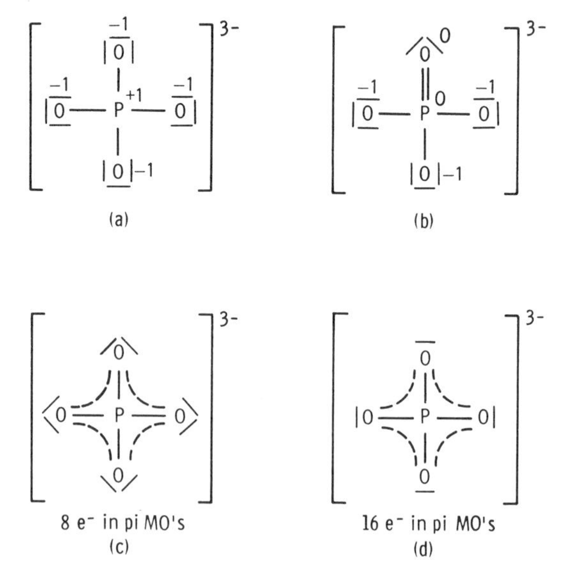

FIGURE 9.14 Electron dash structures for PO_4^{3-} [(a), (b)]. Average electron dash structures for eight (c) and sixteen (d) electrons in the π MOs are also shown.

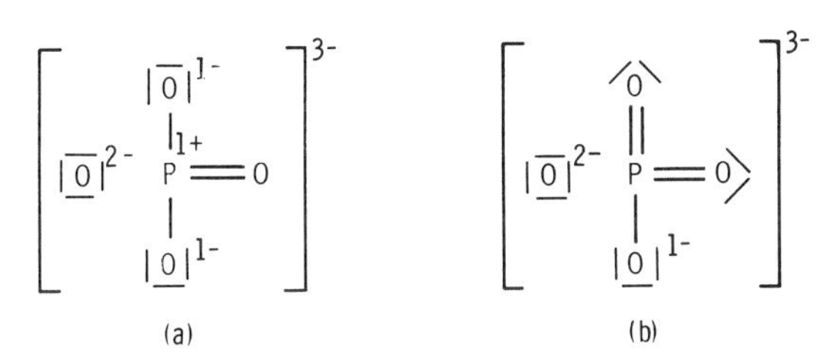

FIGURE 9.15 Two ionic resonance forms of PO_4^{3-}.

have a sufficient number of VAOs on phosphorus, we could incorporate an additional oxygen lone pair into the π system as depicted in Figure 9.14d. By involving all five (P3d) VAOs we will see that this is possible. Although oxygen is more electronegative than phosphorus, the electronegativity difference is not as great as that which occurs in metallic oxides where ionic interactions are of greater importance. Thus the resonance structures shown in Figure 9.15 do not play as important a role in PO_4^{3-}.

Since the experimentally determined P—O distance is shorter than that expected for a single bond, some π bonding undoubtedly occurs. To accommodate π bonding we see that more than a total of four VAOs on phosphorus will be needed. This observation implies that we could use (P3d) VAOs even though their involvement requires an investment of energy.

After reserving the (O2s) VAOs for lone pairs in the resonance structure in Figure 9.14d and drawing the SOs in Figure 9.8, we construct Generator Table 9.4. Because the (P3d) AOs are included in the phosphorus VAO set, we must, when taking linear combinations of SOs, consider the fact that the SOs generated by $dx^2 - y^2$, dxz, dyz duplicate those generated by px, py, pz. For simplicity, we will assume that the π interactions between the phosphorus and oxygens involve the (Pd_2) and (Pd_3) VAO equivalence sets rather than the Pp set. (The subscripts in Pd_2 and in Pd_3 in Table 9.4 denote the number of (3d) VAOs in the equivalence set.) Thus we have omitted the px-, py-, pz-generated O$p\pi$ SOs in Table 9.4. We will also assume that the P—O σ interactions include only the VAOs in the Ps and the Pp equivalence sets. This simple approach ignores contributions to both the π and σ systems from the Pp, Pd_2, and Pd_3 equivalence sets, which indeed do occur. The 24 electrons we have available for the MO system occupy the 9 BMOs of the BMO–ABMO pairs and the 3 f-generated NBMOs, as shown in the MO energy level diagram in Figure 9.16. Thus there are 16 electrons in the π system (10 in π BMOs and 6 in π NBMOs) and 8 electrons in the σ systems. The average bond order is 2.25, which is in accord with the average Lewis structure in Figure 9.14d.

It is instructive at this point to note that the splitting of the triply degenerate π BMO–ABMO pairs in Figure 9.16 is greater than that of the doubly degenerate π BMO–ABMO pairs. This is best visualized by considering an octahedron and its axis system, shown in Figure 8.12, and placing the four oxygen atoms of PO_4^{3-} on triangular faces of the octahedron, as shown in

TABLE 9.4 Generator Table for PO_4^{3-} (24 e^-)

VAO Equivalence Sets	GOs											
	s	px	py	pz	dxy	dz^2	$dx^2 - y^2$	dxz	dyz	$f3'$	$f3''$	$f2''$
$Ps = (P3s)$	n											
$Pp = (P3px), (P3py), (P3pz)$		n	n	n								
$Pd_2 = (P3dxy), (P3dz^2)$					n	n						
$Pd_3 = (P3dx^2 - y^2), (P3dxz), (P3dyz)$							n	n	n			
$Opr = (O_A2pz), (O_B2pz), (O_C2pz), (O_D2pz)$	n	n	n	n								
$Op\pi = (O_A2px), (O_A2py), (O_B2px), (O_B2py), (O_C2px), (O_C2py), (O_D2px), (O_D2py)$					n	n	n	n	n	n	n	n

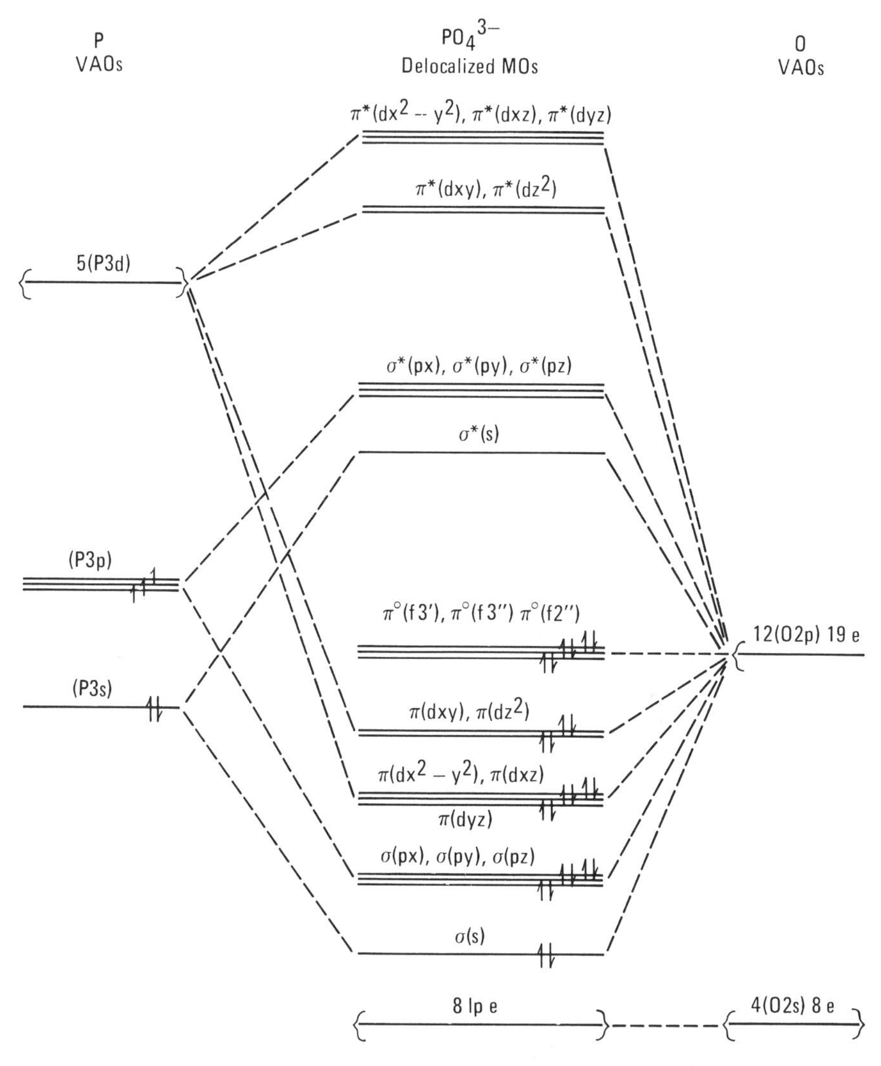

FIGURE 9.16 Delocalized MO energy level for PO_4^{3-}.

Figure 9.17. In the latter figure, the oxygens are at the corners of a tetrahedron which can be inscribed in the octahedron. *Notice that as a result of the axis system we have consistently chosen for the octahedron* (*Figures* 8.12 *and* 9.17) *the axis system in the inscribed tetrahedron in Figure* 9.17 *is rotated by* 45° *around z from that used in our treatments of tetrahedral* P_4 and PO_4^{3-} (*Figure* 9.7b). Because the geometrical relationships of the (d) orbitals are easier to see in the octahedral axis system we have been using, let us retain the axis system in Figure 9.17. From this figure it is easy to see that the π orbital lobes on any given oxygen are closer to the lobes of the (dxy), (dxz),

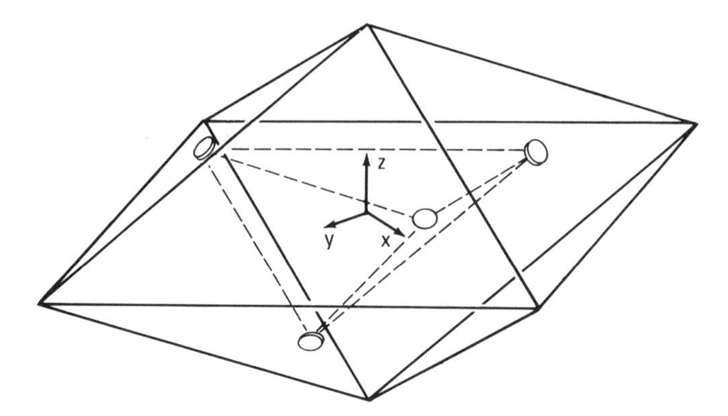

FIGURE 9.17 The tetrahedral arrangement of four PO_4^{3-} oxygens inscribed in an octahedron.

(dyz) partner set than those of the $(dx^2 - y^2)$, (dz^2) partner set, since the lobes of the former are directed to the edges of the octahedron while those of the latter are pointed to its vertices. Thus the center of a triangle is closer to its edges than it is to its vertices. Thus compared to the Pd_2 VAO equivalence set, we expect greater interaction of the Pd_3 VAO equivalence set with the $Op\pi$ VAOs. Note that because of the 45° rotation of the axis system in the tetrahedron in Figure 9.17 compared to Figure 9.7b, the Pd_2 and Pd_3 partner sets exchange their (dxy) and $(dx^2 - y^2)$ GO members.

The σ MOs in Figure 9.16 can be localized analogously to those in CH_4. The bonding π MOs, all being d-generated, cannot be further localized.

9.5. VCl_4^-

Transition metals surrounded by four ligands are either tetrahedral or square planar. Factors influencing the choice of geometry will be discussed later in this section. Let us begin by considering VCl_4^-, which has been determined to be tetrahedral. The delocalized view of this ion can be obtained analogously to that of PO_4^{3-} and the generator table for VCl_4^- given in Table 9.5 is seen to be very similar to Table 9.4. The main difference between these tables is that in contrast to PO_4^{3-}, the $(3d)$ VAOs of vanadium lie below the valence $(4s)$ and $(4p)$ VAOs, whereas in PO_4^{3-} the $(3d)$ valence orbitals lie above the $(3s)$ and $(3p)$ orbitals. Consequently the π ABMOs fall below the σ ABMOs, as shown in Figure 9.18. Thus the additional two electrons in VCl_4^- occupy the $\pi^*(dxy)$ and $\pi^*(dz^2)$ MOs. The presence of two unpaired electrons in this anion, as shown in Figure 9.18, has been verified experimentally.

The localized view of VCl_4^- can be obtained analogously to that of PO_4^{3-}, except that the two unpaired electrons in VCl_4^- as well as the π BMO electrons cannot be localized.

TABLE 9.5 Generator Table for VCl_4^- (26 e^-)

VAO Equivalence Sets	GOs											
	s	px	py	pz	dxy	dz^2	$dx^2 - y^2$	dxz	dyz	$f3'$	$f3''$	$f2''$
$Vs = (V4s)$	n											
$Vp = (V4px), (V4py), (V4pz)$		n	n	n								
$Vd_2 = (V3dxy), (V3dz^2)$					n	n						
$Vd_3 = (V3dx^2 - y^2), (V3dxz), (V3dyz)$							n	n	n			
$Clpr = (Cl_A3pz), (Cl_B3pz), (Cl_C3pz), (Cl_D3pz)$	n	n	n	n								
$Clp\pi = (Cl_A2px), (Cl_A2py), (Cl_B2px), (Cl_B2py), (Cl_C2px), (Cl_C2py), (Cl_D2px), (Cl_D2py)$					n	n	n	n	n	n	n	n

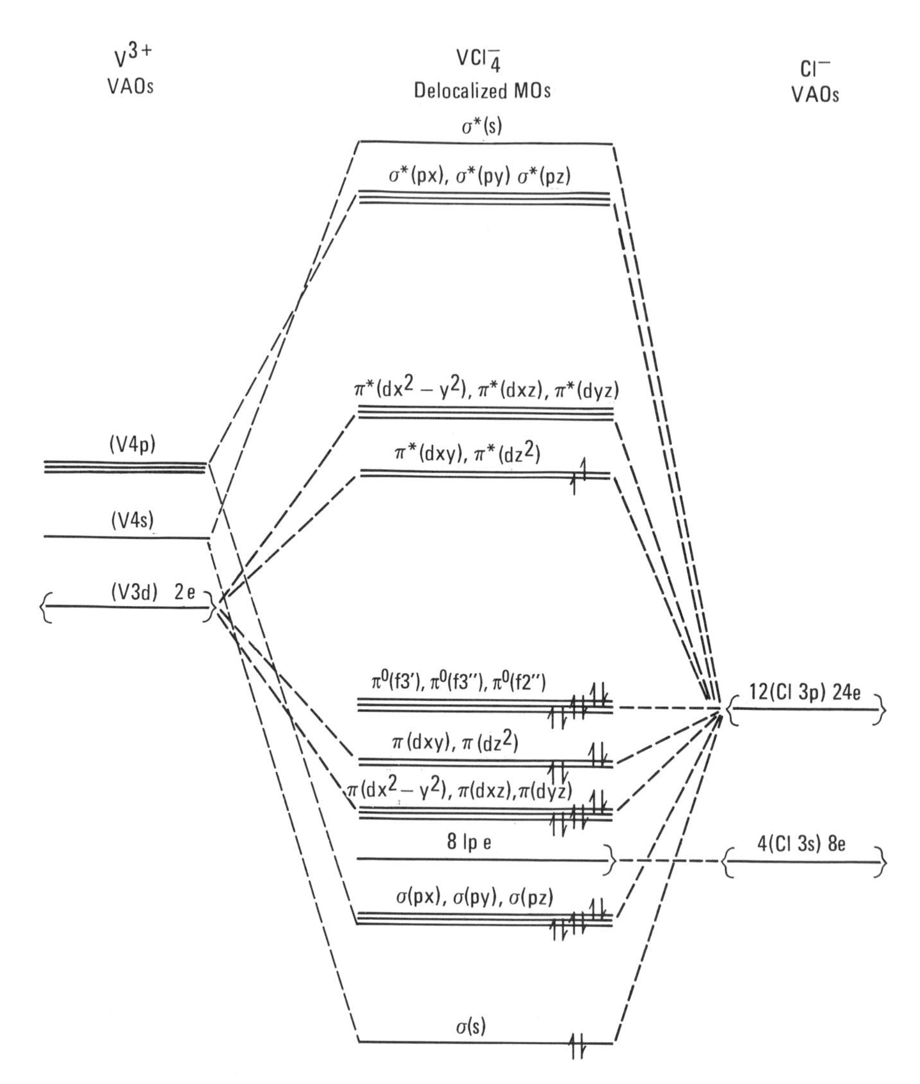

FIGURE 9.18 Delocalized MO energy level diagram for VCl_4^-.

As with octahedral transition metal complexes, a CFT view can also be obtained for other geometries, including tetrahedral complexes such as VCl_4^-. Again the splittings of (*d*) VAOs are considered in this model, and because the (*d*) orbitals are most easily visualized in octahedral symmetry, we again make use of Figure 9.17. Building a model of this figure is the best way of becoming convinced that the (*dxy*), (*dxz*), and (*dyz*) orbitals experience the same environment or field provided by the four negative chloride charges in VCl_4^-. Thus these VAOs are degenerate. Moreover, none of these orbitals have the full symmetry of the tetrahedron. That the ($dx^2 - y^2$) and (dz^2)

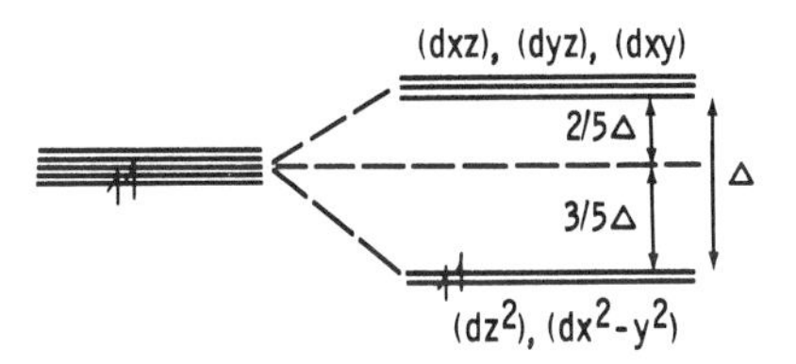

FIGURE 9.19 Splitting of the metal (*d*) VAOs in a tetrahedral crystal field. The axis system employed here is shown in Figure 9.17.

VAOs also feel identical crystal fields can be appreciated from the observation that neither of these AOs contains the full symmetry of the tetrahedron and hence they form a second partner set. Because the lobes of the triply degenerate set are closer to the four ligand charges than the lobes of the doubly degenerate set, the triply degenerate orbitals are raised in energy compared to the doubly degenerate orbitals. The splitting pattern for VCl_4^- which results is that shown in Figure 9.19. Notice that the ordering of the (*d*) VAOs in the CFT model is the same as that of the π-delocalized ABMOs in Figure 9.18, which have mainly vanadium (3*d*) character. The same parallel holds for octahedral CoF_6^{3-} except that the order of the orbitals is inverted and that the doubly degenerate ABMOs for the octahedral case are of the σ instead of the π type. It is interesting to observe here that the differentiation of strong and weak-field complexes from the experimentally measured degree of paramagnetism first occurs at the d^3 configuration for tetrahedral coordination complexes whereas it initially occurs at the d^4 configuration for octahedral ones. Although admittedly somewhat confusing in the present context, d^n is a standard notation that refers to the orbital occupation number and not to the number of orbitals, as in a hybrid set. Tetrahedral complexes show a remarkable tendency to be of the weak-field type because four ligands are unable to split the (*d*) orbitals as strongly as six.

In closing this section, we briefly address the question of why four ligands can frequently surround a transition metal in a square planar array even though there are no lone pairs above and below the plane as in ICl_4^-. A rationale for this observation comes from a comparison of the stabilization energy achieved in a tetrahedral and a square planar crystal field and of the relative sizes of the metal ion and the ligand. Most square planar complexes have d^8 metal configurations. From Figure 9.20 it can be seen that if the crystal field is large enough to overcome the electron pairing energy, the orbital energy stabilization achieved in the square planar configuration is expected to be greater than in the tetrahedral array (in the absence of countervailing effects such as large ligand size which would tend to sterically drive the structure to a tetrahedral configuration). It should be pointed out that in general, it has not been possible to predict geometries of new transition metal complexes from simple considerations as unambiguously as geometries can

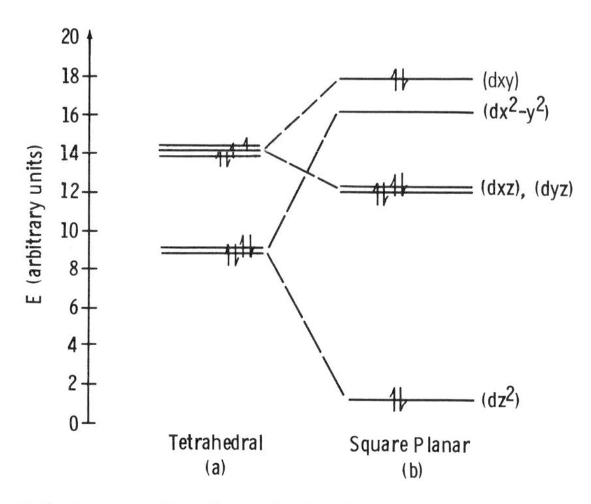

FIGURE 9.20 Relative energies of metal (*d*) VAOs in a tetrahedral (a) and a square planar (b) crystal field.

be predicted for compounds of main-group elements, owing to the variety of effects arising from incomplete $(n - 1)d$ orbitals. For this reason, a great deal of challenging research in this area still remains.

The molecular motions of tetrahedral Z_4 and ZY_4 species can be obtained in the usual way from the appropriate LCAVs represented by the SOs composed of vector-like (*p*) VAOs in Figure 9.8.

Summary

The delocalized and localized bonding views of PnH_3 were seen to be an extension of the considerations employed for generating these views in HChH (Chapter 6). In examining P_4 we saw that generating SOs among tangential VAOs in a tetrahedral geometry with the higher GOs $f1'$ and $f1''$ requires either a knowledge of the angle θ of the conal node, or a prior consideration of the rest of the *f*-generated SOs. After studying the bonding in CH_4, in which no unusual features arose, we found that in PO_4^{3-} the unequal splitting for the triply and doubly degenerate BMO–ABMO pairs generated by the *d* GOs could be rationalized on the basis of geometrical arguments. These considerations were also used in developing the crystal field and ligand field views of VCl_4^-. Finally, some of the factors governing the stability of square planar and tetrahedral transition metal complexes were discussed.

Since no fundamentally new procedures for employing GOs were introduced in this chapter, Table 8.10 is still a valid summary of our recipe for the generation of SOs and SMs.

Exercises

1. Sketch the SOs for PnH_3 in Table 9.1 and also its molecular motions.
2. Draw the localized MO energy level diagrams for (a) CH_4; (b) PO_4^{3-}; (c) VCl_4^-.
3. Develop the delocalized and localized bonding views for (a) $SbCl_3$; (b) the tetrahedral C_4 cluster in the stable tetrahedrane derivative $[CC(CH_3)_3]_4$; (c) $AlCl_4^-$; (d) $TiCl_4$; (e) $PtCl_4^{2-}$ (d^8); (f) the tetrahedral Au_4P^+ cluster in $[(CH_3)_3CAu]_4P^+$.
4. Generate the molecular motions for (a) P_4; (b) AsH_4^+.
5. $Ir_4(CO)_{12}$ consists of a tetrahedral cluster of iridium atoms, each of which is bonded to the carbons of three CO groups. Each member of a set of three CO groups is colinear with a tetrahedral edge of the tetrahedron and can be considered to donate a pair of electrons from an *sp* hybrid on the carbon to an sd^3 orbital on the metal. Such a set is obtained by hybridizing the (4*s*) metal VAO with (3*dxy*), (3*dxz*), and (3*dyz*), and the main lobe of each member of this set points to the corners of a tetrahedron. Develop the delocalized view of the cluster binding.
6. Develop the delocalized view of $CH_3C[Os(CO)_3]_3$ which contains a COs_3 cluster. (Assume all bonds external to the cluster are localized.)
7. CO is a poor Lewis base and BH_3 is not a particularly good Lewis acid. The remarkable stability of the adduct H_3BCO has been attributed to back-donation of electron density from B–H σ BMOs into the π^* MOs of CO. Using GOs between the boron and carbon, sketch the MOs expected from such an interaction and show from an MO energy level diagram why this interaction is expected to lead to increased stability.
8. Calculational evidence supports a strong measure of back-donation from filled metal (*d*) VAOs to σ^* P–F MOs in zero-valent metal PF_3 complexes with virtually no involvement of (*d*) VAOs on phosphorus. Using GOs, sketch the MOs expected from the appropriate metal (*d*) VAOs and the σ^* PF_3 MOs in octahedral $Cr(PF_3)_6$.
9. From the information given in Table 1 of Appendix II.B, calculate the θ values of the ($d0$), ($f0$), ($g0$), ($g1'$), ($g1''$), ($g2'$), ($g2''$), ($h0$), ($h1'$), ($h1''$), ($h2'$), ($h2''$), ($h3'$), and ($h3''$) AOs. Verify your values where appropriate with those given in Figure 2.6.
10. By drawing the appropriate SOs, show that there are (3*d*) VAOs on phosphorus that are of the correct symmetry to participate in σ bonding in PO_4^{3-}.
11. Deep red tetrahedral Na_4^{3+} ions have been observed by x-ray crystallography in a certain zeolite mineral. Develop a delocalized view for this ion.
12. The smallest tetrahedral oxo-anion so far seems to be the NO_4^{3-} anion observed by x-ray crystallography in Na_3NO_4. Develop a delocalized bonding view of this ion.
13. The indium atom in the anion of the salts K, Rb and $Cs[In(CH_3)_4]$ is tetrahedral. Develop a delocalized and localized view of the InC_4^- part of this ion.
14. Develop delocalized views of (a) $TeCl_3^+$; (b) CF_3; (c) $GeCl_3$, all of which are pyramidal.

CHAPTER 10

Bipyramidal and Related Molecules

In this chapter we will consider molecules having trigonal and pentagonal bipyramidal symmetry. The tetragonal bipyramid appears to have been skipped here but you can see that this geometry is the same as an octahedron if all six vertex atoms are identical. Molecules having trigonal bipyramidal (tbp) symmetry include the cluster anion $B_5H_5^{2-}$ and PF_5, depicted in Figures 10.1a and b, respectively. Other examples of trigonal bipyramidal molecules are PCl_5(gas), AsF_5, and $SbCl_5$. Related to PF_5 are SF_4 and BrF_3 (Figures 10.1c and d), which can be viewed as having, respectively, one and two vertex atoms in the equatorial plane of the tbp replaced by lone pairs. Additional examples of this type of molecule are SeF_4, $TeCl_4$, PCl_4^-, AsF_4^-, $SbCl_4^-$, ClF_3, and XeF_3^+. Of pentagonal bipyramidal symmetry are $B_7H_7^{2-}$ and IF_7 (Figures 10.1e and f). Several metal fluoride complexes also appear to possess this geometry [e.g., MF_7^{3-} (M = Zr, Hf, Nb, U) and ReF_7]. Species of geometrical order higher than pentagonal bipyramidal do not appear to have been characterized.

For each bonding scenario discussed in this chapter, you should use the appropriate column in Table 10.7 as a guide for repeating the steps in the text using the Homework Drawing Board in Node Game.

10.1. BrF_3

This volatile interhalogen compound, formed by direct combination of the elements, has been determined to have the structure shown in Figure 10.1d. The electron dash structure in Figure 10.1d is analogous to the dash structure of XeF_2 in Figure 10.1g. We therefore expect the five electron pairs to point toward the corners of a trigonal bipyramid in the localized view. We could have chosen to treat XeF_2 as a tbp molecule with three vertex atoms missing, but we have already discussed it as a linear molecule in Chapter 4. It should be recognized from our treatment of XeF_2 that the localized description of BrF_3 does not necessitate the inclusion of (*d*) orbitals on the bromine. The electronic structures of XeF_2 and BrF_3 are quite similar except that one

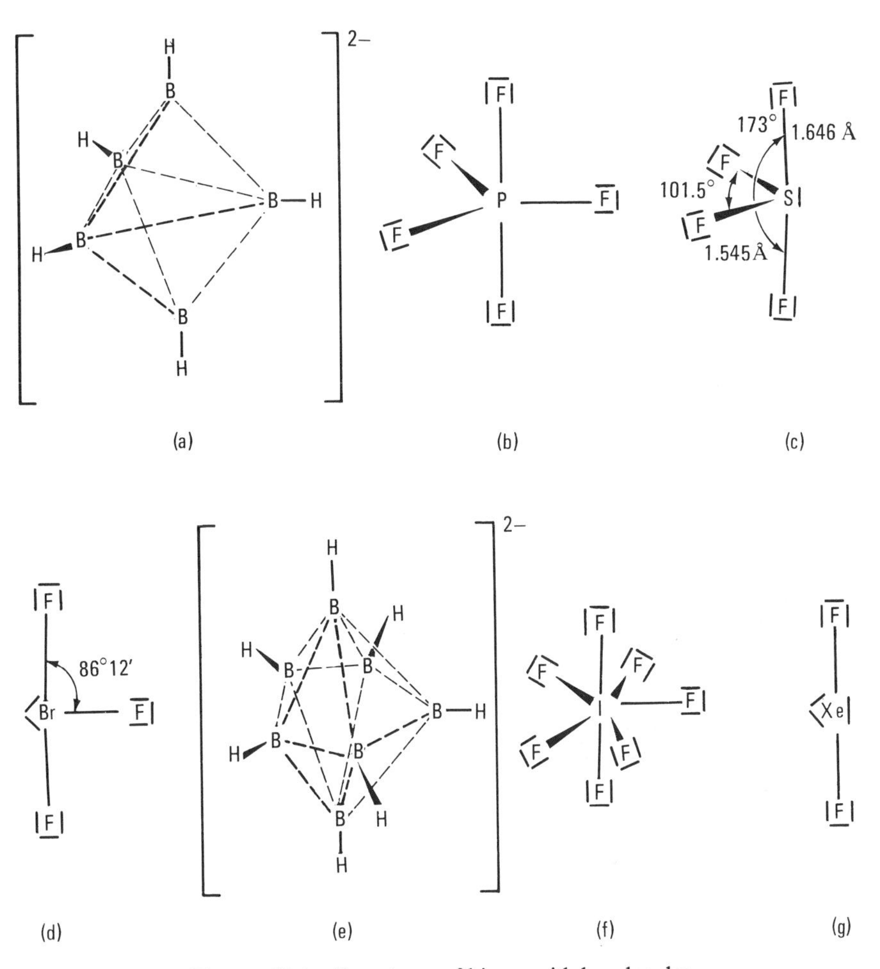

FIGURE 10.1 Structures of bipyramidal molecules.

lone pair in the equatorial plane of XeF_2 (as shown in Figure 10.1g) is replaced by a bond pair in a Br—F link (Figure 10.1d). Recall also that it is a general rule of VSEPR theory that in the localized view, five electron pairs arrange themselves in a trigonal bipyramidal array with lone pairs (if any) occupying equatorial positions. Notice that the less-than-90° $F_{ax}BrF_{eq}$ bond angles are a consequence of the VSEPR rule that bond-pair lone-pair repulsions are less than lone-pair lone-pair repulsions.

For convenience, we will choose the coordinates shown in Figure 10.2 (which, for simplicity, do not include the slight bending of the vertical molecular axis away from 180°). From the symmetry of the molecule it should be clear that the fluorine sites are not equivalent. This site inequivalence also induces *chemical* inequivalence in these atoms because their valence electronic environments are different. Thus F_A and F_B are equivalent to each

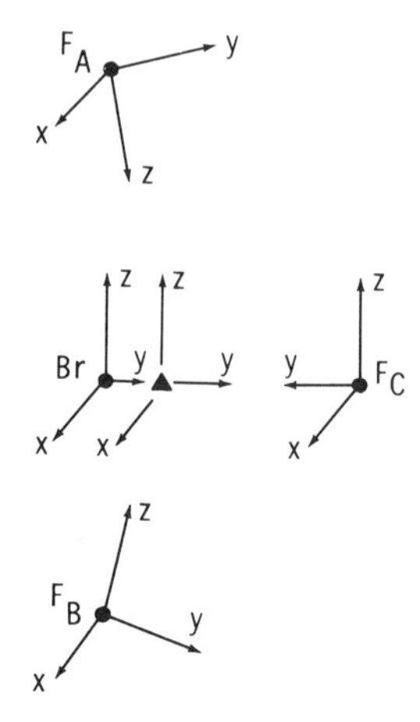

FIGURE 10.2 Axis system for BrF_3.

other but inequivalent with respect to F_C. We assign lone pairs to the (4*s*) and (4*px*) VAOs on bromine, the (2*s*), (2*px*), and (2*pz*) VAOs on F_C, and the (2*s*), (2*px*), and (2*py*) VAOs on F_A and F_B. [The (Br4*s*) lone pair must actually possess some (Br4*px*) and (Br4*py*) character for the two bromine lone pairs to be sterically active.] From the remaining VAOs we construct Generator Table 10.1 in the usual way by using GOs to draw the SOs for diatomic BrF_C and for the linear triatomic F_ABrF_B portions of the molecule (Figure 10.3). The MOs can then be drawn from the information in the GO columns of Table 10.1.

Since there are three electron pairs for the five MOs of this molecule, two MOs will be unoccupied. These two will obviously be the *s*- and *pz*-generated ABMOs. Therefore two of the three *s*-generated MOs are occupied. We now will establish that the SO $\tilde{\sigma}(s|Fpz)$ in Figure 10.3b can be taken to be an occupied NBMO. As explained earlier, we can always choose an orbital pair so that the third SO will not contribute. There are three choices. First, if we choose $\sigma(s|Brpy)$ (Figure 10.3a) to be excluded as the NBMO, we must linearly combine the SOs in Figures 10.3b and c to provide a BMO and an

TABLE 10.1 Generator Table for BrF_3 ($6\ e^-$)

VAO Equivalence Sets	GOs *s*	GOs *pz*
Br*py* = (Br4*py*)	n	
Br*pz* = (Br4*pz*)		n
F*pz* = (F_A2*pz*), (F_B2*pz*)	n	n
F*py* = (F_C2*py*)	n	

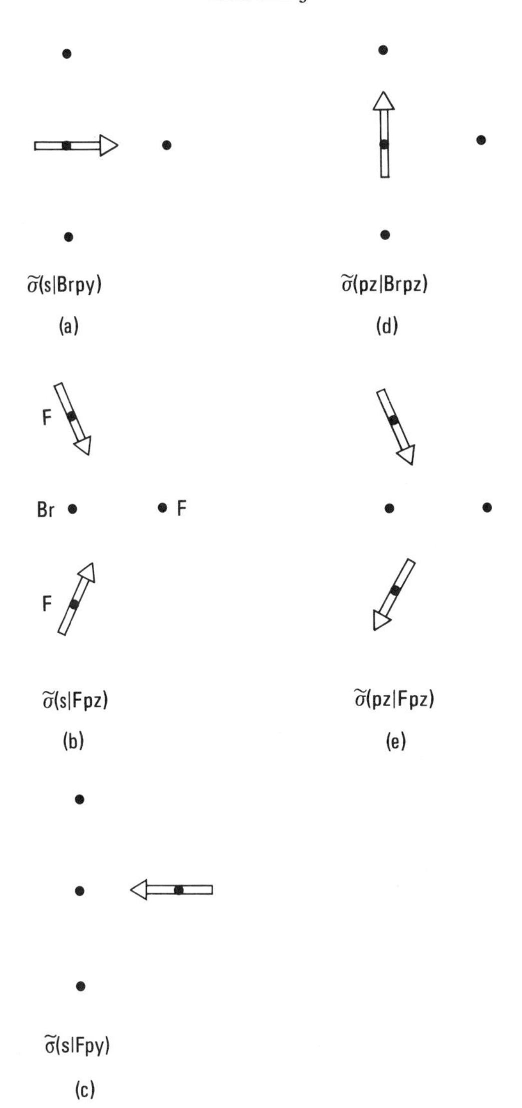

FIGURE 10.3 SOs of BrF_3 as given in Table 10.1. The GOs are not shown.

ABMO. These SOs involve orbitals on nonadjacent atoms, however, which essentially do not interact. Second, if we choose to exclude the SO in Figure 10.3b as the NBMO, linear combination of the remaining SOs in Figures 10.3a and d clearly give rise to a BMO–ABMO pair. Third, if we choose to exclude the SO in Figure 10.3c, linear combination of the SOs in Figures 10.3a and b would give rise to a BMO–ABMO pair, but the resulting three-center bond, along with the three-center BMO obtained by linearly combining the *pz*-generated SOs in Table 10.1, would result in a bond order of 1.0 in the triatomic part of the molecule and a bond order of zero in the diatomic part. This choice, like the first one, is obviously bad and so let us see what the consequences of the second choice are. By placing the NBMO in the triatomic part of the molecule and the BMO–ABMO pair in the diatomic part, we insure that the bond order in the former moiety is 0.5 and 1.0 in the latter. This makes sense because the BrF_A and BrF_B bonds are known to be longer (weaker) than the (stronger) BrF_C bond.

Of the 28 valence electrons in the molecule, only 6 have not been assigned as lone pairs, and they occupy the MOs $\sigma(s)$, $\sigma^0(s)$, and $\sigma(pz)$. The energy level diagram in Figure 10.4 shows the bonding parallel between BrF_3 and XeF_2, which we made at the beginning of this section. Thus there are only two bonding electrons delocalized over three centers (F_A, Br, and F_B), as was the case in XeF_2, and an equatorial lone pair in XeF_2 is replaced by an occupied BMO and BrF_3. This figure is also consistent with the observation that the

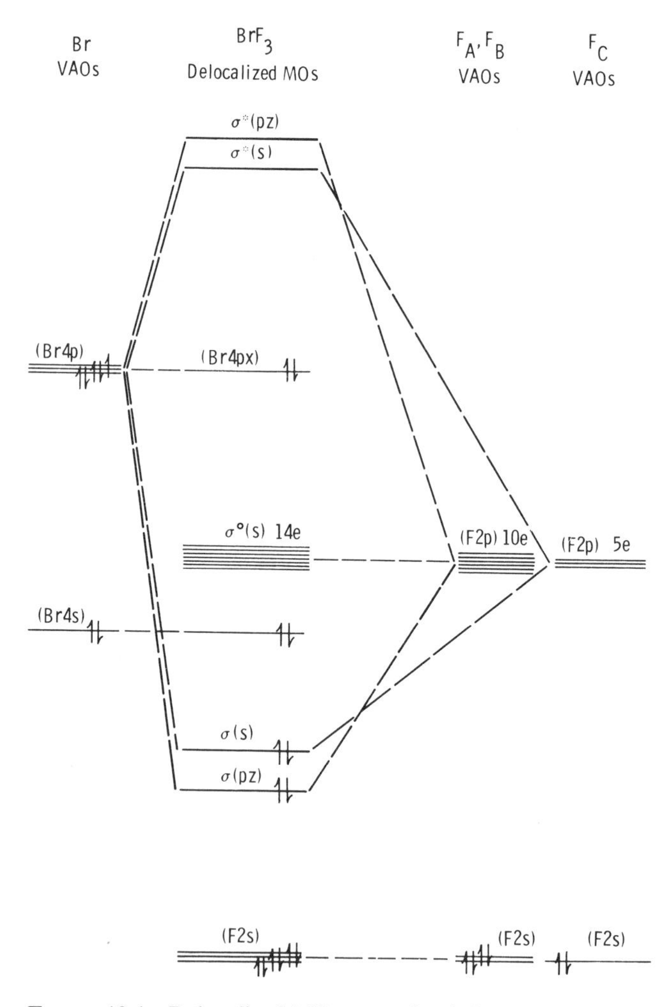

FIGURE 10.4 Delocalized MO energy level diagram for BrF_3.

BrF_A and BrF_B bond lengths are equal but longer than BrF_C. The former are three-center two-electron bonds (bond order = 0.5) while the latter is a two-center two-electron bond (bond order = 1.0).

You can see from the delocalized bond picture that the bonding in the diatomic part is already localized. The only further localization required is that which we can do for the triatomic part and also the lone pairs. The delocalized σ bonding system along the F_A—Br—F_B axis can be localized into two half-bonds as was the case for XeF_2. The lone pairs on bromine can be localized into (*spx*) VAO hybrids and those on the fluorines can be localized into (sp^2) VAO hybrids. The localized MO diagram is shown in Figure 10.5.

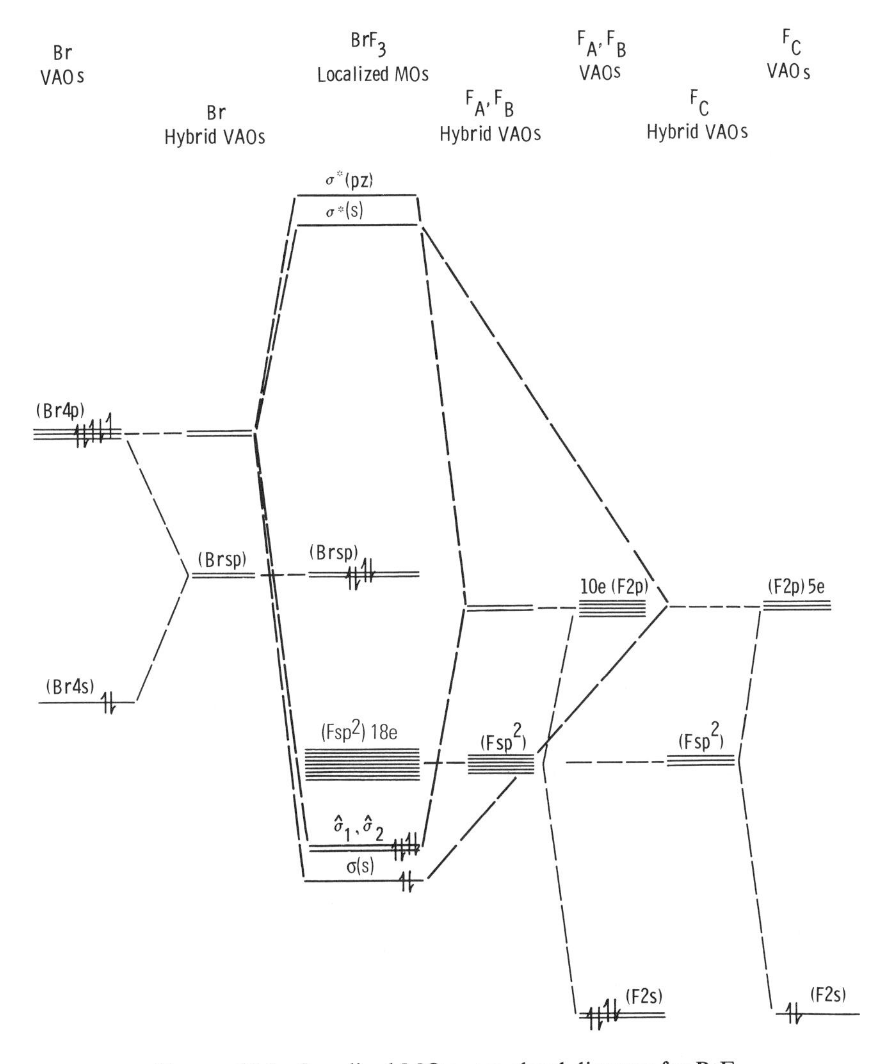

FIGURE 10.5 Localized MO energy level diagram for BrF_3.

Although the generation of the molecular motions of BrF_3 follows our usual recipe, a somewhat unusual feature emerges from that process because of the molecule's low symmetry. From the GO-generated SMs drawn in Figure 10.6 and summarized in Generator Table 10.2, we see that the *s* GO (Figure 10.6a) calls in SMs from three AV equivalence sets. Interestingly, *py* generates the same set of SMs plus an additional one (Figure 10.6b). Thus the column entries headed by the less inclusive *s* GO in Table 10.2 is enclosed in parentheses and can be disregarded. Also, *dyz* (Figure 10.6e), $dx^2 - y^2$

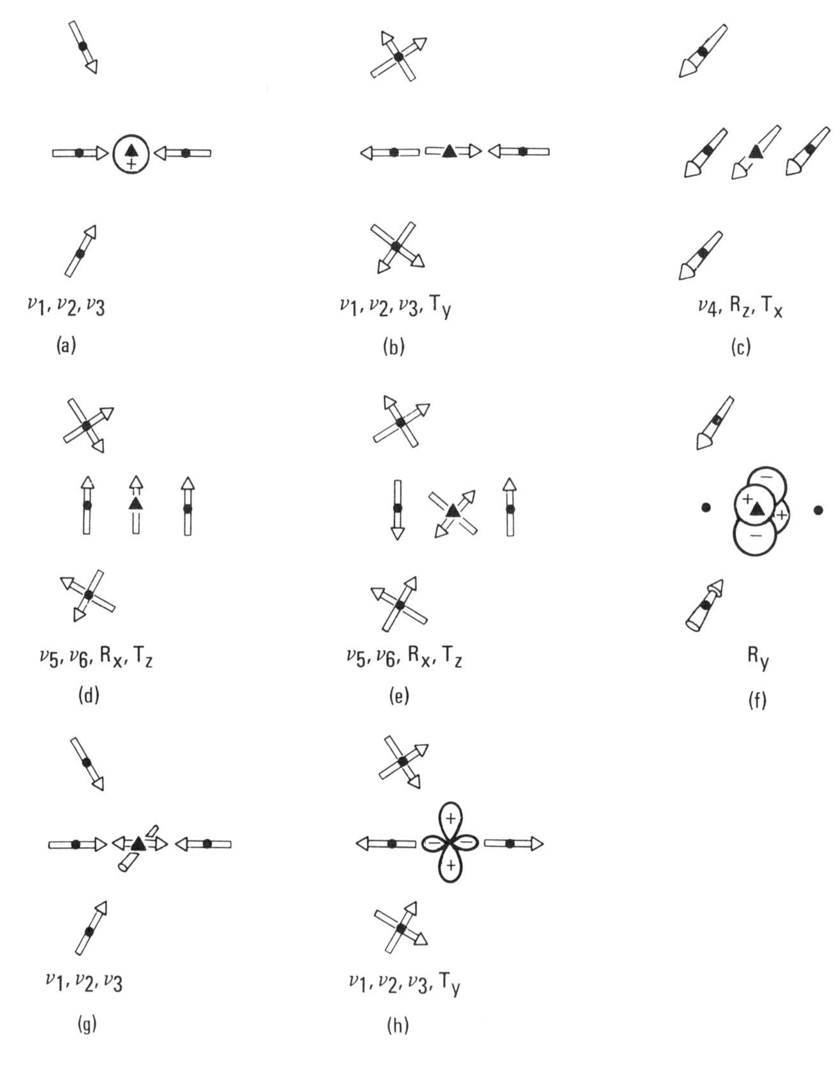

FIGURE 10.6 SMs for T-shaped ZY_3.

TABLE 10.2 Generator Table for T-Shaped ZY_3

	GOs							
AV Equivalence Set[a]	*s*	*px*	*py*	*pz*	dz^2	dx^2-y^2	*dxz*	*dyz*
Z*rx*		m						
Z*ry*	(m)		m		(m)	(m)		
Z*rz*				m				(m)
Y*xa*		m					m	
Y*ya*			m	m	(m)			(m)
Y*za*	(m)		m	m	(m)	(m)		(m)
Y*xe*		m						
Y*ye*	(m)		m		(m)	(m)		
Y*ze*				m				(m)

[a] Here "a" denotes the axial Y atoms and "e" denotes the equatorial Y atom.

(Figure 10.6g) and dz^2 (Figure 10.6h) generate the same SMs as *pz*, *s*, and *py*, respectively, and can therefore be ignored. It is clear from Figure 10.6b that the translation T_y must arise from an AV equivalence set combination involving Y*ya* + Z*ry* − Y*ye* + Y*za*. T_z and R_x in Figure 10.6d clearly arise from linear combinations of the type Z*rz* + Y*ze* − Y*za* + Y*ya* and Z*rz* − Y*ze* + Y*ya* − Y*za*, respectively. As expected from the $3n-6$ rule, six vibrational modes are generated.

10.2. SF_4

Sulfur tetrafluoride is a gas formed in the oxidative fluorination of SCl_2. In its electron dash structure in Figure 10.1c sulfur (like bromine in BrF_3) is presumed to be hypervalent since it has an expanded octet involved in σ bonds and lone pairs. This electron dash structure permits the formal charges on the atoms to be zero. Our experience with isoelectronic BrF_3 and XeF_2 suggests that in the localized view of SF_4 one lone pair will be in the equatorial plane of a tbp electron pair arrangement. The experimentally determined structure in Figure 10.1c parallels that of BrF_3, with the longer bonds bent away from 180° because of dominating lp–bp repulsions. The large electronegativity difference between sulfur and fluorine gives rise to strong ionicity in the S–F bonds.

Using the axis system in Figure 10.7, in which we ignore the slight axial angular distortion, the electron dash structure in Figure 10.1c indicates that we can assign the F_A, F_B lone pairs to the (F2*s*), (F2*px*), and (F2*py*) VAOs and the F_C, F_D lone pairs to (F2*s*), (F2*px*), and (F2*pz*) VAOs. The sulfur lone pair is assigned to the (S3*s*) VAO. In constructing generator Table 10.3 with the aid of the pictorial representations of the SOs in Figure 10.8, similarities to the generator table and SOs for BrF_3 are revealed. Again there are three SOs

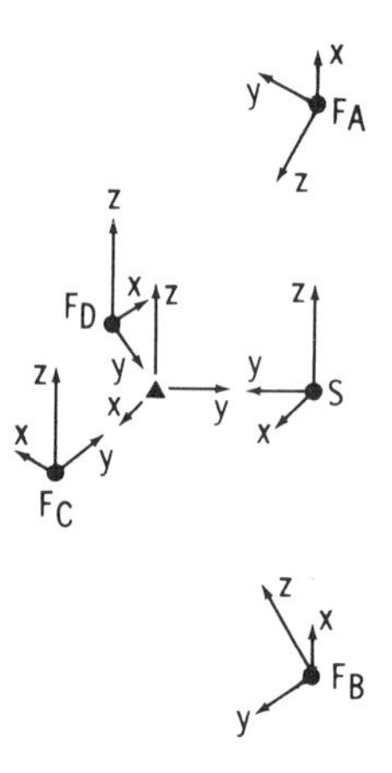

FIGURE 10.7 Axis system for SF_4. The x axes on F_A and F_B are to be viewed as coming out of the plane of the paper.

generated by s. Like BrF_3 (and several other molecules that we have considered), SF_4 is an example of a system for which two SOs interact mainly with each other but not (significantly) with a third. We then again expect an s-generated BMO–NBMO–ABMO set of MOs, with the NBMO consisting mainly of the Fpz SO because eventually we want less bond order in the axial triatomic part than in the equatorial part. That way, our bonding scheme will reflect the experimental fact that the axial bonds are longer (i.e., weaker) than the stronger equatorial ones. Consequently, the bonding and antibonding linear combinations arise from the (S3py) VAO and the Fpy SO. A second BMO–ABMO pair for SF_4 results from linearly combining $\tilde{\sigma}(px|\text{F}py)$ and (S3px) (Figures 10.8d and e). The BMO–ABMO pair that stems from linear combinations of the pz-generated SOs (Figures 10.8f and g) is analogous to the one we formulated earlier for the axial triatomic part of BrF_3. We see then that SF_4 has two triatomic parts: a bent one in the equatorial plane and a linear one along the z axis. Drawing a qualitative energy level diagram is

TABLE 10.3 Generator Table for SF_4 (8 e^-)

	GOs		
VAO Equivalence Sets	s	px	pz
Spx = (S3px)		n	
Spy = (S3py)	n		
Spz = (S3pz)			n
Fpz = (F_A2pz), (F_B2pz)	n		n
Fpy = (F_C2py), (F_D2py)	n	n	

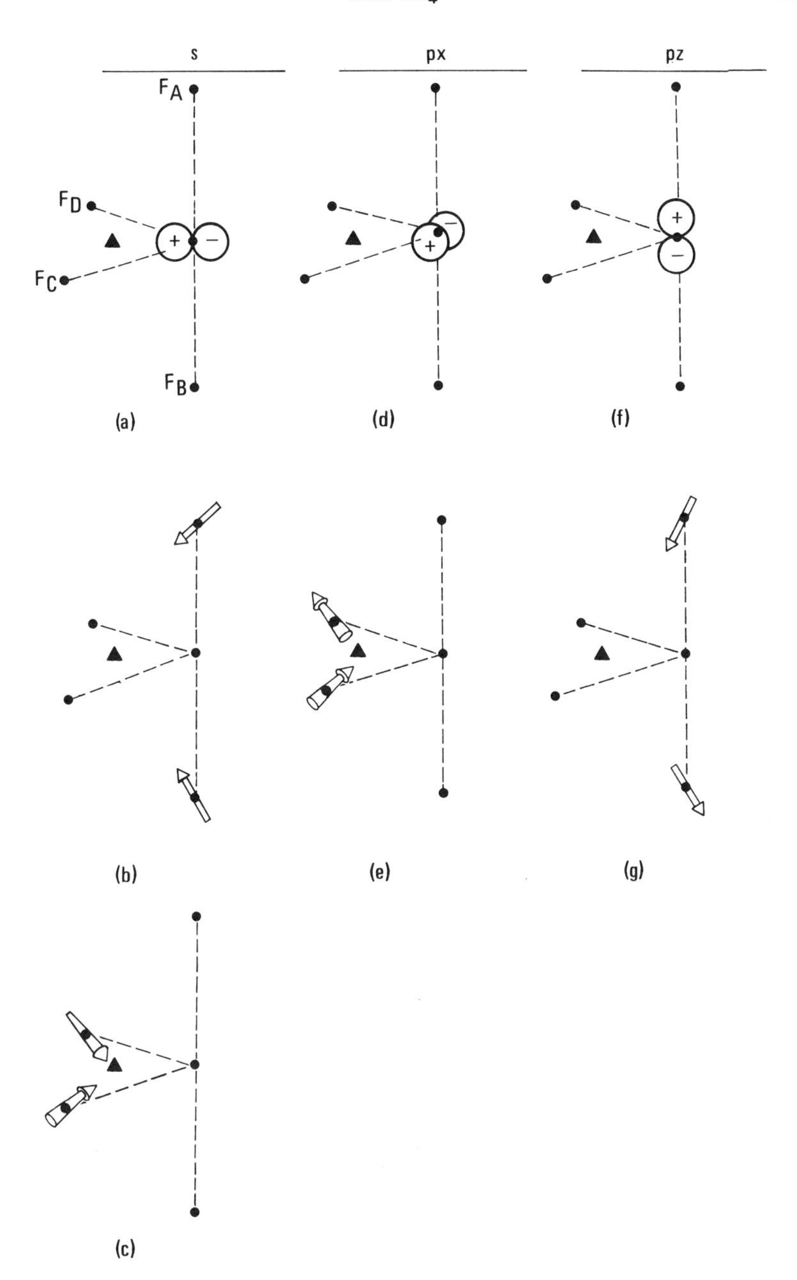

FIGURE 10.8 SOs for SF_4. The GOs are not shown in these drawings.

straightforward and is left as an exercise. All three of the BMOs are filled, giving average bond orders of 1.0 and 0.5 for the shorter equatorial bonds and the longer axial bonds, respectively.

Localization of the occupied MOs associated with the axial triatomic part of the molecule is analogous to that in BrF_3 and XeF_2, and we will not discuss this further. Localization of the occupied MOs of the equatorial triatomic moiety also poses no new considerations since this can be accomplished in an analogous manner to the localization of the occupied σ bonding MOs of H_2O.

Generating the molecular motions of a species having the geometry of SF_4 is done in the usual way and this is left as an exercise.

10.3. $B_5H_5^{2-}$

This anion is another example of an electron-deficient system. Anions of the type $B_nH_n^{2-}$ where $n = 6$–12 are well known and all have been found to possess *closo* (i.e., closed) configurations. Thus, for example, the geometries of $B_6H_6^{2-}$, $B_7H_7^{2-}$, and $B_{12}H_{12}^{2-}$ are octahedral, pentagonal bipyramidal, and icosahedral, respectively. It is not unreasonable to expect that the as yet uncharacterized $B_5H_5^{2-}$ ion will have tbp symmetry (Figure 10.1a) since $B_3C_2H_5$, an isostructural isoelectronic analogue, is known.

The B—H links in $B_5H_5^{2-}$ are two-center two-electron bonds and as usual we reserve the appropriate (Bsp^{α}) hybrid on each boron for bonding to a terminal hydrogen. Note that α will undoubtedly be different for the apical and equatorial boron VAOs. As you can see in the axis system of Figure 10.9, the inwardly directed orthogonal hybrid VAOs divide into two sets, the apical hybrids (B*sp*a-in) and the equatorial hybrids (B*sp*e-in). Furthermore, the two (B2*px*) and (B2*py*) VAOs, which are directed tangentially around the trigonal bipyramidal framework, also form apical and equatorial VAO sets.

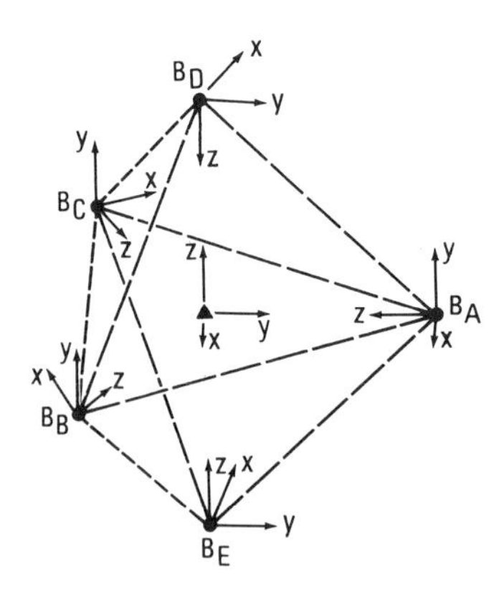

FIGURE 10.9 Axis system for $B_5H_5^{2-}$.

Since ten electrons are in the localized B—H bonds, there are 12 electrons left to bind the boron cluster.

Using the axis system shown in Figure 10.9, we can construct Generator Table 10.4 with the help of Figure 10.10. The GOs *px* and *py* in Table 10.4 are seen to each generate three SOs. Notice that these GOs and the corresponding SOs (and MOs) they generate are degenerate in the triangular symmetry in which they reside (see Chapter 5.1). By linearly combining the SOs generated by each of the GOs (*s*, *px*, *py*, and *pz*) 4 BMOs can be obtained which will house 8 of the 12 electrons that bind the borons together. Linear combinations of the antibonding and nonbonding SOs generated by *dxz* and *dyz* GOs give rise to two degenerate BMOs and two degenerate ABMOs. The two BMOs are occupied by the remaining four electrons. From the energy level diagram in Figure 10.11 you can see that the overall B—B bond order is 2/3 since there are nine edges on a trigonal bipyramid.

Although it is clear from Table 10.4 that there should be a total of six BMO–ABMO pairs and three NBMOs, the ordering among the BMOs and the ABMOs cannot be predicted without sophisticated calculations and no order should be inferred from the spacings shown in the energy level diagram in Figure 10.11. You can see from this diagram, however, that the 12 electrons available for boron cluster binding fill all of the BMOs but none of the NBMOs. It is because of the empty NBMOs that this ion is electron deficient.

Hybridization of the GO set associated with the BMOs of Figure 10.11 leads to an sp^3d^2 hybrid set. When such a set contains a *dxz* and a *dyz* orbital, the main lobes of its members point to the vertices of a trigonal prism. In Figure 10.12 is shown how a trigonal prism can be inscribed inside a trigonal bipyramid. Recall that in the previous chapter's discussion of the localized view of P_4 we saw that if an sp^3d^2 hybrid set contains a *dxy* and dz^2 orbital, then the main lobes of the set point to the vertices of an octahedron. Thus the directionality of the hybrids in Figure 10.12 tells us that the six BMO electron pairs should be localized at the edges of the tbp, as shown.

There arises a question regarding the choice of boron VAO orientation that will accommodate our localization scheme. This question can be resolved

TABLE 10.4 Generator Table for B_5 Cluster in $B_5H_5^{2-}$ (12 e^-)

	GOs						
VAO Equivalence Sets	*s*	*px*	*py*	*pz*	*dyz*	*dxz*	*f*3′
B*spa*-in = (B_D*spi*), (B_E*spi*)	n			n			
B*pxya* = (B_D2*px*), (B_E2*px*), (B_D2*py*), (B_E2*py*)		n	n		n	n	
B*spe*-in = (B_A*spi*), (B_B*spi*), (B_C*spi*)	b	a	a				
B*pxe* = (B_A2*px*), (B_B2*px*), (B_C2*px*)		b	b				a
B*pye* = (B_A2*py*), (B_B2*py*), (B_C2*py*)				b	a	a	

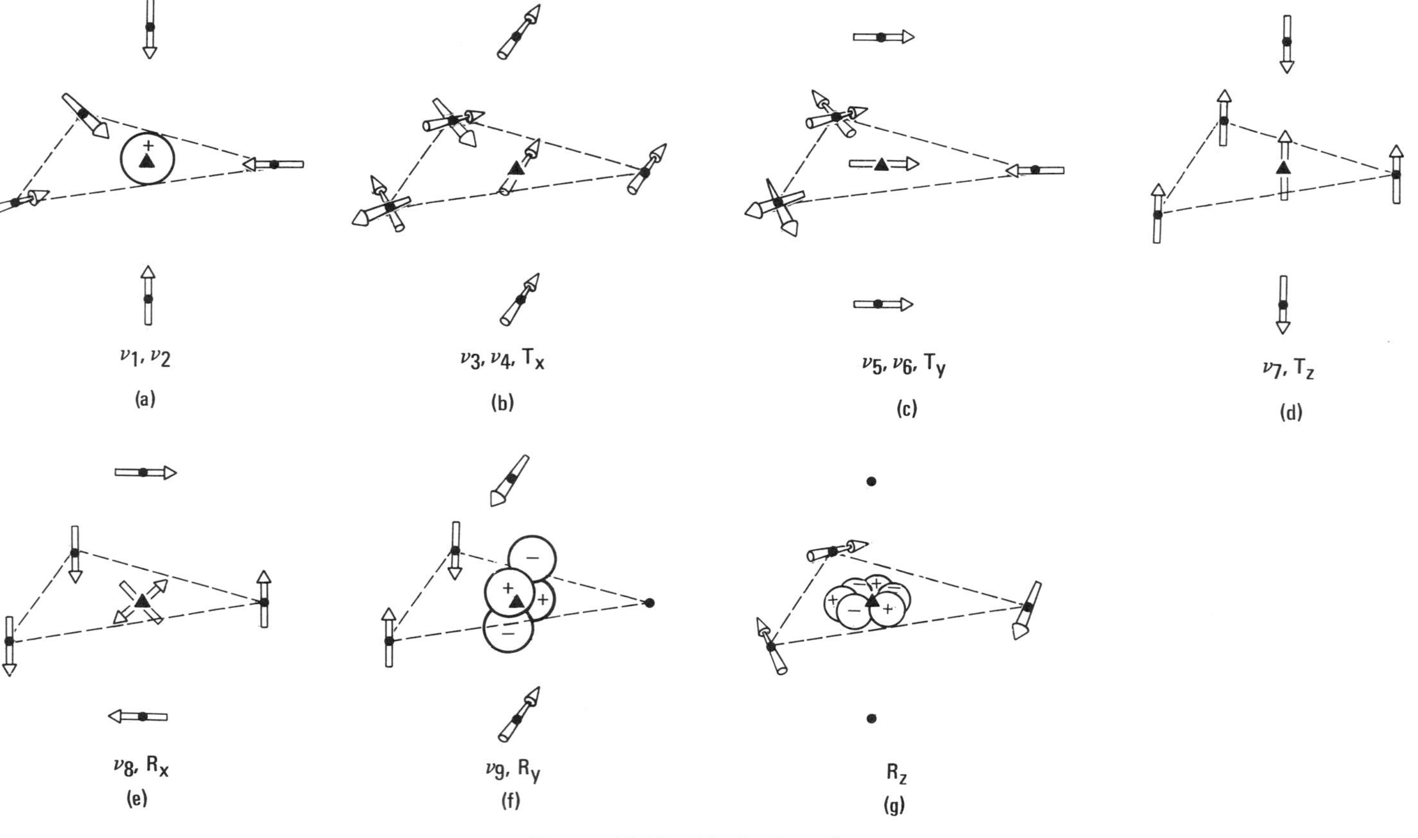

FIGURE 10.10 SOs for $B_5H_5^{2-}$.

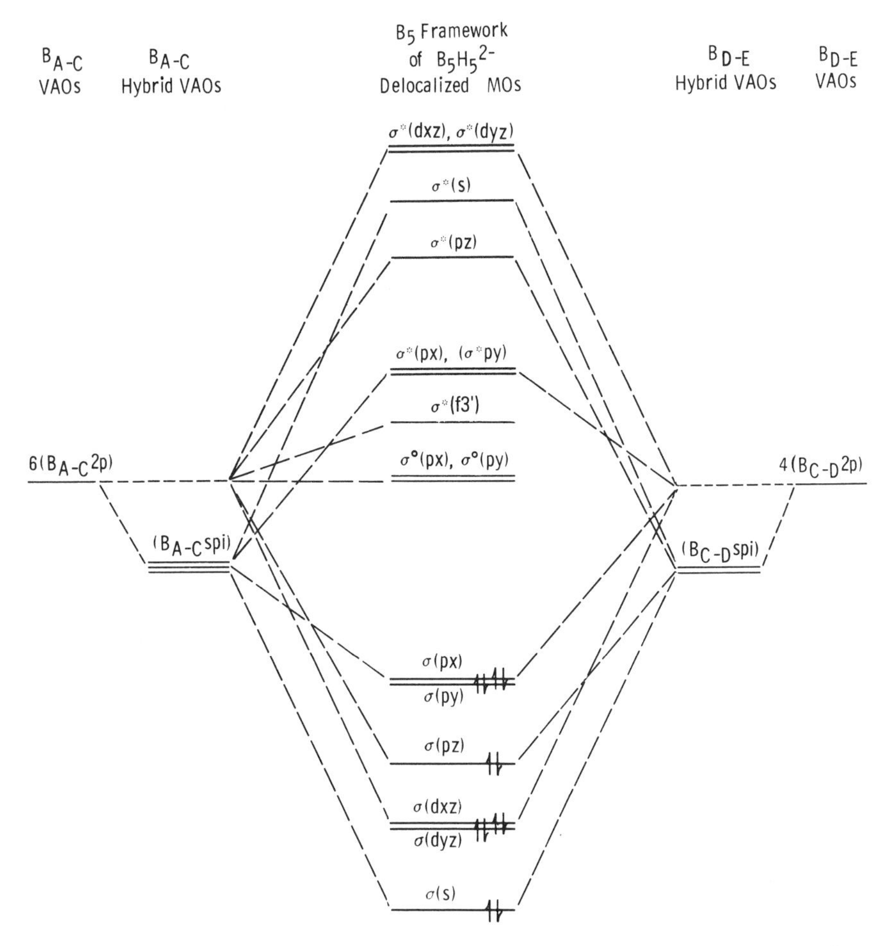

FIGURE 10.11 Delocalized MO energy level diagram for B_5 framework of $B_5H_5^{2-}$.

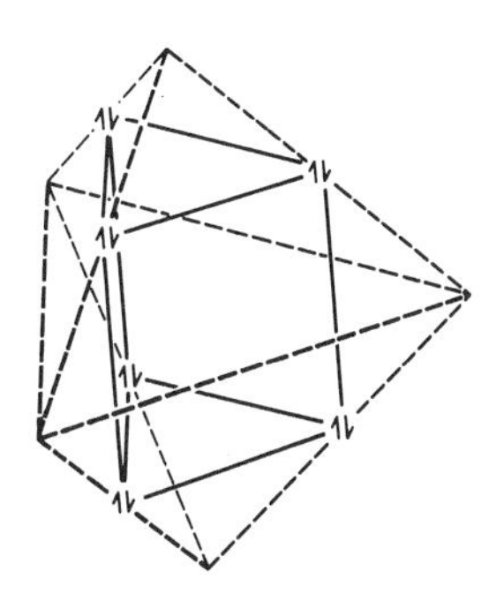

FIGURE 10.12 Localized bond pair positions in $B_5H_5^{2-}$ indicated by the $sp^3dxzdyz$ GOs.

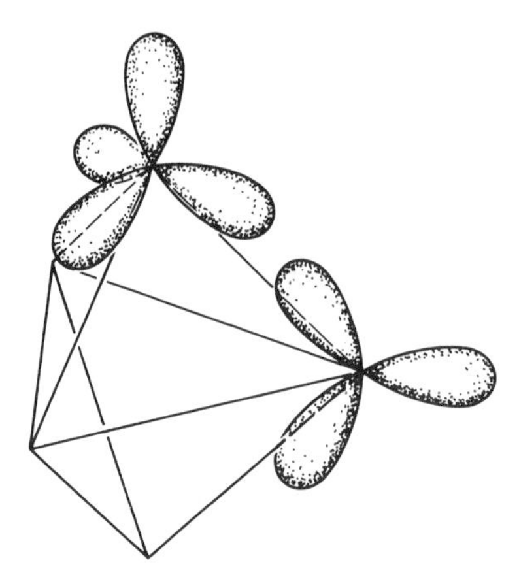

FIGURE 10.13 Sketch showing directions of apical sp^3 and equatorial sp^2 VAOs conveniently oriented for localized two-center bonds in $B_5H_5{}^{2-}$.

by considering the apical boron VAOs to consist of a set of approximately (sp^3) hybrids, one of which is used to make a two-center two-electron bond with a hydrogen, while the other three are directed over the three edges of the trigonal bipyramid as shown in Figure 10.13. To avoid bent bonds with the equatorial borons, we may wish to place more (p) character in the edge-oriented lobes of the apical boron hybrids, which would bring them more parallel to the edges of the trigonal bipyramid. Of course more (s) character must then be moved into the hybrids pointed toward the apical hydrogens. By assuming that the VAOs on the equatorial borons consist of (sp^2) orbitals, as schematically depicted in Figure 10.13 (in which the outwardly directed hybrids are each used to make a two-center two-electron bond with an equatorial hydrogen), we can make two-center two-electron bonds between pairs of equatorial and apical boron hybrids as shown in the localized energy level diagram in Figure 10.14. In this energy level diagram the delocalized ABMOs shown in Figure 10.11 are omitted because we have changed the hybridization on the borons and therefore it would not be straightforward to construct in Figure 10.14 the dashed-line connections to the ABMOs in Figure 10.11. Recall also that the localization of ABMOs is not very meaningful anyway.

It should be noted that the orientation of the sp^2 VAO hybrids on the equatorial borons requires that there be three boron ($2px$) VAOs (one from each boron) oriented tangentially around the equatorial plane. These ($2px$) VAOs form the delocalized NBMOs and the $f3'$-generated ABMO of the molecule. Thus the localized bond picture we have developed permits electron pair binding between equatorial and apical pairs of borons but not between equatorial borons even though they would presumably be within bonding distance of one another.

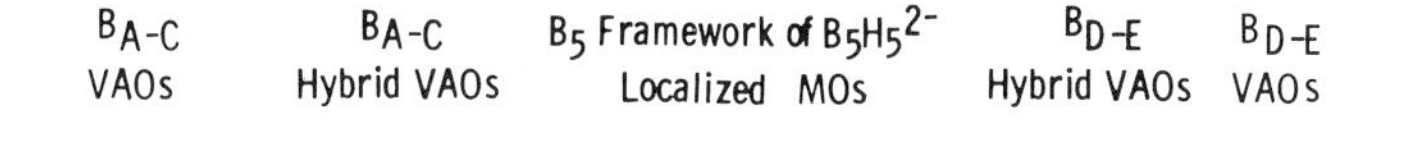

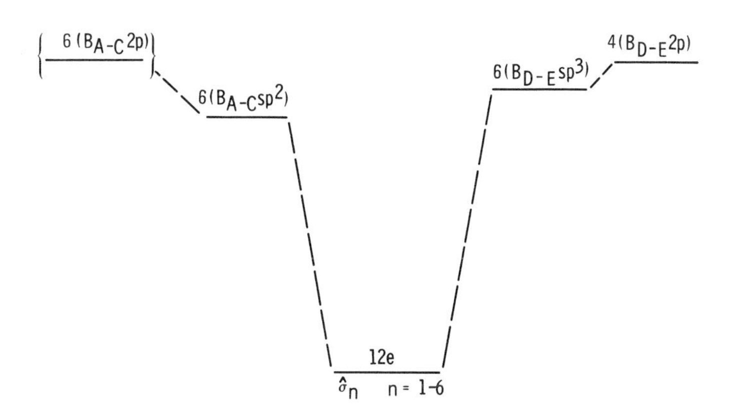

FIGURE 10.14 Localized MO energy level diagram for the B_5 framework of $B_5H_5^{2-}$. The delocalized NBMOs and ABMOs are not shown.

The interesting and useful principle of *isolobality* is briefly discussed here. Notice that the B-H fragment we used to build the $B_5H_5^{2-}$ ion has the same set of VAOs on the boron as we have on the carbon in a C-H fragment. We say, therefore, that the B-H and C-H fragments are *isolobal.* The isolobality principle can be used to predict that we could perhaps substitute a C-H fragment or two for one or two B-H fragments in $B_5H_5^{2-}$. Since a carbon atom has four valence electrons instead of three as does boron, we could expect the negative charges on the $B_5H_5^{2-}$ to be reduced by one with each sequential C-H substitution, in order to keep the total electron count the same. Indeed, $B_3C_2H_5$ exists (see exercise 8 at the end of the chapter).

10.4. PF_5

Fluorination of phosphorus produces PF_5, a colorless, stable gas. The electron dash structure (Figure 10.1b) is similar to the geometrical arrangement of the five electrons pairs around the central atom in XeF_2, BrF_3, and SF_4. The expected trigonal bipyramidal structure of this hypervalent molecule has been confirmed by electron diffraction experiments and the axial bond lengths are found to be slightly longer than the equatorial ones. Both types of bonds contain strong ionic character.

After assigning the lone pairs to fluorine $(2s)$, $(2px)$, and $(2py)$ VAOs, Generator Table 10.5 can be constructed with the help of SOs generated by GOs and the use of the coordinate system in Figure 10.9. This table reflects a fact we will use later, namely, that s and dz^2 generate the same SOs. As in

TABLE 10.5 Generator Table for PF_5 (10 e^-)

	GOs				
VAO Equivalence Sets	*s*	dz^2	*px*	*py*	*pz*
P*s* = (P3*s*)	n	(n)			
Ppa = (P3*pz*)					n
Ppe = (P3*px*), (P3*py*)			n	n	
F*pz*a = (F_D2pz), (F_E2pz)	n	(n)			n
F*pz*e = (F_A2pz), (F_B2pz), (F_C2pz)	n	(n)	n	n	

BrF_3 and SF_4, the apical fluorines and the equatorial fluorines in PF_5 interact with the central atom but not with each other. Since PF_5 has ten valence electrons and a total of nine MOs according to Table 10.5, we expect that the one ABMO corresponding to each of the GOs will be unoccupied and that the *s*-generated NBMO will be occupied. The SO to choose as the NBMO in our linear combinations generated by the *s* GO is F*pz*a because that way the bond order will be 1.0 in the equatorial plane, and 0.5 in the axis. As we saw with BrF_3 and SF_4, such a bond order distribution reflects the experimental result predicted by VSEPR considerations that the axial bonds are longer (weaker) than the stronger equatorial ones. These bond orders are seen to arise in the following way. The BMOs of the BMO–ABMO pairs generated by *px*, *py*, and *pz* are occupied. The BMO $\sigma(pz)$ forms a three-center bond in the linear triatomic part of PF_5, $\sigma(px)$ and $\sigma(py)$ form four-center bonds in the tetra-atomic equatorial plane, and $\sigma(s)$ gives a 4-center BMO there as well. We then have a bond order of 1.0 in each of the P—F equatorial links and a bond order of 0.5 in each P—F axial link. Drawing the delocalized MO energy level diagram is left as an exercise.

In obtaining a localized bond picture we are faced with the problem that two occupied MOs are generated by the same GO, namely (*s*). This situation sometimes occurs for electron-rich systems where NBMOs and/or ABMOs are occupied. If we choose to localize only the occupied BMOs, we must deal with an sp^3 GO set. To inscribe the tetrahedral symmetry of such a GO set inside the trigonal bipyramidal PF_5 is awkward at best. If we include the GO of the occupied NBMO (namely, *s*) we face the strange prospect of forming hybrid GOs with the *sspxpypz* set, which includes two *s* GOs. We could circumvent this problem by recalling that the *s* GO is not the only generator of the first column in Table 10.5. In fact, any GO with axial symmetry about the *z* axis, which is also symmetric on reflection through the *xy* plane gives rise to the same SOs. In addition to *s*, this set includes dz^2, *g*0, and still higher GOs, all of which are characterized by having either zero conal nodes in the case of *s*, or an even number of conal nodes (two for dz^2, four for *g*0, etc.). On the basis of the nodal structure of the orbitals, it is in fact easy to see that the BMO and the NBMO are most naturally associated with the *s* GO and the dz^2 GO, respectively. With this assignment of the GOs to the MOs, the localized isoenergetic MOs are formed from a set of sp^3d GOs. The orbitals of this set are not all spatially equivalent but consist of two equivalent GOs directed oppositely along the z axis and three equivalent GOs in the equatorial plane. Having done this, we must realize that by introducing the dz^2 GO, we have automatically assumed that a corresponding $3dz^2$ VAO is generated on phosphorus; a VAO we purposely did not include in the delocalized view because d VAOs lie too high in energy to participate appreciably in bonding. (Recall our discussion of localized bonding in XeF_2 in Chapter 4.) By using a ($3dz^2$) *phosphorus* VAO, it is not surprising that the bond order in our localized picture of PF_5 is 1.0 in the axial part of the molecule instead of the fractional bond order of 0.5 we saw in the delocalized view.

By creating an isoenergetic set of sp^3d GOs, we find that the five localized MOs σ_n ($n = 1 - 5$) are also at the same energy. Because the axial and equatorial bond lengths of PF_5 are different, however, we might expect that their localized bond energies are also slightly different. We could incorporate this difference into our (unrealistic) localized view by creating an *spxpy* GO set for the bonds in the equatorial plane and a $pzdz^2$ GO set for the bonds along the three-fold axis. This provides us with three isoenergetic localized equatorial MOs at a somewhat lower energy than the two isoenergetic localized axial MOs lying at a higher energy. The best localized view of molecules such as PF_5 involves localizing the equatorial portion using sp^3 GOs and partially localizing the axial part as we did for XeF_2, for example.

The molecular motions of Z_5 and ZY_5 species having tbp symmetry are obtained by our usual route, as shown in Figure 10.10 for Z_5. Generating the motions of tbp ZY_5 is left as an exercise.

10.5. IF_7

Heating fluorine with iodine gives IF_7. The electron dash structure shown in Figure 10.1f indicates that there are seven bond pairs around iodine. From VSEPR considerations, these seven bond pairs are predicted to arrange themselves into a pentagonal bipyramidal geometry and this has been experimentally verified for IF_7. Since the equatorial bond pairs are spanned by smaller angles (72°) than the 90° angles separating the axial and equatorial bond pairs, the axial I—F links are expected from VSEPR considerations to be shorter than the equatorial I—F bonds. This has also been verified experimentally for these (quite polar) linkages.

After assigning lone pairs to the fluorine ($2s$), ($2px$), and ($2py$) VAOs, we use GOs to pictorially generate SOs with the remaining VAOs (see axis system in Figure 10.15) and construct Generator Table 10.6. Like PF_5, IF_7 is an example of a molecule in which the apical fluorines and the equatorial fluorines interact with the central atom to slightly different extents, and do not interact with each other. Thus, as before, we must identify a BMO–NBMO–ABMO set from the MOs we create from the s-generated SOs in Table 10.6. Unlike

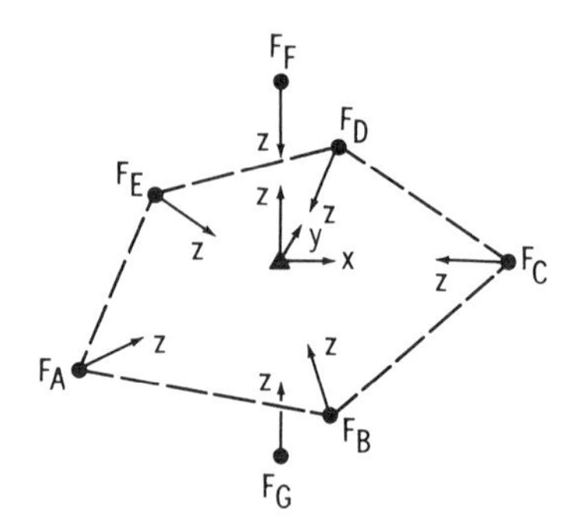

FIGURE 10.15 Axis system for IF_7.

TABLE 10.6 Generator Table for IF_7 (14 e^-)

	GOs					
VAO Equivalence Sets	*s*	*px*	*py*	*pz*	*dxy*	$dx^2 - y^2$
I*s* = (I5*s*)	n					
Ipe = (I5*px*), (I5*py*)		n	n			
Ipa = (I5*pz*)				n		
F*pe* = (F_A2*pz*), (F_B2*pz*), (F_C2*pz*), (F_D2*pz*), (F_E2*pz*)	n	n	n		n	n
F*pa* = (F_F2*pz*), (F_G2*pz*)	n			n		

our procedure with PF_5, we will choose the *s*-generated SO in the equatorial plane of IF_5 to be the NBMO because we want a higher bond order in the axial I—F bonds than in the equatorial ones to be consistent with the VSEPR prediction that the equatorial bonds are longer (weaker). Further examination of this table tells us that in addition to the BMO and the NBMO arising from the *s*-generated SOs, we have three BMOs that come from the linear combination of the pairs of *p*-generated SOs. The SOs generated by *dxy*- and $dx^2 - y^2$ are both NBMOs, since the equatorial fluorines are not close enough to each other to interact. The seven electron pairs available for bonding are therefore housed in the four BMOs (two in the equatorial plane, two in the axis) and the three NBMOs (two in the equatorial plane, one in the axis). The bond order along the axis of IF_7 is 1/2 plus 1/7, and that in the equatorial plane is 2/5 plus 1/7.

In generating a localized view of IF_7 we face a problem similar to the one we encountered in PF_5, namely, that two of the seven occupied delocalized MOs of IF_7 are generated by *s*. From the fact that the ($5dz^2$) orbital on iodine points toward the axial fluorines and the ($5dx^2 - y^2$) and (5*dxy*) orbitals are in the same plane as the equatorial fluorines, we might, *just for the sake of argument*, conclude that the GO hybrid set which would produce seven energy-equivalent localized MOs for these seven electron pairs is sp^3d^3.

TABLE 10.7 A Useful Summary of Procedures for Generating Pictorial Representations of Delocalized and Localized Bonding Views of a Molecule and for Visualizing its Vibrational Modes

Delocalized View	Localized View	Normal Modes
1. Decide on best electron dash structure using lowest possible formal charges and drawing resonance forms. In such forms, some lone pairs may be involved in a π system.	1. Decide on which set(s) of doubly occupied delocalized MOs to localize.	1. Identify the motional atomic vector (AV) equivalence sets and list them in the generator table.
2. Establish the molecular geometry from VSEPR theory if possible. For transition metal complexes and boron hydrides, other considerations apply.	2. Hybridize the GOs which give rise to the sets(s) in step 1.	2. Using GOs, draw the symmetry motion (SMs).
3. After preassigning lone pairs and localized bond pairs to (*s*) and suitably oriented (*p*) VAOs, identify all the VAO equivalence sets (hybrids may also be used) and list them in the generator table.	3. Using the hybridized GOs, draw the SOs. *Treat groups of inequivalent peripheral atoms separately.*	3. Fill in the rest of the generator table (i.e., the GOs and m designations).
4. Using GOs located at a molecular center or centroid, sketch all the SOs, recalling that #SOs = #VAOs. *Treat groups of inequivalent peripheral atoms separately.*	4. Sketch the MOs, being aware of the consequences of a central atom on which the full set of SOs (hybridized VAOs) is not generated by the hybridized GO set. If necessary, rehybridize peripheral atom VAOs and rotate hybrid GOs to locate bond pairs between atoms.	4. Take appropriate linear combinations of SMs generated by the same GO and draw all the molecular motions.
5. Fill in rest of generator table (i.e., the necessary GOs and the n, a, or b designations). Establish partner GO and SO sets.	5. Draw the resulting MO energy level diagram and occupy the appropriate levels with electrons.	5. Identify the normal vibrational modes.
6. Take linear combinations of SOs generated by the same GO and draw all the MOs.	6. Verify that the bond order per link is the same as in the delocalized view.	
7. Draw the resulting MO energy level diagram and occupy the appropriate levels with electrons.		
8. Deduce the bond order per link.		

From the geometry of IF_7 we see that all the members of this hybrid set are not spatially equivalent. (Recall that a similar inequivalence of the orbitals occurred in the sp^3d set we temporarily used for the localized view of PF_5.) This would allow us to create two hybrid sets, namely sp^2d^2 and *pd* for the equatorial and axial regions, respectively. As we saw from our discussion of the localized view of PF_5, however, the inclusion of *d* VAOs in delocalized views of expanded-octet molecules is unrealistic and should be avoided. The main reason for including them in our discussions is to demonstrate that if such higher VAOs did participate in bonding, they could easily be incorporated into the GO approach to creating localized and delocalized MOs schemes.

Summary

The geometries of all the molecules examined in this chapter contained two inequivalent sets of peripheral atoms: axial and equatorial. We treated these sets separately, both in the delocalized and localized bonding schemes we generated. In our discussions of the delocalized views of BrF_3, SF_4, PF_5 and IF_7, we learned that choosing the essentially NBMO from a set of three SOs generated by the same GO involves the experimentally justified decision to place the NBMO along the axial part of all of these molecules except IF_7. We also came to appreciate once again why the inclusion of *d* VAOs in expanded-octet (hypervalent) molecules should be avoided and why localized views of electron-deficient molecules are not very informative.

In Table 10.7 is an amended summary of our procedures for using GOs.

Exercises

1. Develop the localized view of SF_4.
2. Draw delocalized MO energy level diagrams for (a) PF_5; (b) IF_7.
3. Sketch the molecular motions for (a) XeF_3^+; (b) $SbCl_5$; (c) SF_4.
4. From a consideration of the delocalized MO energy level diagram for $B_5H_5^{2-}$ (Figure 10.11) comment on the feasibility of reducing this ion to the as yet unrealized $B_5H_5^{4-}$. Describe its expected magnetic properties.
5. Give the Lewis structure of F_3ClO_2, remembering to make the formal charges as low as possible. Sketch the structure of the molecule. (Hint: Both oxygens are in identical geometrical environments.) Justify your structure on the basis of VSEPR arguments and the relative sizes of the peripheral atoms.
6. Develop the delocalized bonding view of Sn_5^{2-}.
7. The neutral species PF_4 is known and it has the structure shown below. Rationalize this structure on VSEPR grounds and compare it with that of SF_4.

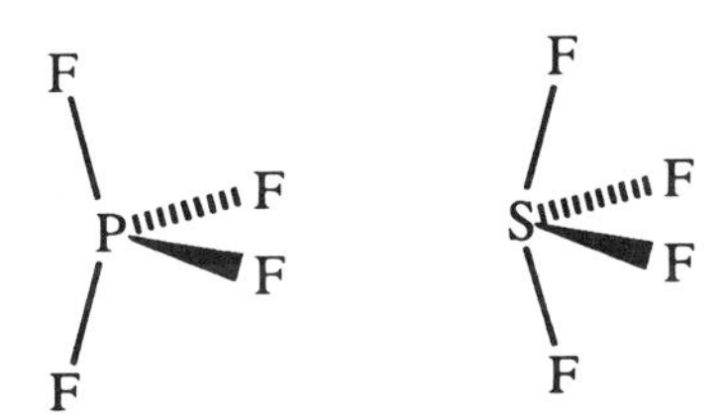

8. Develop the delocalized view of $B_3C_2H_5$. The carbons lie on the axis of a trigonal bipyramid and the borons are in the equatorial plane.

9. Develop a delocalized view of (a) the B_7 cluster in pentagonal bipyramidal $(BH)_7{}^{2-}$ and the B_5C_2 cluster in $(BH)_5(CH)_2$ in which one carbon lies above and one lies below the pentagonal B_5 plane.

10. Develop delocalized views of (a) $TeF_7{}^-$ and (b) $XeF_7{}^+$, both of which are pentagonal bipyramidal. Are they isoelectronic?

11. Develop a delocalized view for $XeF_3{}^+$.

12. Develop a delocalized view of the [*cyclo*-$C_5(CH_3)_5Sn]^+$ cation, which has a pentagonal pyramidal structure with the tin atom at the unique vertex.

13. The trigonal bipyramidal gold cluster in $[(H_3PAu)_5C]^+$ contains a central carbon surrounded by gold atoms bound to one another. Develop a delocalized view of this ion.

Chapter 11

Prismatic Molecules

Prismatic species without central atoms are rare. The Te_6^{4+} ion is an example of a trigonal prism and prismane (Figure 11.1a) is the as yet unknown parent compound of several characterized derivatives which have been found to contain a trigonal prismatic arrangement of carbon atoms. Cubane (Figure 11.1b) has been synthesized by photolytically dimerizing cyclobutadiene that had first been stabilized in a metal complex. Cubic and trigonal prasmatic metal cluster compounds are known, and a pentagonal prismatic $(RSn)_{10}$, where R is a bulky benzene derivative, has also been reported.

Examples of prismatic molecules containing central atoms are "arene" organometallic "sandwich" compounds of which many are known. The arene ligands are typically cyclopentadienide, benzene, or cyclooctatetraenide, and two of these planar moieties "sandwich" a metal atom between them. In Figure 11.2 are shown two conformations of *bis*-cyclopentadienide metal complexes, which have been experimentally verified. There is also structural evidence that the hydrogens on the C_5 rings are bent slightly away from the molecular center and out of the carbon planes. Since we are concerned with prismatic rather than antiprismatic molecules in this chapter, we will restrict our attention to the pentagonal prismatic conformation in Figure 11.2a. *Bis*-arene complexes are known for a variety of transition metals [including lanthanides and actinides in the case of $[(\eta^8\text{-}C_8H_8)_2M]^x$ and metal oxidation states from +1 to +4 can be stabilized in such organometallic compounds. The η^n symbol denotes that *n* ring atoms of the ligand are bonded to the central atom. It might be noted that the full set of such ring atoms do not always bond to a metal atom, in which case the complex is no longer a true "sandwich" compound. Cubic ions such as $U(NCS)_8^{4-}$, UF_8^{3-}, and NpF_8^{3-} are also known.

As in previous chapters, you should repeat the steps for each bonding situation in the discussions using the Homework Drawing Board in Node-Game. Use the appropriate column in Table 11.4 as a guide. Note that the bonding in $(\eta^5\text{-}C_5H_5)_2Fe$ is included in Practice Exercise 5 of Node-Game.

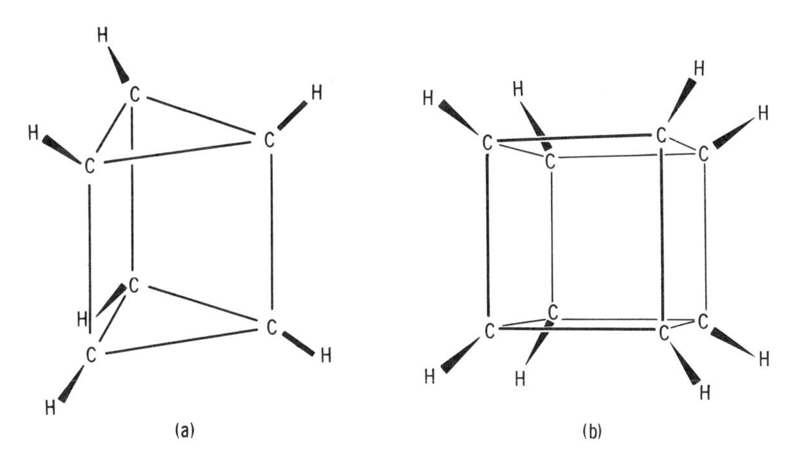

FIGURE 11.1 Prismane (a) and cubane (b).

11.1. C_6H_6 (Prismane)

The electron dash structure for prismane is shown in Figure 11.1a. Since there are thirty valence electrons for a total of the 15 C—C and C—H links, this carbon cluster system is not electron deficient. The C—C bonds are strained, however, because of the nontetrahedral carbon angles. This molecule is therefore prone to bond-breaking reactions and is unstable with respect to thermal rearrangement to benzene, which is unstrained and is stabilized by resonance.

To obtain a delocalized view of this molecule, we first localize the prismane C—H bond pairs in carbon (Csp^{α}) VAO hybrids. The VAOs on a given carbon atom which are used to bond the carbon framework must, of course, be orthogonal to sp^{α}. In the spirit of previous examples, we could pick an orthogonal (Csp) hybrid oppositely directed from (Csp^{α}) and toward the molecular center indicated by the triangle in Figure 11.3b. The remaining

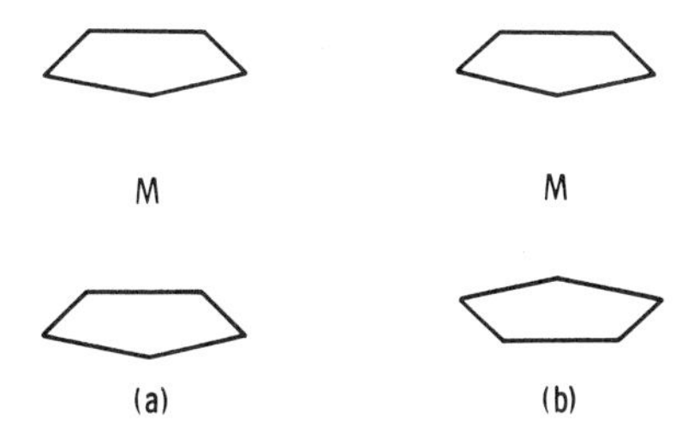

FIGURE 11.2 Eclipsed (a) and staggered (b) conformations of *bis*-cyclopentadienide metal complexes.

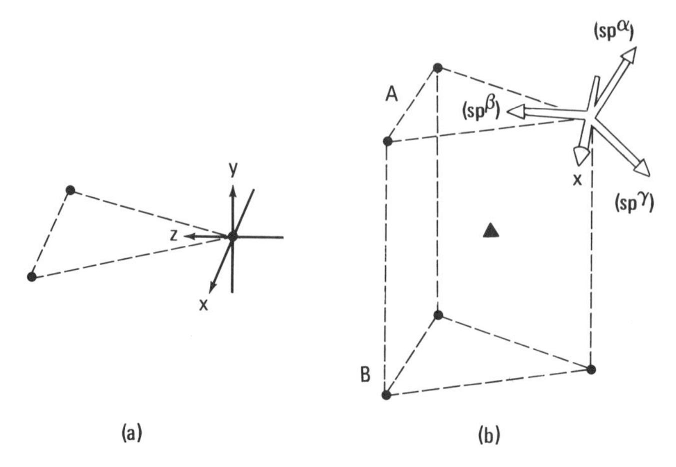

FIGURE 11.3 Axis system for the top triangle of prismane (a). In (b) is shown a conveniently used configuration of VAOs on one of the carbons of prismane.

two (2*p*) VAOs could then lie at 90° to one another in a plane perpendicular to the two hybrids. However, because we can make good use here of what we learned about the bonding in triangular species in Chapter 5, it is convenient to pick one VAO to lie in the plane of the triangle (see Figure 11.3b), just as we did in the case of N_3^+. The remaining VAO, sp^{γ}, is now determined. It lies more or less (but not exactly) along the C—C link connecting the lower and the upper triangle. It is slanted slightly with respect to the vertical C—C bonds toward the outside of the trigonal prism at an angle that depends on the orientation of the C—H bond.

If we were to follow our usual procedure, we would use GOs at the center of the C_6 framework of prismane to generate six MOs in each of the VAO sets Csp^{β}, Csp^{γ}, and Cpt. Although this approach is valid, it is more convenient to treat prismane as being composed of two triangular C_3H_3 fragments, which we label A and B in Figure 11.3b. Separating a molecule into components is especially useful when the components possess MO systems that can be individually treated. A prismatic molecule is in this sense analogous to the bonding treatment we used for a diatomic molecule, except that instead of generating SOs in VAOs of a pair of atoms, we will generate SOs in MOs of a pair of polygonal fragments.

Our set of VAOs for each of the C_3H_3 moieties of prismane is only slightly different from the set for the N_3^+ or cyclopropenium ions discussed in Chapter 5. By our usual procedure we can deduce the MOs for each of the C_3 halves of prismane and they are listed as MO equivalence sets in Table 11.1. For the sake of argument, assume that $\alpha > \beta > \gamma$. Then the energy levels for the hybrid VAOs of these fragments can be arranged as shown in Figure 11.4. We now place GOs at the molecular center between identical pairs of C_3 fragment MOs as shown in Figure 11.5. These C_3 fragment MOs [σ_A(s),

TABLE 11.1 Generator Table for the C_6 Cluster Framework of Prismane (18 e^-)

Equivalence Sets	GOs							
	s	px	py	pz	dxz	dyz	$f3''$	$g3''$
$C_3sp^\beta a = \sigma_A(s), \sigma_B(s)$	b			u				
$C_3sp^\beta b = \sigma_A^*(px), \sigma_B^*(px), \sigma_A^*(py), \sigma_B^*(py)$		u	u		a	a		
$C_3sp^\gamma a = \pi_A(pz), \pi_B(pz)$	b			u				
$C_3sp^\gamma b = \pi_A^*(dxz), \pi_B^*(dxz), \pi_A^*(dyz), \pi_B^*(dyz)$		u	u		a	a		
$C_3pta = \sigma_{pA}(px), \sigma_{pB}(px), \sigma_{pA}(py), \sigma_{pB}(py)$		b	b		u	u		
$C_3ptb = \sigma_{pA}^*(f3''), \sigma_{pB}^*(f3'')$							u	a

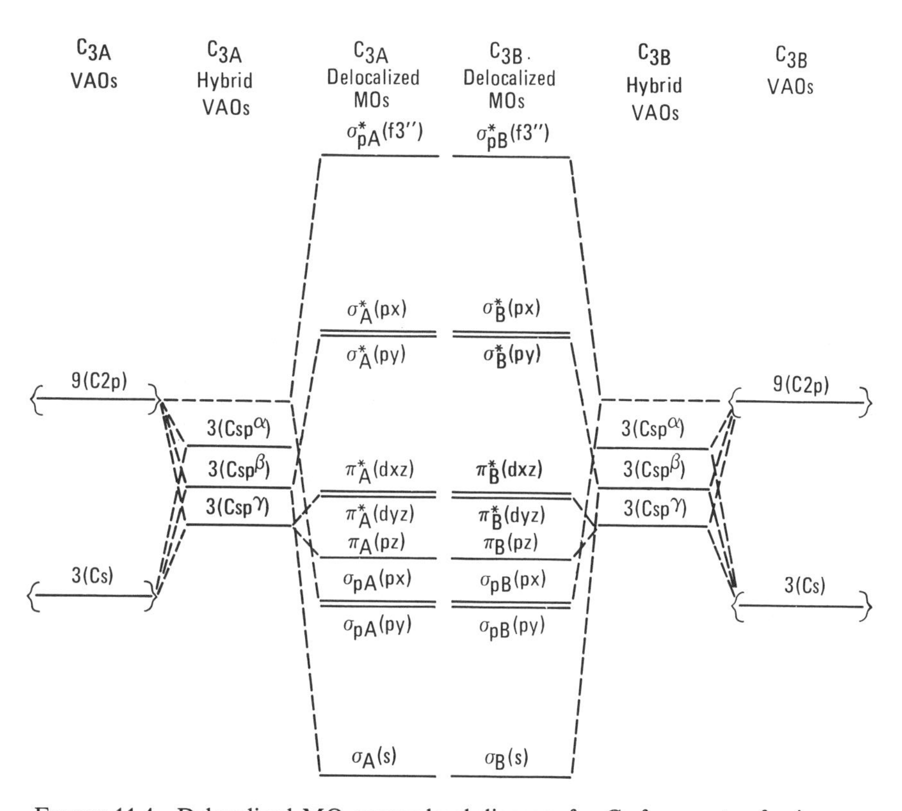

FIGURE 11.4 Delocalized MO energy level diagram for C_3 fragments of prismane. Note that the (sp^α) hybrid VAOs on each fragment are assumed to form localized two-center MOs with the hydrogens.

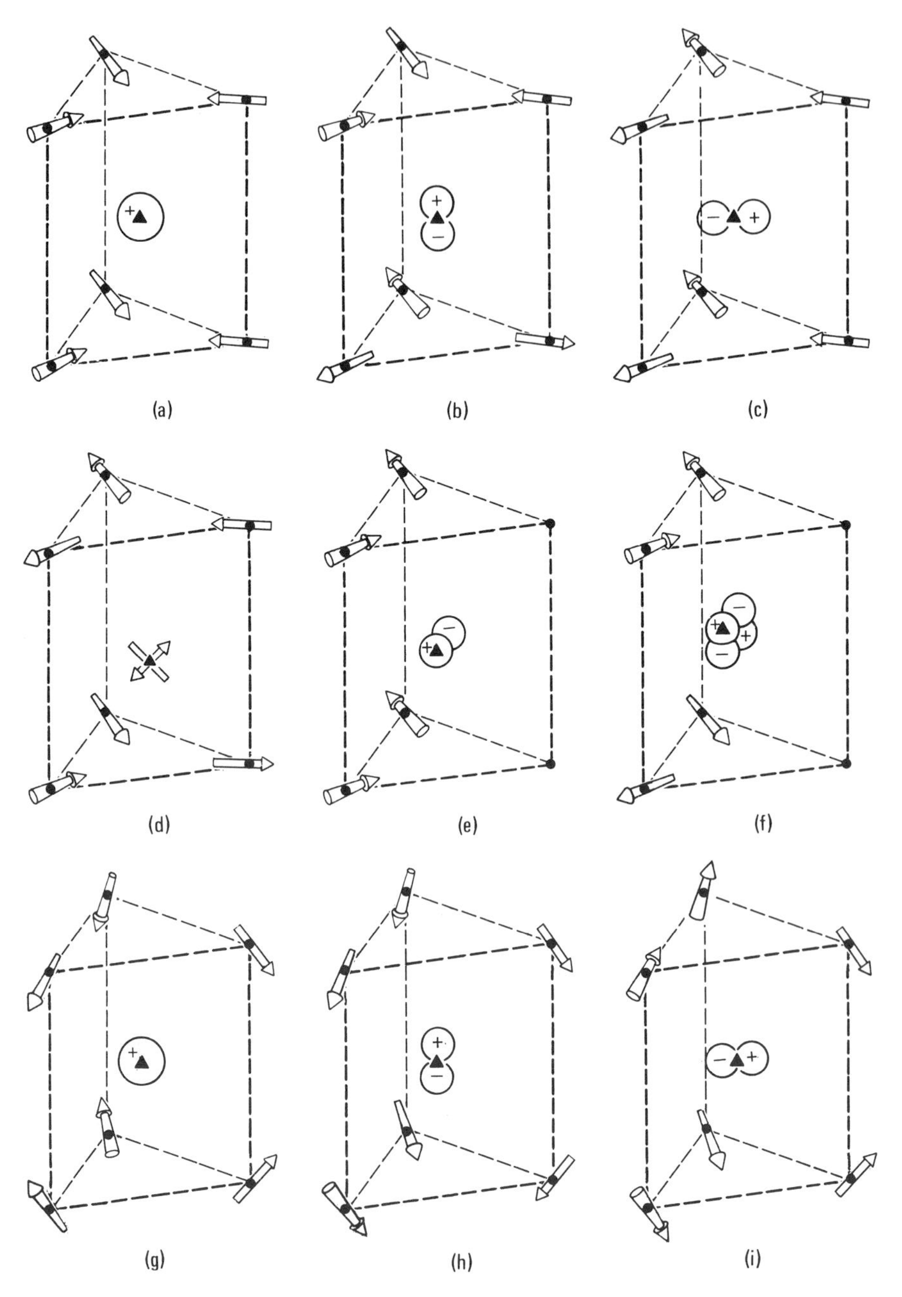

FIGURE 11.5 Sketches of GO-generated linear combinations of C_3 fragment MOs in prismane. The C_3 fragment MOs are composed of linear combinations of carbon (sp^{β}) [(a)–(f)], (sp^{γ}) [(g)–(l)], and ($2px$) [(m)–(r)] VAOs.

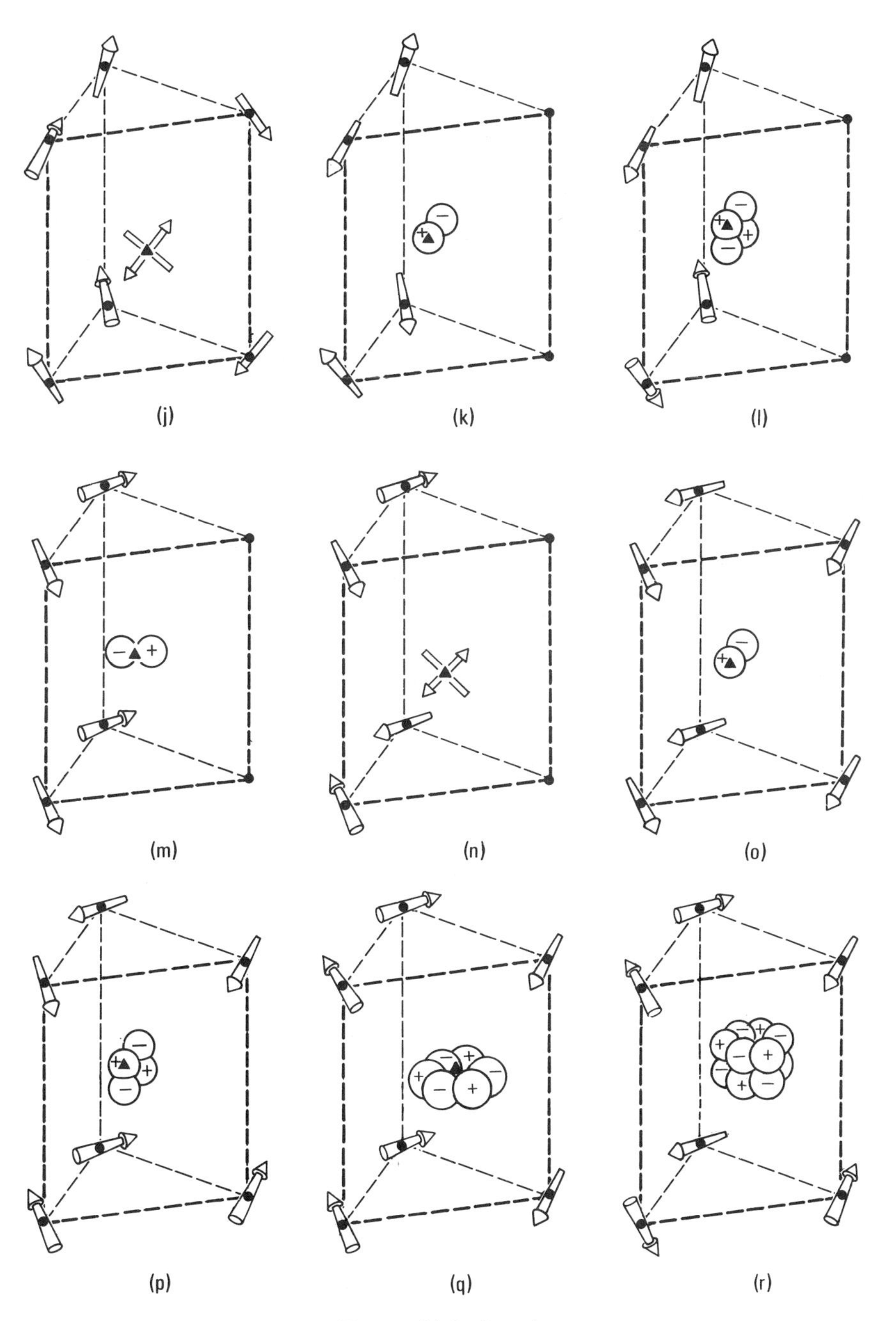

FIGURE 11.5 (cont.)

$\sigma_B(s)$, etc., in Table 11.1 and Figure 11.4] are generated analogously to the way we generated the MOs of N_3^+, for example. The identical pairs of C_3 fragment MOs contained in the equivalence sets of Table 11.1 are thus seen to give rise to pairs of linear combinations which are either bonding or antibonding along the vertical edges of the prism (Figures 11.5a, b; c, d; e, f; g, h; etc.). These pairs are designated in the horizontal rows of Table 11.1 according to their *overall* bonding (b), antibonding (a), or unclear bonding (u) nature. Thus in Figures 11.5d, f, j, l, and r we can see that interactions within the C_3 moieties as well as the interactions between the two C_3 fragments are antibonding and so they are identified as such in Table 11.1. Similarly, the MOs in Figures 11.5a, g, m, and o are clearly BMOs. While the remaining MOs are clearly either bonding or antibonding between pairs of fragment MOs, their overall bonding or antibonding natures are not obvious because it is very difficult to gauge the relative importance of intra- versus inter-fragment interactions owing to the different orbital overlap angles. The splittings between all the pairs of fragment MOs, as indicated by the MOs identified in the rows of Table 11.1, are depicted in Figure 11.6a. From the columns in Table 11.1 we also see that additional splittings can occur owing to identical symmetries of the MOs (i.e., MOs generated by the same GO) and these are depicted in an approximate way in Figure 11.6b.

It seems appropriate that we fill the four completely bonding MOs and leave the five antibonding MOs empty. It is true, of course, that the two *s*-generated BMOs and the two *dxz*- and *dyz*-generated ABMOs will change their relative energies as a result of interaction, but they will still retain their overall bonding or antibonding character.

On the basis of our straightforward Lewis structure, we would expect to find five additional BMOs from among the nine "u"-type MOs in Table 11.1. We have no choice but to designate as BMOs a degenerate pair of "u" MOs from the *px*, *py* columns of Table 11.1 and another such pair from the *dxy*, *dyz* columns. (We could, of course, have designated two pairs of "u" MOs from the *px*, *py* columns of Table 11.1 as BMOs, but that would presume that there are no BMOs among *dxz*-, *dyz*-generated MOs and this does not seem likely.) The remaining BMO must be either the $f3''$-generated MO or a linear combination of the *pz*-generated MOs. The choice is not obvious. Let us assume, however, that by linear combination of the *pz*-generated MOs we drive the energy of one of them well below the $f3''$-generated MO as shown in Figure 11.6b.

The occupied MOs in Figure 11.6 are generated by a set of GOs that includes two *s*, two *px*, two *py*, one *pz*, one *dxz*, and one *dyz* GO. In the prismane geometry, the dz^2 GO generates identically with *s* and so we can replace one of the *s* GOs by a dz^2 GO. Similarly, one of the *px* and *py* GOs can be replaced by an $f1'$ and $f1''$ GO, respectively. This gives rise to an $sp^3d^3f^2$ hybrid GO set, whose main lobes point toward the midpoints of the edges of a trigonal prism (i.e., sp^2 toward the vertical edges and pd^3f^2 toward the upper and lower edges). The nine localized bonds are then centered

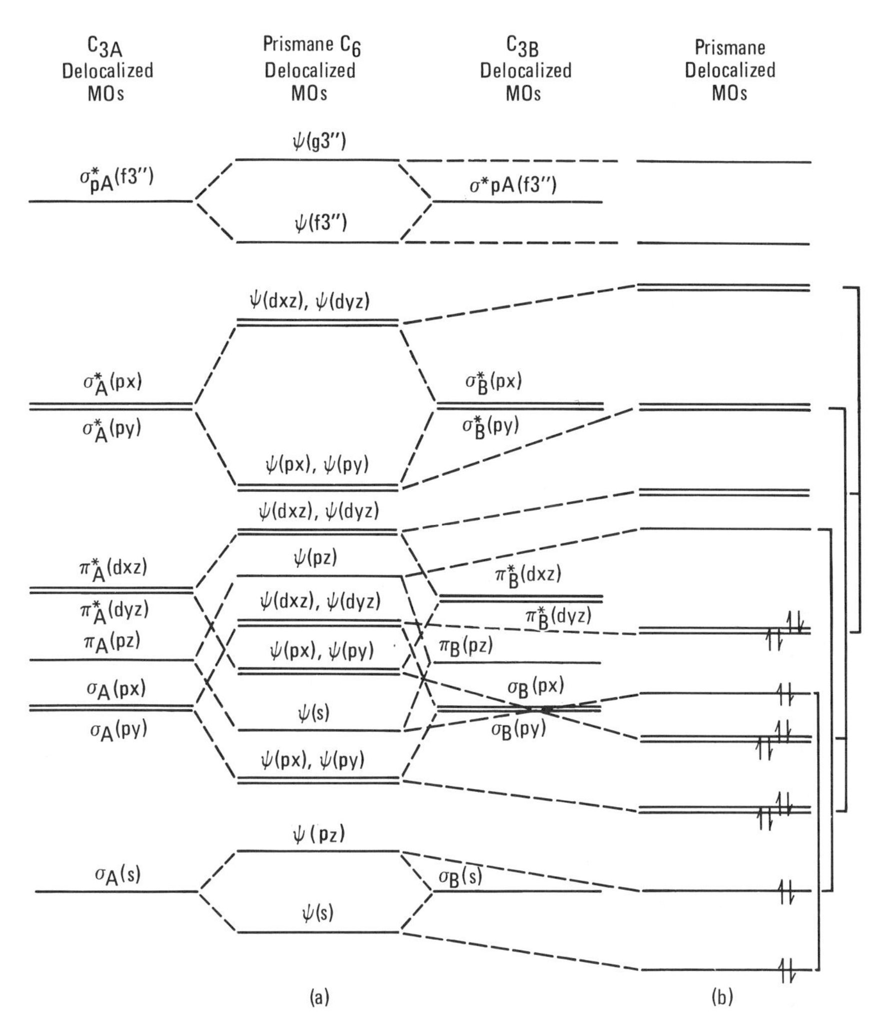

FIGURE 11.6 Delocalized MO energy level diagram for prismane. In (a) are shown the splittings indicated by the entries in the rows of Table 11.1 and in (b) are depicted further splittings of levels generated by the same GOs.

between each pair of carbon atoms with appropriate contributions from the Csp^{β}, Csp^{γ}, and Cpt VAOs on each carbon. We can also rehybridize these carbon VAOs to produce two identical hybrids pointing their major lobes over the edges of the triangular rings, and a unique hybrid directing its main lobe over a vertical edge of the prismatic C_6 cluster. To the extent that the C—C bonds in the rings are not equivalent to those between the two rings, the energies of the two types of localized MOs will be different. As in the delocalized bonding picture, the bond order is 1.0. Note that the presence of more than one identical GO for generating the occupied delocalized MOs we wish to localize does not alter the bond order derived from the delocalized view. Unlike the analogous situation in PF_5 and IF_7, there is no central atom in prismane to respond to the additional GOs.

The molecular motions of prismane are linear combinations of the molecular motions of the C_6 prism and the H_6 prism generated by the same GO. The molecular motions of a prism are generated in the same way as the MOs were in Figure 11.5, except that z in the axis system in Figure 11.3a could be pointed toward the center of gravity (i.e., the center of the molecule) with the assumption that the C—H bonds also lie along this axis.

11.2. C_8H_8 (Cubane)

The electron dash structure represented by Figure 11.1b indicates that there are sufficient valence electrons to achieve a bond order of 1.0 in all the links. If we treat this molecule analogously to prismane we can see that each of the three VAO sets Csp^{β}, Csp^{γ}, and Cpt in a C_4H_4 moiety gives rise to four MOs that have the same symmetry characteristics as the radial, π, and tangential sets, respectively, in Te_4^{2+} (Chapter 7). The MOs of the frameworks of the two C_4 fragments A and B are grouped into eight valence orbital equivalence sets in Table 11.2. From each identical pair of C_4 fragment MOs is generated a pair of SOs (Figure 11.7) that can be linearly combined. And these combinations are identified in the rows of Table 11.2. These MOs are, of course, further split by the fact that in most cases one GO generates more than one SO in the columns of this table. As is also seen from this table we have 9 clearly bonding MOs and so only 3 more are needed to house the 24 electrons in the C_8 cluster. It seems reasonable to assume that linear combination of the pz- and dxy-generated "u" orbitals could give rise to one BMO each. The third BMO is assumed to be the $g4''$-generated "u" MO since the remaining columns all provide clearly antibonding MOs.

The occupied delocalized MOs are generated by a set of GOs, which includes two s, three px, three py, one pz, one dxy, one $dx^2 - y^2$, and one $g4''$. Substituting a dz^2 GO for an s, and $f1'$, $f1''$ and $h1'$, $h1''$ for two px, py pairs, the GO set becomes $sp^3d^3f^2h^2g$ which, upon hybridization, produces GOs whose main lobes point to the midpoints of the edges of a cube. In these regions are concentrated the electron pairs that are contained in the localized bonds. The localized bonds are made up of appropriate contributions from the Csp^{β}, Csp^{γ}, and Cpt VAOs on each of the eight carbons. Alternatively, we can rehybridize these VAOs in a manner analogous to that used for a localized view of prismane.

If we make use of the isolobality principle learned in the previous chapter, we might have expected that by analogy to $B_6H_6^{2-}$ (Chapter 8), octahedral $C_6H_6^{4+}$ should exist. However, if it does, it is expected to be quite unstable (owing to its high positive charge) and therefore to have a strong tendency to

TABLE 11.2 Generator Table for C_8 Framework of Cubane ($24\ e^-$)

VAO Equivalence Sets	GOs											
	s	px	py	pz	dxy	dxz	dyz	$d(x^2-y^2)$	$f2'$	$f2''$	$g4''$	$h4''$
$C_4sp^{\beta}a = \sigma_A(s), \sigma_B(s)$	b			u								
$C_4sp^{\beta}b = \sigma_A^0(px), \sigma_B^0(px), \sigma_A^0(py), \sigma_B^0(py)$		b	b			a	a					
$C_4sp^{\beta}c = \sigma_A^*(dxy), \sigma_B^*(dxy)$					u					a		
$C_4sp^{\gamma}a = \pi_A(pz), \pi_B(pz)$	b			u								
$C_4sp^{\gamma}b = \pi_A^0(dxz), \pi_B^0(dxz), \pi_A^0(dyz), \pi_B^0(dyz)$		b	b			a	a					
$C_4sp^{\gamma}c = \pi_A^*(f2''), \pi_B^*(f2'')$					u					a		
$C_4pta = \sigma_{pA}^0(px), \sigma_{pB}^0(px), \sigma_{pA}^0(py), \sigma_{pB}^0(py)$		b	b			a	a					
$C_4ptb = \sigma_{pA}(dx^2-y^2), \sigma_{pB}(dx^2-y^2)$								b	u			
$C_4ptc = \sigma_{pA}^*(g4''), \sigma_{pB}^*(g4'')$											u	a

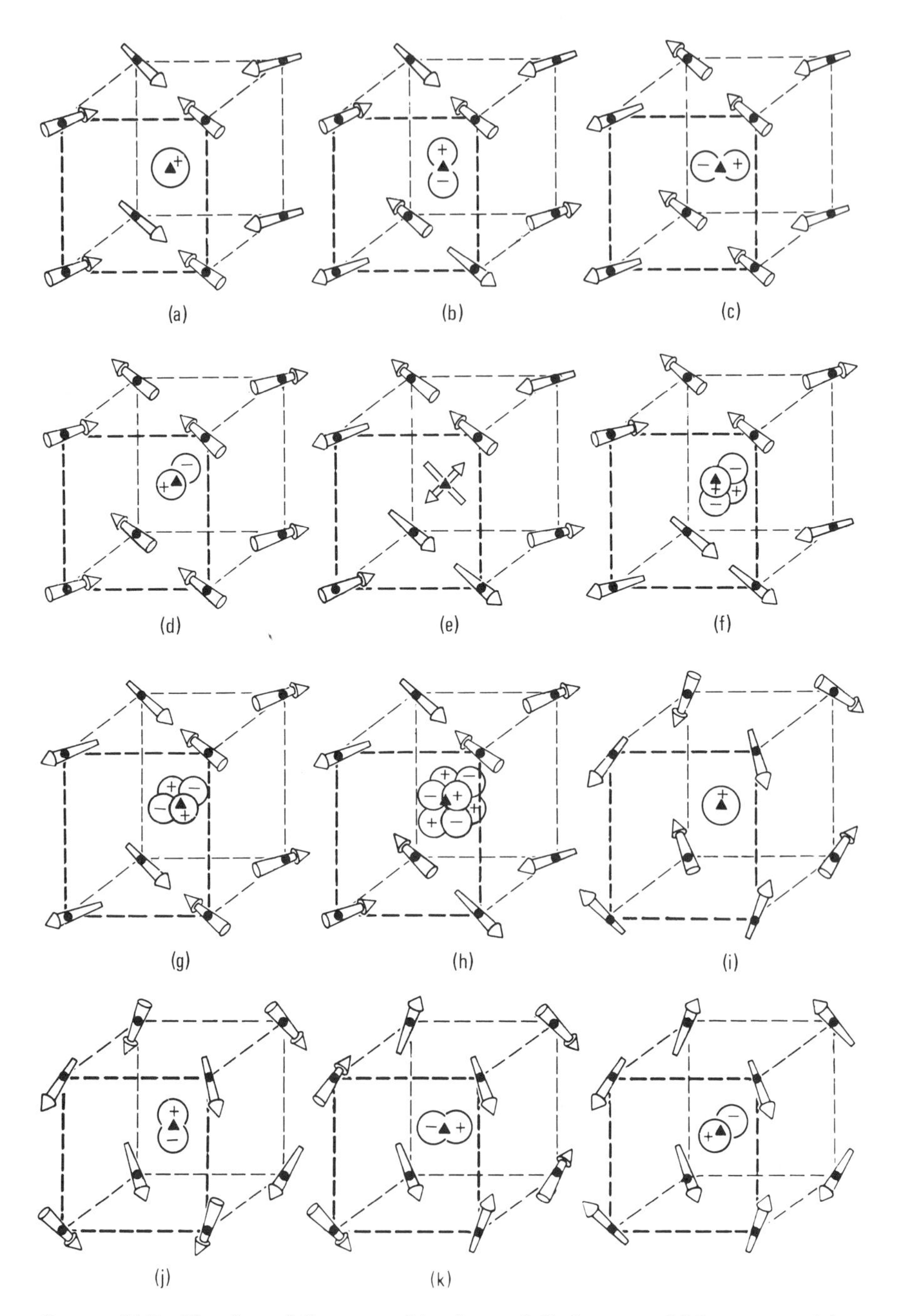

FIGURE 11.7 Sketches of linear combinations of C_4 fragment MOs generated by GOs in cubane. The C_4 fragment MOs are composed of linear combinations of carbon (sp^{β}) [(a)–(h)], (sp^{γ}) [(i)–(p)], and ($2px$) [(q)–(x)] VAOs.

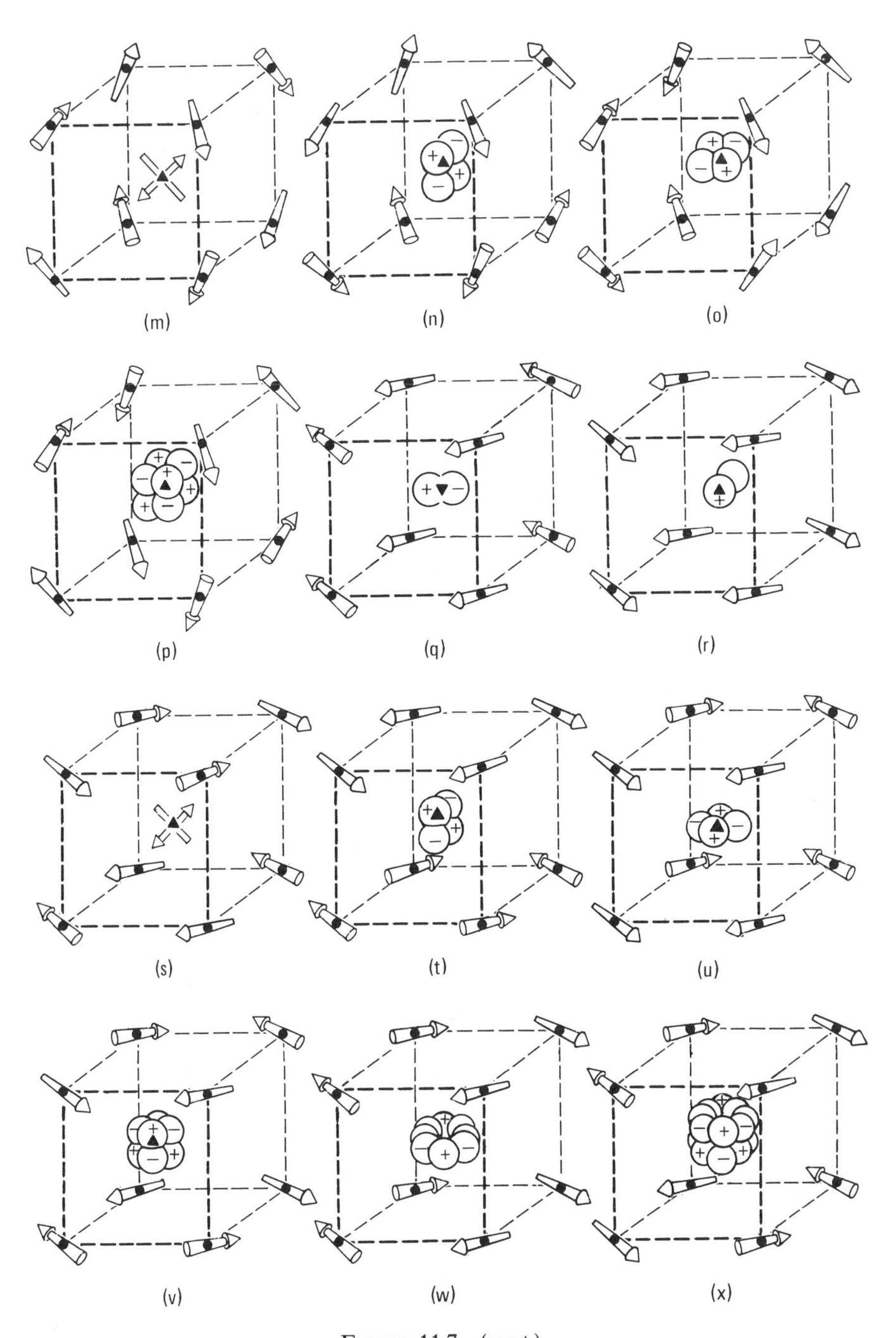

FIGURE 11.7 (cont.)

be an oxidizing agent and pick up four electrons to form prismane prior to rearranging to the much more stable aromatic benzene molecule. We conclude, then, that although some predictions of the isolabality principle work (e.g., $B_5H_5^{2-}$ correctly predicts $B_3C_2H_5$, Chapter 10) other factors can destroy a prediction.

11.3. $(\eta^5\text{-}C_5H_5)_2Fe$

Ferrocene was the first "sandwich" compound to be synthesized. This orange sublimable crystalline organometallic is obtained from the reaction of NaC_5H_5 and $FeCl_2$ are and it is stable in excess of 500°C!

Sandwich compounds generally contain aromatic rings such as $C_5H_5^-$, C_6H_6, and $C_8H_8^{2-}$ as ligands. The two rings may be eclipsed (i.e., lie one on top of one other) or staggered (i.e., one ring rotated so that its atoms lie between those of the other). For treatment of the bonding in sandwich molecules, we will assume that the rings have an eclipsed conformation.

The electron dash structure of ferrocene depicted in Figure 11.8 in its eclipsed form suggests that each ligand anion contributes six π electrons for bonding to the metal. The iron atom can then be considered as an Fe^{2+} ion, having a $3d^6$ VAO configuration. These six iron VAO electrons are shown as iron lone pairs in Figure 11.8. Notice that the eighteen electrons around the iron (6 lone pairs and 12 bond pairs) provide an electron configuration for this atom which is identical to that of the inert gas Kr at the end of the periodic row in which we find iron. Eighteen-electron metal configurations play an important role in the stability of many complexes including organometallic ones. Complexes that have this electronic configuration are said to obey the "18-electron" or "effective atomic number" (EAN) rule. Examples of species that fall into this category generally have ligands of high field in which π backbonding from the metal to empty ligand orbitals can occur (Chapter 8). Examples of such ligands are CO, CN^-, $P(OR)_3$, and in the present case, $\eta^5\text{-}C_5H_5^-$. In ferrocene as with the CoF_6^{3-} ion, we defer the question of orbital assignment of the transition metal lone pairs until later. With the aid of the axis system shown in Figure 11.9a we construct the SOs

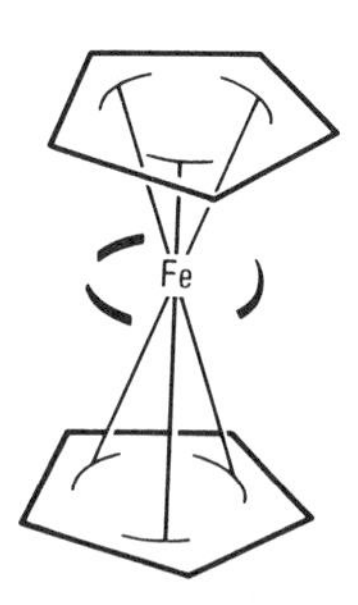

FIGURE 11.8 Electron dash structure for ferrocene. The curved wedges around Fe denote lone electron pairs.

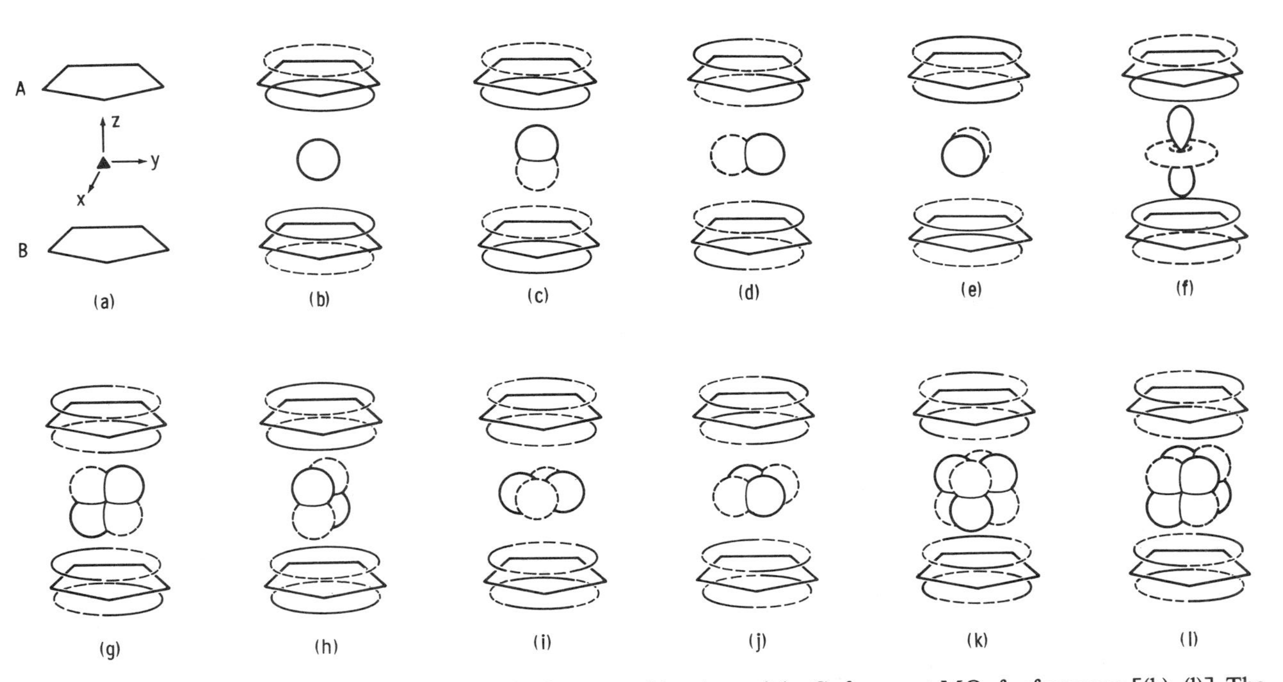

FIGURE 11.9 Axis system for ferrocene (a) and the linear combinations of the C_5 fragment MOs for ferrocene [(b)–(l)]. The dotted and solid lines in the ligand π MOs and GOs denote regions in which ψ is negative and positive, respectively.

depicted in Figure 11.9b–l. Notice that we use an abbreviated way of depicting the $C_5H_5^-$ MOs in which dashes represent negative lobes of Cpz VAOs making up an MO and solid lines represent positive Cpz VAO lobes in these MOs. This convention is then also used in the GOs shown in Figure 11.9. The GOs allow us to generate SOs from identical pairs of $C_5H_5^-$ delocalized π MOs. They also permit us to generate the (Fe3d), (Fe4s) and (Fe4p) VAOs on iron. Notice in Figure 11.9 that the s and dz^2 GOs generate the same SO on the $C_5H_5^-$ ligands. As we saw in Chapter 7, the five $C_5H_5^-$ π MOs consist of one BMO, a degenerate pair of BMOs, and a degenerate pair of ABMOs. Therefore these MOs group themselves into three MO equivalence sets. From the SOs depicted in Figure 11.9, generator Table 11.3 is constructed. In this table we recognize from the pentagonal prismatic symmetry of the molecule that the iron VAO equivalence sets Feda, Fedb, and Fepa contain VAO partners, while the VAO equivalence sets Fedc, Fes, and Fepb contain only one VAO each.

The occupation of the MOs in ferrocene can be arrived at beginning with the assumption that seven BMOs are obtained upon linear combination of each of the SO pairs generated by the same GO in Table 11.3. We can expect at least one BMO from the linear combination of the SO's in the $s(dz^2)$-generated SOs, and in this MO we place two more electrons. The remaining two electrons in this diamagnetic compound must also be placed in an MO arising from this set, because if we place them in the degenerate $f2'$, $f2''$-generated ABMOs, these electrons would be unpaired. What about the last occupied MO? Notice that the $\pi_A(pz)$, $\pi_B(pz)$ BMO's (which employ carbon (2pz) VAOs) lie lower than any of the iron VAOs (i.e., the principal quantum numbers are $n = 2$ and $n = 3$ and 4, respectively). Since (Fe3d) lies below (Fe3s), we expect the (Fe3d) orbital to interact more strongly with the ligand

TABLE 11.3 Generator Table for Ferrocene (18 e)

	GOs									
VAO Equivalence Sets[a]	$s(dz^2)$	px	py	pz	dxy	dx^2-y^2	dxz	dyz	$f2'$	$f2''$
Feda = (Fe3dxy), (Fe3dx^2-y^2)					n	n				
Fedb = (Fe3dxz), (Fe3dyz)							n	n		
Fedc = (Fe3dz^2)	n									
Fes = (Fe4s)	n									
Fepa = (Fe4px), (Fe4py)		n	n							
Fepb = (Fe4pz)				n						
Cpπa = $\pi_A(pz)$, $\pi_B(pz)$	n			n						
Cpπb = $\pi_A(dxy)$, $\pi_B(dxy)$ = $\pi_A(dyz)$, $\pi_B(dyz)$		n	n		n	n				
Cpπc = $\pi_A^*(f2')$, $\pi_B^*(f2')$ = $\pi_A^*(f2'')$, $\pi_B^*(f2'')$							n	n	n	n

[a] The abbreviation Cp is commonly used for the cyclopentadienide ion.

$\pi_A(pz)$, $\pi_B(pz)$. Thus (Fe4*s*) becomes the largely nonbonding MO in the linear combinations, and it is in this orbital of largely lone pair character that we place the last pair of electrons as shown in the MO energy level diagram of Figure 11.10.

Before going on to the localized view of ferrocene, repeat the delocalized view we have just developed using the appropriate practice exercise in Node Game.

The Lewis structure in Figure 11.8 suggests that we should localize three bond pairs between each of the ligands and the metal. The GOs that generate the six lowest BMO's in Figure 11.10 are *s* (or dz^2), *px*, *py*, *pz*, *dxz*, and *dyz*. When hybridized, the members of the GO hybrid set point their main lobes

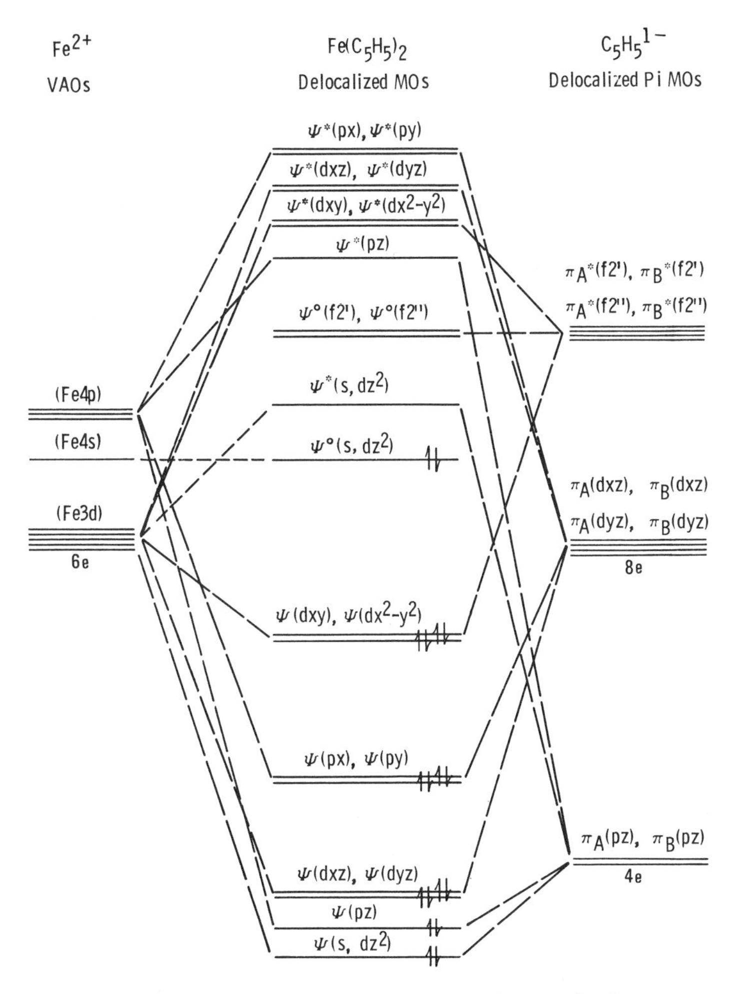

FIGURE 11.10 Delocalized MO energy level diagram for ferrocene.

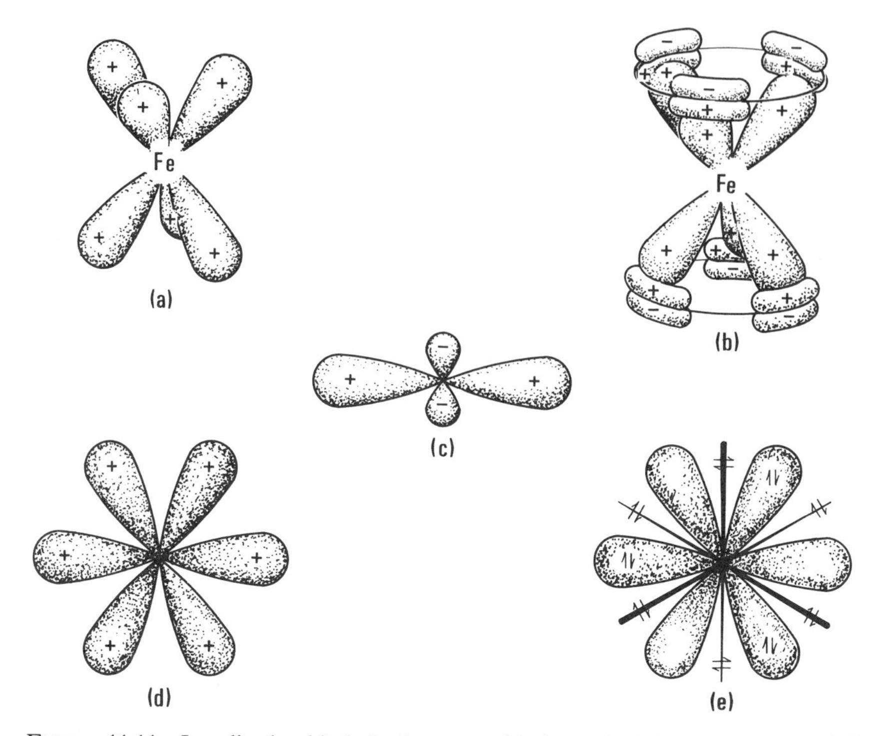

FIGURE 11.11 Localized orbitals in ferrocene: (a) the main lobes of a set of sp^3d^2 (p^3d^3) hybrids; (b) the formation of localized bonds via the localized π MOs of the C_5 fragments; (c) an (sd^2) hybrid; (d) the orientation of the main lobes of such a hybrid set; (e) and orientation of the lone pair lobes in the ($Fesd^2$) set relative to the bond pairs, as viewed down the z axis of the molecule. The heavy and light wedges represent bond pair orbital lobes pointed above and below the plane of the paper, respectively.

to the corners of a trigonal antiprism as shown in Figure 11.11a. The localized MOs generated by these GOs consist essentially of the corresponding iron hybrid VAOs and the localized ligand MOs concentrated in regions of the rings toward which the GOs point (see Figure 11.11b). Here the ligand contributions are the localized MOs discussed for $C_5H_5^-$ in Chapter 7.

If we use a dz^2 GO for our localized bonds, the GOs associated with the remaining three delocalized MOs in Figure 11.10 are s, dxy, and $dx^2 - y^2$. Hybridizing these GOs leads to three hybrid GOs in the equatorial plane. A top view of one such hybrid is shown in Figure 11.11c and a top view of the main lobes of all of them is given in Figure 11.11d. The corresponding localized MOs resulting from superposing the highest three delocalized MOs will consist of the identical (sd^2) iron hybrids and certain contributions from the ligand ions. The bonding character of $\psi(s)$, $\psi(dxy)$ and $\psi(dx^2 - y^2)$ will be evenly distributed over these three localized MOs. They can be described

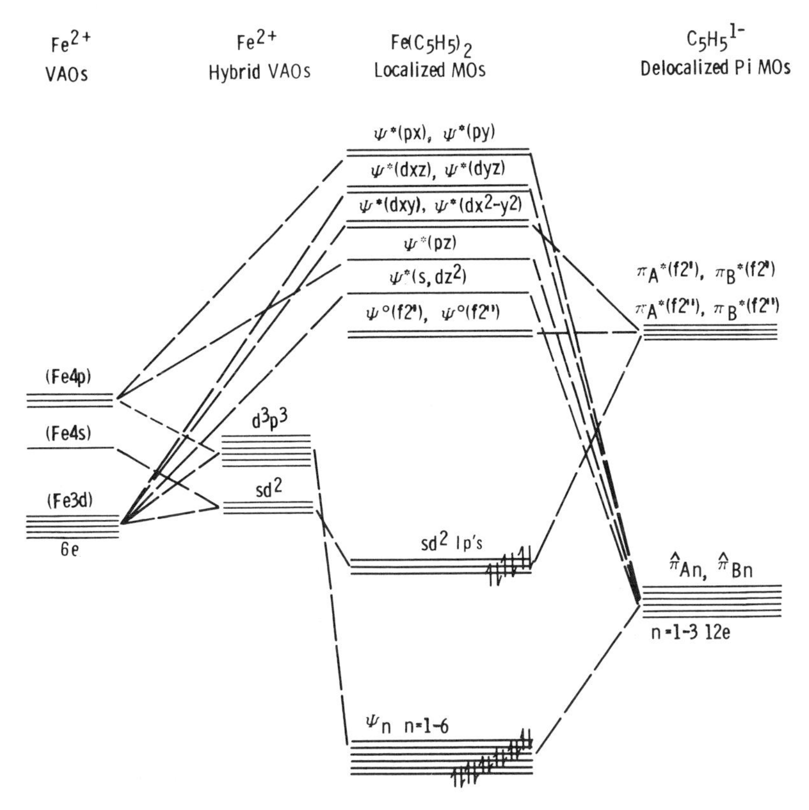

FIGURE 11.12 Localized MO energy level diagram for ferrocene.

as being largely iron lone pairs that are somewhat delocalized, in a bonding fashion, to the two ligand rings. These six lone pair-like lobes can be oriented by the appropriate hybridization anywhere in this plane. On the basis of VSEPR considerations, however, one might expect the lone pair lobes to be staggered with respect to the antiprismatic bonding lobes in order to be maximally localized as shown in Figure 11.11e. However, influences beyond the scope of this book may dominate that would alter this conclusion. A localized energy level diagram is given in Figure 11.12.

Summary

In this chapter the technique of treating prismatic molecules such as prismane, cubane, and ferrocene as having pairs of regular polygonal moieties possessing valence MOs was developed. In addition ferrocene possesses a central atom with three "lone pairs" that we found could be housed in three metal (sd^2) hybrids.

In Table 11.4 is an updated version of our procedures for using GOs.

TABLE 11.4 A Useful Summary of Procedures for Generating Pictorial Representations of Delocalized and Localized Bonding Views of a Molecule and for Visualizing its Vibrational Modes

Delocalized View	Localized View	Normal Modes
1. Decide on best electron dash structure using lowest possible formal charges and drawing resonance forms. In such forms, some lone pairs may be involved in a π system. 2. Establish the molecular geometry from VSEPR theory if possible. For transition metal complexes and boron hydrides, other considerations apply. 3. After preassigning lp's and localized bp's to (s) and suitably oriented (p) VAOs, identify all the VAO equivalence sets (hybrids may also be used) and list them in the generator table. *For groups of atoms (e.g., polygons) valence MO equivalence sets are used.* 4. Using GOs located at a molecular center or centroid, sketch all the SOs, recalling that #SOs = #VAOs. Treat groups of inequivalent peripheral atoms separately. 5. Fill in rest of generator table (i.e., the necessary GOs and the n, a, or b designations). Establish partner GO and SO sets. 6. Take linear combinations of SOs generated by the same GO and draw all the MOs. 7. Draw the resulting MO energy level diagram and occupy the appropriate levels with electrons. 8. Deduce the bond order per link.	1. Decide on which set(s) of doubly occupied delocalized MOs to localize. 2. Hybridize the GOs which give rise to the sets(s) in step 1. 3. Using the hybridized GOs, draw the SOs. Treat groups of inequivalent peripheral atoms separately. 4. Sketch the MOs, being aware of the consequences of a central atom on which the full set of SOs (hybridized VAOs) is not generated by the hybridized GO set. If necessary, rehybridize peripheral atom VAOs and rotate hybrid GOs to locate bond pairs between atoms. 5. Draw the resulting MO energy level diagram and occupy the appropriate levels with electrons. 6. Verify that the bond order per link is the same as in the delocalized view.	1. Identify the motional atomic vector (AV) equivalence sets and list them in the generator table. 2. Using GOs, draw the symmetry motion (SMs). 3. Fill in the rest of the generator table (i.e., GOs and m designations). 4. Take appropriate linear combinations of SMs generated by the same GO and draw all the molecular motions. 5. Identify the normal vibrational modes.

EXERCISES

1. Draw the MOs of a C_3 triangular fragment of prismane and draw the localized MO energy level diagram for prismane.

2. Draw the molecular motions of the C_6 cluster of prismane assuming that the z axis in Figure 11.3a points to the center of gravity of the C_6 cluster. Hint: It is possible to take linear combinations of SMs of like symmetry of two identical halves of a molecule to obtain the molecular motions. As a simple example, treat Te_4^{2+} as a pair of diatomic molecules and take linear combinations of their SMs generated by the same GO.

3. In molecules such as prismane and cubane, pure (Csp^3) hybrids cannot be used to make localized bonds. Comment on this statement and qualitatively compare more realistic hybrid sets in these two molecules. Comment on the possibility of direct head-on overlap of adjacent carbon hybrids along the edges of these molecules as opposed to angular overlaps.

4. Draw delocalized and localized MO energy level diagrams for cubane.

5. What is the bond order in ferrocene? Compare your answer with the one implied by its Lewis structure.

6. Generate the delocalized and localized views of Te_8^{8+} assuming that it has a cubic molecular structure.

7. The XeF_8^{2-} ion is known and in $(NO_2)_2[XeF_8]$ this ion possesses a square antiprismatic array of fluorines with a Xe atom at the center. Generate the delocalized view and comment on the orbital occupation in the MO energy level diagram.

8. On the basis of the delocalized MO diagram for $[Cr(C_6H_6)_2]^{1+}$ in the eclipsed conformation give a reason why this ion is reasonably stable in spite of its nonconformance to the 18-electron rule.

9. The complex $[U(C_8H_8)_2]^0$ with two unpaired electrons can be synthesized by simply heating finely divided uranium metal with cyclooctatetraene. (a) What is the formal oxidation state of the metal? (b) Develop the delocalized MO picture assuming that the uranium (5f), (6d), (7s), and (7p) VAOs all participate in MO formation with the ligand orbitals.

10. The $Rh_6C(CO)_{15}^{2-}$ ion consists of a trigonal prismatic array of metal atoms with a carbon atom in the center. Each rhodium bears a terminal CO group and three CO groups that bridge an edge as indicated in the partial structure below. Each metal atom can be considered to be bound by two electrons from a terminal CO. Each bridging CO is bound to a metal by one carbon electron and one metal electron. The Rh—C links can be assumed to use a set of roughly tetrahedrally oriented hybrids obtained by mixing the (5s), (4dxy), (4dyz), and (4dxy) VAOs on rhodium. Develop the delocalized bonding view for the Rh_6C cluster system, assuming that all the metal atoms are close enough to interact with one another.

11. The cubic cluster $Ni_8(PC_6H_5)_6(CO)_8$ has the structure indicated in partial form above. Each nickel binds to a phosphorus using one phosphorus electron and one

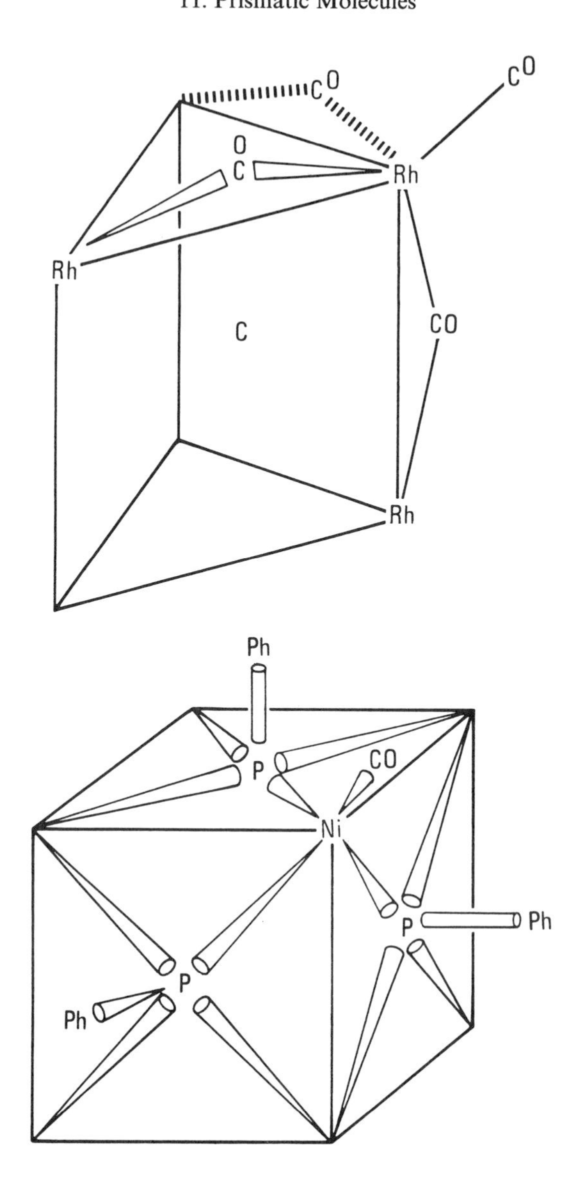

nickel electron and the terminal CO binds with two carbon electrons. Each P_3NiC moiety is assumed to employ a set of tetrahedrally deployed sd^3 metal hybrids obtained by mixing the nickel (4*s*), (3*dxy*), (3*dxz*), and (3*dyz*) VAOs. Develop a delocalized bonding view of the Ni_8 cluster.

12. Develop a delocalized view of the tin cluster in pentagonal prismatic $(ArSn)_{10}$ where Ar = the bulky benzene derivative below.

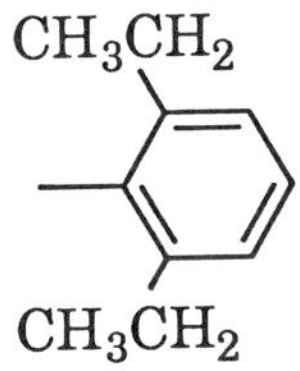

CH_3CH_2

CH_3CH_2

13. Develop delocalized views for the eclipsed conformation of $(\eta^5\text{-}C_5H_5)_2M$ where M = (a) Cr, (b) Ni and comment on their magnetic properties.

14. Repeat problem 13 for the "triple-decker" sandwich $[(\eta^5\text{-}C_5H_5)Ni(\eta^5\text{-}C_5H_5)Ni(\eta^5\text{-}C_5H_5)]^{1+}$ in which the central $\eta^5\text{-}C_5H_5^-$ is bonded on both sides to an Ni atom. Treat the molecule in three sections: the central $\eta^5\text{-}C_5H_5$, the two Ni^{2+} atoms and then the two outside $\eta^5\text{-}C_5H_5^-$ ligands.

15. The LiH_6^+ ion should be trigonal prismatic according to calculations. Develop a delocalized view for this hypothetical cation.

16. Develop a delocalized view for the square antiprismatic form of $B_8H_8^{2-}$.

17. Which of the following species do not obey the 18-electron rule? (a) $Cu(NH_3)_6^{2+}$; (b) $Fe(CO)_5$; (c) $Fe(H_2O)_6^{2+}$, (d) $Mn(CN)_6^{2+}$?

CHAPTER 12

Polymers

When one atom or a group of atoms repeats many times in the form of a chain or "back-bone," we refer to the substance as a *polymer*. The repeating units in polymer backbones can be made up of a single type of atom, such as carbon or sulfur, or they can be composed of more than one type of atom, such as carbon and silicon or of phosphorus and nitrogen. Depending upon their composition and length (which can be many thousands of units long) polymers can display varying degrees of toughness, pliability, heat stability, and resistance to air oxidation. They can also be insulators, conductors, semiconductors, or superconductors. They can be neutral or bear charges. Due to the ever expanding utility of polymers in our lives, it is useful to understand some aspects of the bonding in their backbones.

In this section, we shall examine only homonuclear polymers with main group elements (carbon) as the backbone (i.e., organic polymers). Since polymers are quasi-infinite in their extension, we shall see that carbon-based polymers are one-dimensional analogues of both graphite (which is two-dimensional) and diamond (which is three-dimensional).

12.1. σ-Bonded (Nonconjugated) Polymers

The prototype for the entire class of nonconjugated organic polymers is polyethylene (formed from ethylene, $CH_2{=}CH_2$) represented in Figure 12.1. This polymer is used to make a variety of products ranging from milk bottles to space vehicle components. Each carbon in this polymer is tetrahedrally coordinated by two hydrogen atoms and two carbon atoms. Since there is a negligible energy barrier for rotation about each C–C bond, these polymer chains can be curled and even spiraled instead of adopting the planar carbon backbone as depicted in Figure 12.1. Nevertheless, due to the nearly tetrahedral C–C–C bond angle required at each carbon atom, the repeat unit for the polyethylene polymer is the ethylene unit itself (the blocked segment in Figure 12.1).

Polyethylene represents one substitutional derivative for the general polymer $—[CR_1R_2—CR_1R_2]—$, where $R_1 = R_2 = H$. If each carbon atom is unsymmetrically coordinated (i.e., $R_1 \neq R_2$) various stereochemical possibilities exist. If the substitution patterns in adjacent repeat units adopt the same atomic configuration, the pattern is called *meso*, while patterns with opposite configurations are termed *racemic*. Polymers with long meso sequences are called *isotactic*, and those with long racemic sequences are *syndiotactic*. If meso and racemic patterns occur randomly between repeating units along the polymer chain, the polymer is called *atactic*. Although these different types of *tacticity* can lead to different structural, electronic, and chemical behavior, a basic description of their electronic structure and chemical bonding lies fundamentally with the carbon backbone. With this in mind, we will now examine the orbital diagram for the polyethylene prototype —[CH_2—CH_2]—.

The Lewis structure of polyethylene is also represented in Figure 12.1, if we assign every line and wedge as a single chemical bond. By constructing Csp^3 hybrid VAOs and using H1*s* VAOs, two Csp^3 hybrid VAOs will form bonding and antibonding two-center, two-electron MOs for each C—H bond. (We could, of course, recast these MOs into *localized* two-center, two-electron bonds if we wished to do so.) The remaining two Csp^3 hybrid VAOs can then be used for bonding along the carbon chain. To see how we use these orbitals, let us focus on a single repeat unit (i.e., a single CH_2—CH_2 group or CH_2 dimer unit). Since there are four VAOs, the GO approach to this molecular fragment also produces four MOs generated by the 1*s*, 2*px*, 2*py*, and 3*dxy* GOs shown in Figure 12.2.

Two of the GOs in the repeat unit generate a strongly bonding and antibonding MO pair between the carbon atoms in the unit, while the remaining two GOs overlap weakly within the repeat unit but are oriented to overlap strongly with corresponding GOs in adjacent units. This effect between repeat units is called *dispersion* and contributes to the fact that extended structures (e.g., polymers and solids) do not show discrete electronic states but have *bands* of electronic states. To construct orbitals for the polymer, we will use a C_{16} chain. For convenience we will imagine a Cartesian coordinate system in the center of Figure 12.3a such that the x axis is parallel to the chain and the y axis is coplanar with the carbon backbone and each H—C—H plane. Therefore, the Csp^3 hybrid VAOs used for bonding the backbone lie in the xy plane.

Figure 12.3 illustrates the MOs generated by the indicated GOs in the *s*-generated bonding combination of Csp^3 hybrid VAOs within each repeating dimer unit (Figure 12.2a). The eight dimer units in the C_{16} chain contain eight GOs that increase in nodal complexity according to the sequence: 1*s*, 2*px*, 2*s*, 3*px*, 3*s*, 4*px*, 4*s*, 5*px*. This construction makes use of both planar and spherical nodes (shown as lines and partial circles, respectively) in the GOs: the planar node (the *yz*-plane) bisects the C_{16} chain and the spherical nodes are required to intersect C—C contacts between adjacent repeat units. For each GO, there are eight *intraunit* ("within-dimer-unit") σ bonding contacts, modulated by seven weaker *interunit* ("between-dimer-unit") π-antibonding

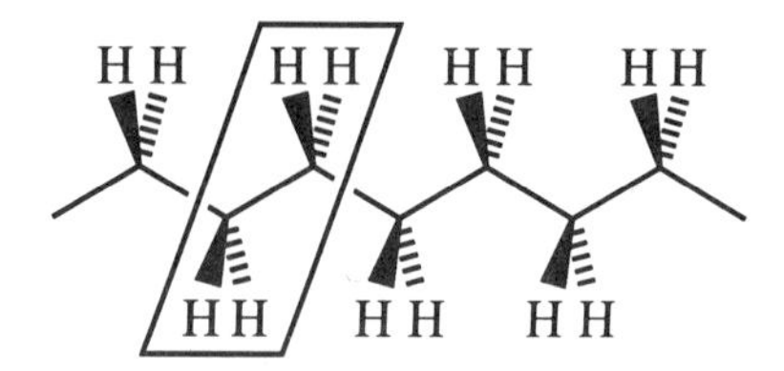

FIGURE 12.1 Polyethylene fragment.

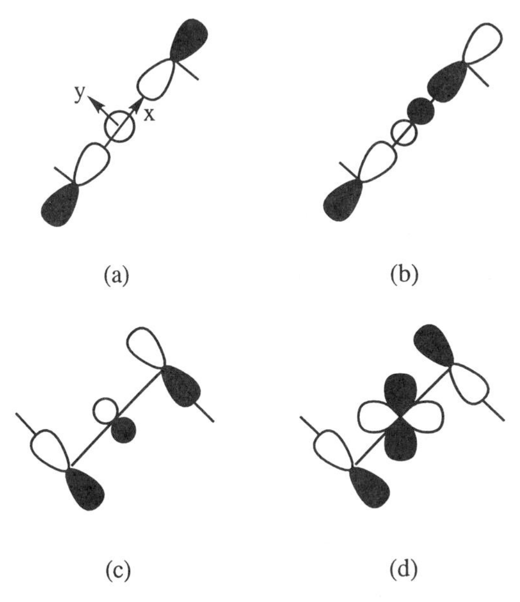

FIGURE 12.2 MOs in a polyethylene dimer fragment generated by a $1s$ (a), $2px$ (b), $2py$ (c) and $3dxy$ (d) GO. The darkened and undarkened lobes represent negative and positive wave function signs, respectively.

interactions. The energy of each of these orbitals is influenced most strongly by the *intraunit* sigma orbital overlap, and together they form a C—C σ-bonding energy band. Generator Table 12.1 summarizes these conclusions drawn from Figure 12.3.

Figure 12.4 illustrates the various GOs for the antibonding $2px$ GO-generated combination of the Csp^3 hybrid VAOs within each repeating dimer unit (Figure 12.2b). Again, the eight dimer units in the C_{16} chain lead to eight GOs that increase in nodal complexity according to the assignment: $5s$, $6px$, $6s$, $7px$, $7s$, $8px$, $8s$, and $9px$. This construction also makes use of both planar and spherical nodes. The planar node (the yz-plane) bisects the C_{16} chain and spherical nodes are required to intersect C—C contacts *within adjacent repeat units*. For each GO there are eight antibonding sigma contacts, modulated by seven weaker π orbital interactions. Again, the energy

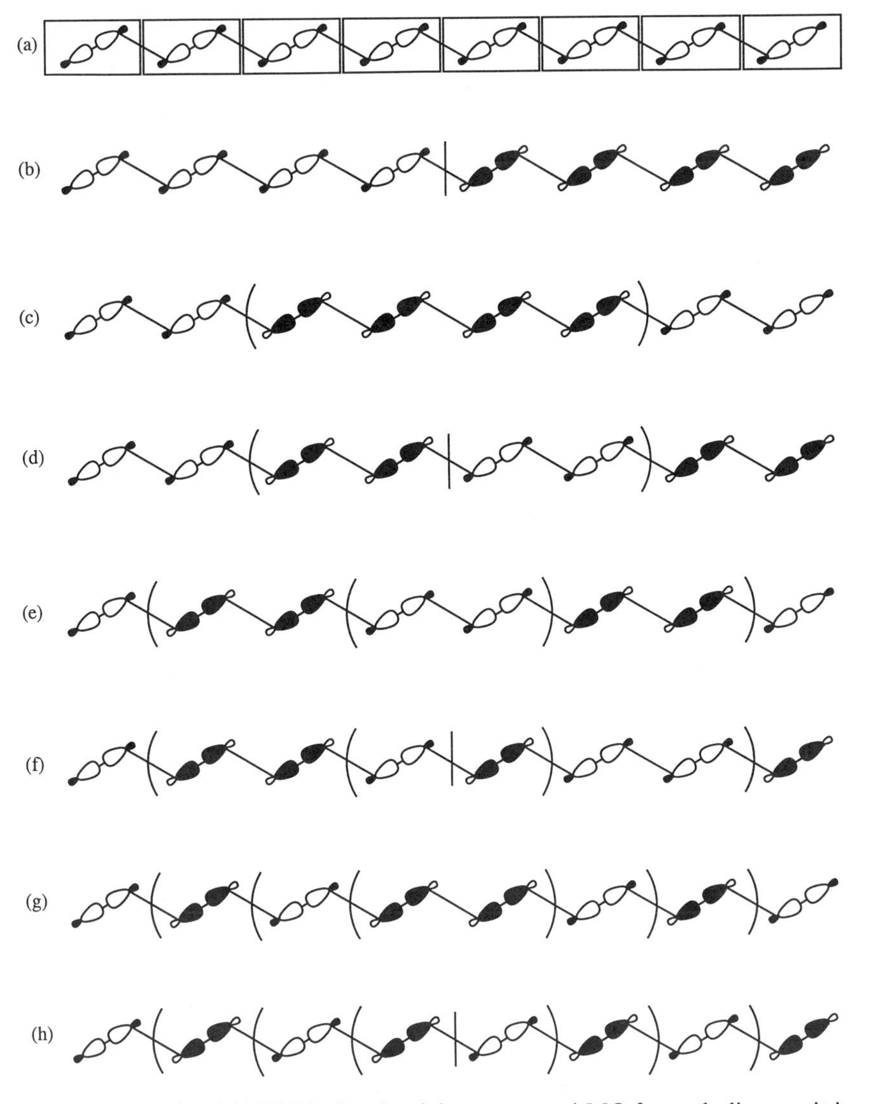

FIGURE 12.3 The eight BMOs for the eight *s*-generated MO for each dimer unit in a C_{16} fragment of polyethylene wherein the C_{16} BMOs are generated by the 1*s* (a), 2*px* (b), 2*s* (c), 3*px* (d), 3*s* (e), 4*px* (f), 4*s* (g), and 5*px* (h) GOs.

TABLE 12.1 Generator Table for the Eight *s*-Generated MOs for Each Dimer Unit in a C_{16} Fragment of Polyethylene

	GOs							
AV Equivalence Set	1*s*	2*s*	2*px*	3*s*	3*px*	4*s*	4*px*	5*px*
16 C*sp*^{3}	b	b	b	b	b	b	b	b

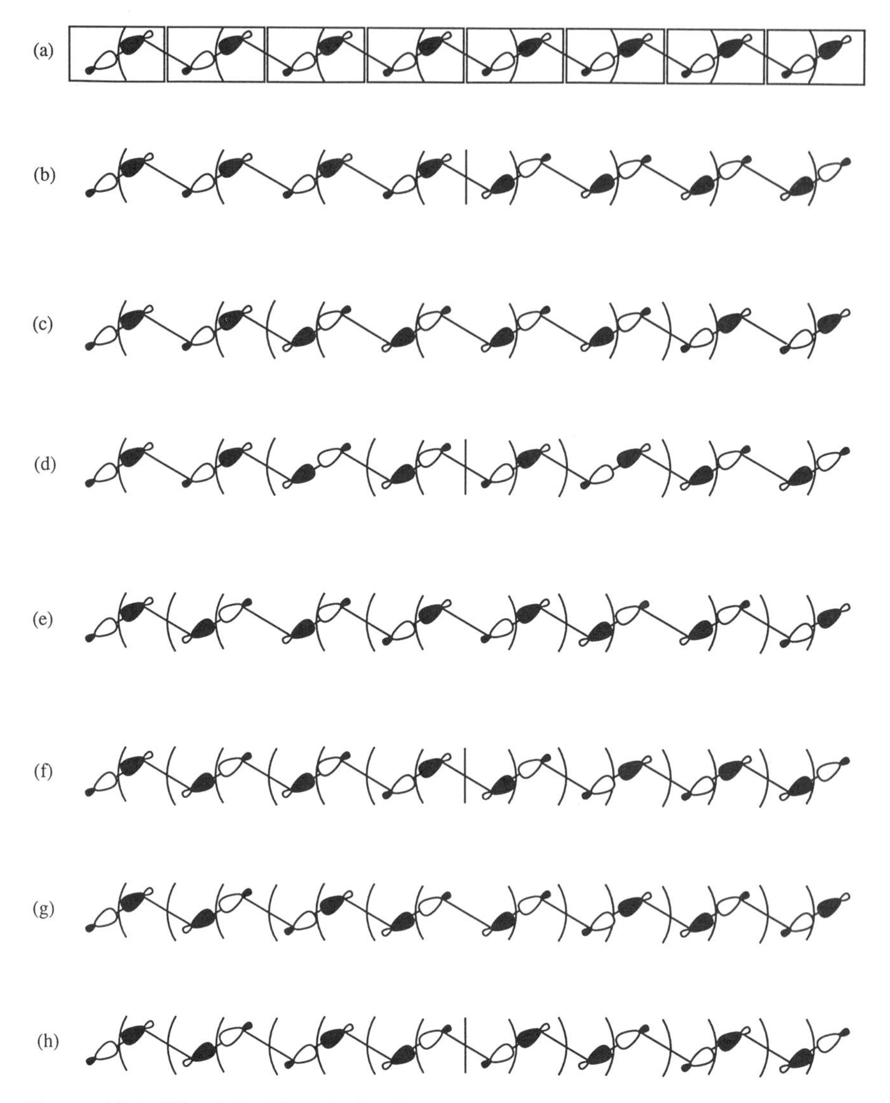

FIGURE 12.4 The eight ABMOs for the eight 2*px*-generated MO for each dimer unit in a C_{16} fragment of polyethylene wherein the C_{16} ABMOs are generated by the 5*s* (a), 6*px* (b), 6*s* (c), 7*px* (d), 7*s* (e), 8*px* (f), 8*s* (g), and 9*px* (h) GOs.

TABLE 12.2 Generator Table for the Eight 2*px*-Generated MOs for Each Dimer Unit in a C_{16} Fragment of Polyethylene

	GOs							
Equivalence Set	5*s*	6*s*	6*px*	7*s*	7*px*	8*s*	8*px*	9*px*
16 Csp^3	a	a	a	a	a	a	a	a

of these orbitals is influenced mostly by the intraunit orbital overlap as recorded in generator Table 12.2.

Figure 12.5 illustrates the C_{16} backbone MOs obtained from the $2py$-generated MO of each dimer unit (Figure 12.2c). These eight MOs show stronger interunit orbital overlap than intraunit overlap. These MOs follow a sequence similar to those we used for the sets of GOs in Figures 12.3 and

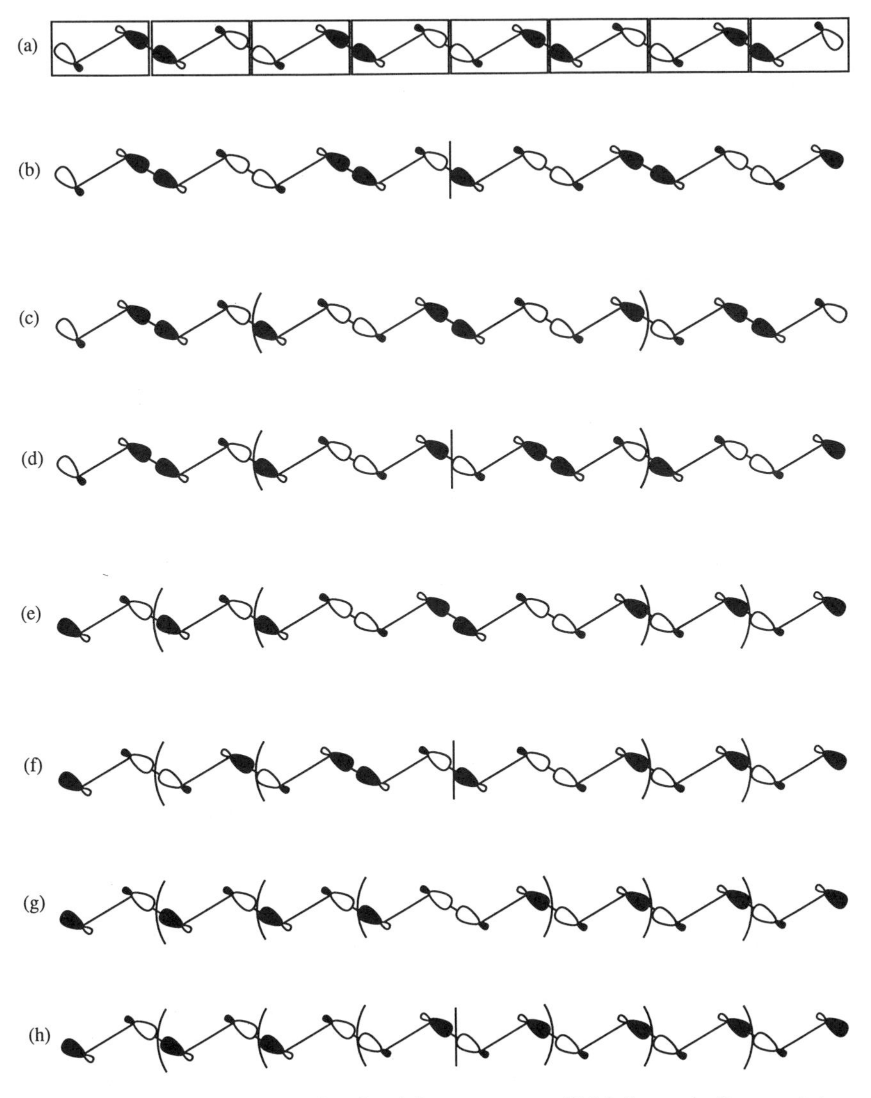

FIGURE 12.5 The eight MOs for the eight py-generated MO for each dimer unit in a C_{16} fragment of polyethylene generated by the $1s$ (a), $2px$ (b), $2s$ (c), $3px$ (d), $3s$ (e), $4px$ (f), $4s$ (g), and $5px$ (h) GOs. Four of the MOs (a)–(d) are BMOs and four (e)–(h) are ABMOs.

TABLE 12.3 Generator Table for the Eight $2py$-Generated MOs for Each Dimer Unit in a C_{16} Fragment of Polyethylene

	GOs							
Equivalence Set	$1s$	$2s$	$2px$	$3s$	$3px$	$4s$	$4px$	$5px$
16 Csp^3	b	b	b	a	b	a	a	a

12.4, and the placement of the nodal surfaces is dictated by the same criteria we employed for the previously used GO sets. Notice that four of the MOs (Figures 12.5a–d) are overall C—C bonding, while the other four (Figures 12.5e–h) are net C—C antibonding. (See generator Table 12.3.) Eight more MOs can be generated from the $3dxy$-generated MO (Figure 12.2d) of each dimer, and these are shown in Figure 12.6. Thus we obtain an additional set of MOs, half of which are C—C bonding, and half are C—C antibonding (Table 12.4). The qualitative electronic structure of polyethylene, depicted in Figure 12.7, contains localized orbitals assigned to bonding and antibonding C—H interactions and it shows two weakly delocalized bands; one assigned to bonding and the other to antibonding C—C interactions. There are two components to the C—C σ and σ^* bands: one is intradimer and the other is interdimer. As Figure 12.2 shows, the orbitals in (a) and (b) have strong intradimer σ overlap but weak interdimer π overlap. The opposite is true for orbitals in (c) and (d). Since band dispersion arises from interdimer overlap, there are different "bandwidths" for the two components of the C—C σ and σ^* bands. This qualitative picture suggests that in polyethylene there is a relatively large separation between the highest occupied MO (HOMO) and lowest unoccupied MO (LUMO) in the bands. Because this energy gap between the HOMO and LUMO in Figure 12.7 is comparatively large, polyethylene is generally classified as an insulator because it takes considerable energy to promote electrons from the HOMO to the LUMO wherein electron conduction takes place. Yet this energy is smaller than the corresponding $\pi \rightarrow \pi^*$ transition energy in ethylene from which polyethylene is made.

Cross-linking is an important process in polymer chemistry because it contributes to the structural and mechanical properties of numerous polymers by linking polymer chains together in locations along their backbones. Sulfur is a common cross-linking agent due to its tendency to form two bonds, thus enabling it to link chains by $C(S)_xC$ bonds via replacement of a hydrogen on a carbon in each chain. In principle, we could also imagine many polyethylene chains bound directly to one another by C—C bonds to form more complex polymeric materials. If this process is continued to a quasi-infinite extent, a diamond crystal would result. The electronic structure of diamond is composed of two bands separated by a gap of 5.4 eV. The C—C σ-bonding band is fully occupied, and the σ-antibonding band is completely empty.

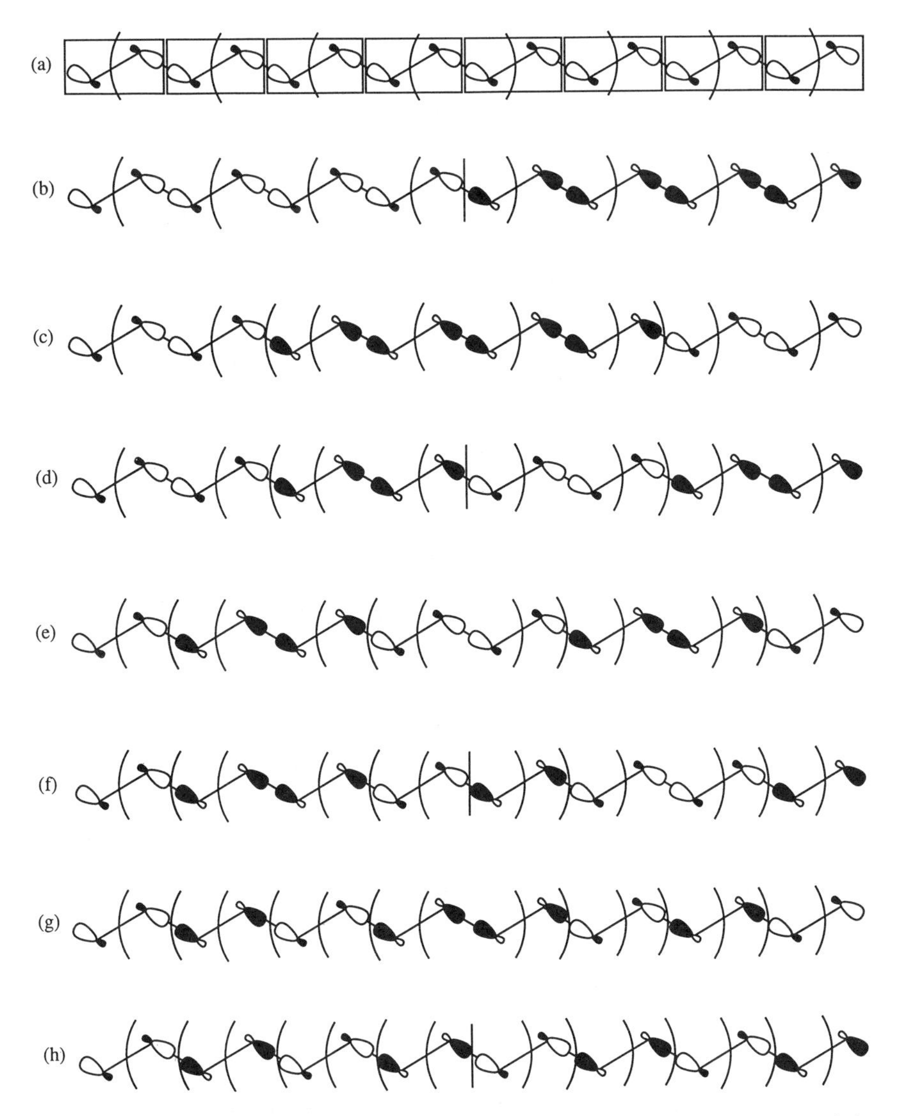

FIGURE 12.6 The eight MOs for the eight *dxy*-generated MO for each dimer unit in a C_{16} fragment of polyethylene generated by the 5*s* (a), 6*px* (b), 6*s* (c), 7*px* (d), 7*s* (e), 8*px* (f), 8*s* (g), and 9*px* (h) GOs. Four of the MOs (a)–(d) are BMOs and four (e)–(d) are ABMOs.

TABLE 12.4 Generator Table for the Eight 3*dxy*-Generated MOs for Each Dimer Unit in a C_{16} Fragment of Polyethylene

	GOs							
Equivalence Set	5*s*	6*s*	6*px*	7*s*	7*px*	8*s*	8*px*	9*px*
16 Csp^3	b	b	b	a	b	a	a	a

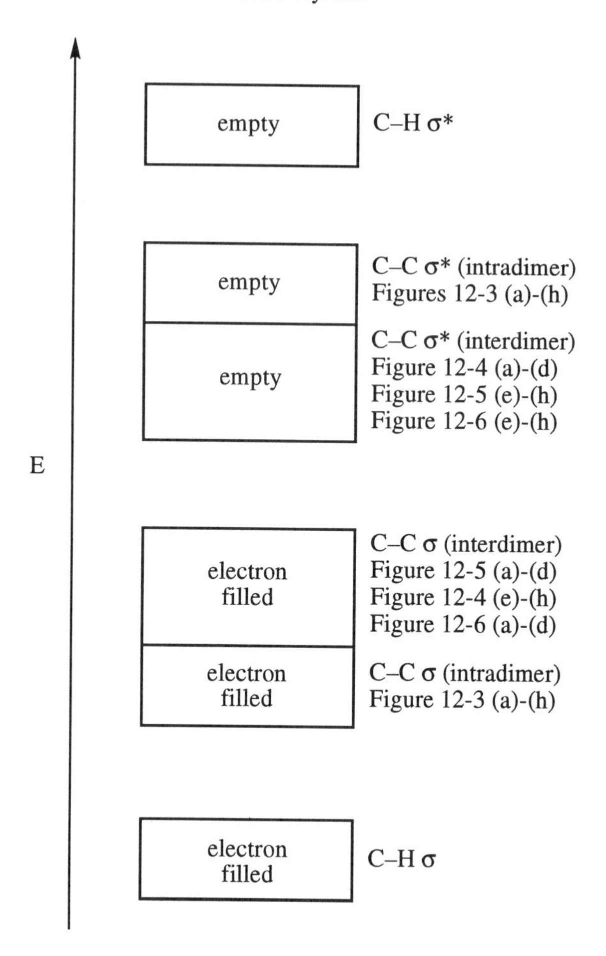

FIGURE 12.7 Energy level diagram for the delocalized MOs of a C_{16} fragment of polyethylene.

12.2. π-Bonded Conjugated Polymers

Nonconjugated polymers (i.e., the C=C bonds do not alternate with C—C bonds) offer many applications as structural materials, but they are generally electrical insulators. Their chemical bonding can be explained using either a delocalized or a localized MO picture. Conjugated organic polymers (i.e., possessing alternating C=C and C—C bonds) have potential metallic characteristics owing to the presence of delocalized electrons in their π orbital bands. Fragments of such polymers have recently been shown to function as single "molecular wires" for electrical conduction. Although different Lewis structures such as those shown in Figure 12.8 can be used to describe a localized bonding view, delocalized MOs have been extremely useful in eluci-

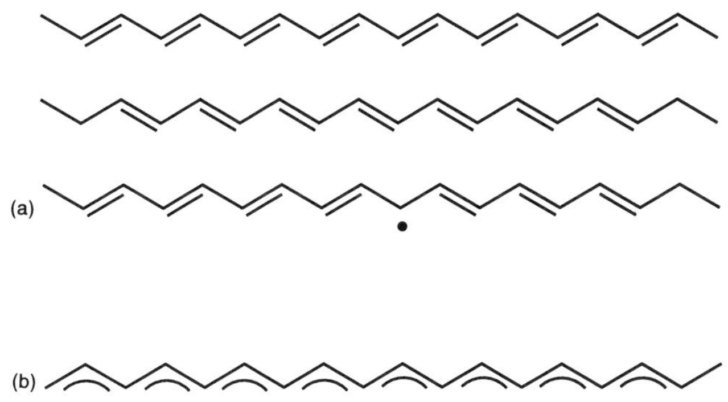

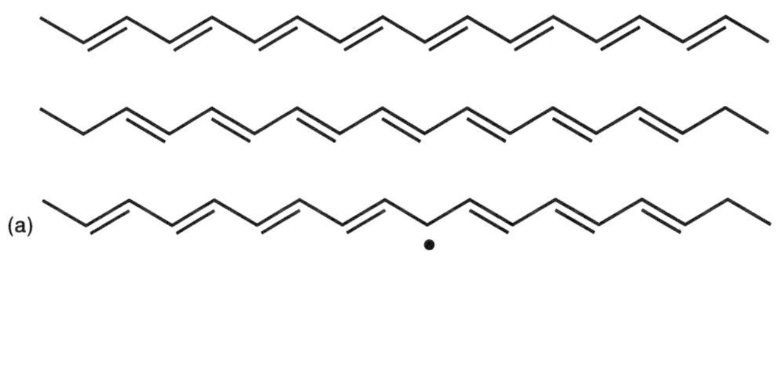

FIGURE 12.8 The localized (a) and delocalized (b) Lewis structures of a conjugated polymer.

dating the band structure of many of these types of polymers. In this section, we will examine a delocalized bonding view of the simplest of all conjugated organic polymers, namely, polyacetylene —[CH=CH]— (produced from acetylene, HC≡CH; see Figure 12.9a) and then move on to polyacene, which is a one-dimensional polymer produced by a condensation of benzene rings (Figure 12.9b). Along our way to the electronic structure of polyacene, we will consider some well-known polycyclic aromatic hydrocarbons (PAHs) such

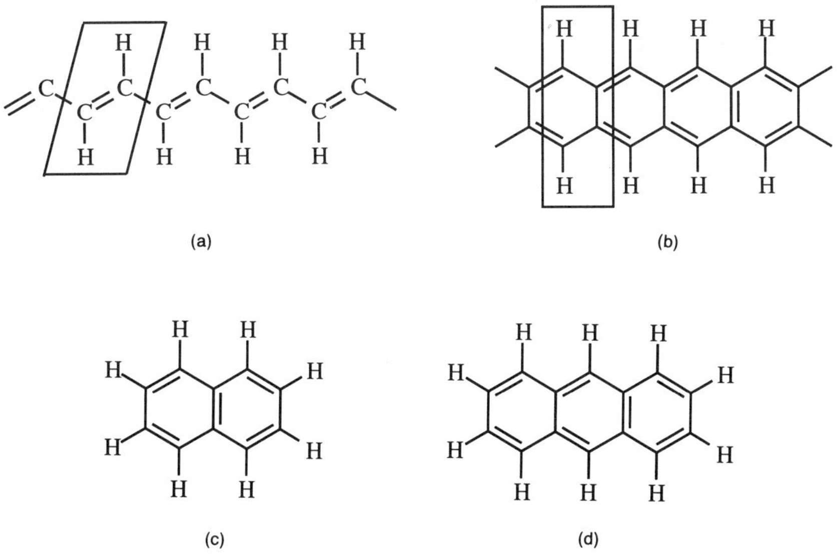

FIGURE 12.9 (a) Polyacetylene fragment; (b) polyacene fragment; (c) naphthalene; (d) anthracene.

as naphthalene (Figure 12.9c) and anthracene (Figure 12.9d). Molecules of this type have been found in meteorites originating beyond our solar system as well as in grilled foods. Finally, we will examine two-dimensional condensations on benzene rings that lead to the ultimate structure, graphite, a quasi-infinite two-dimensional honeycomb sheet of fused benzene rings.

Polyacetylene is a semiconducting conjugated polymer that can be doped with Br_2, for example, to cause metallic conductivity. This polymer is therefore being intensely studied for its applications in polymer batteries and solar cells. Undoped polyacetylene can show two of several Lewis structures, shown in Figure 12.8, which are consistent with its semiconducting behavior. The delocalized view (Figure 12.8b) might be expected to give metallic conduction, but this does not take place until polyacetylene is doped. Other Lewis structures that include radical sites or charged sites are also possible, and these have been examined to account for some of the unusual electronic behavior of polyacetylene. In this section, we will treat polyacetylene as a uniform planar zig-zag chain of (CH) groups. By analogy with polyethylene, each carbon atom is assigned three sp^2 hybrid VAOs plus one *pz* VAO, and the hydrogen has its 1*s* VAO. The repeat unit contains two CH groups as shown in Figure 12.9a. Since we have treated a very similar type of σ-bonded fragment when we discussed the bonding pattern in polyethylene, we will highlight only the π MOs originating from the C2*pz* VAOs in polyacetylene.

The essential difference between the π bonds and the σ bonds in the polyacetylene chain is the fact that C2*pz* – C2*pz* orbital overlaps (i.e., perpendicular to the carbon plane) are identical within repeat units and between repeat units, whereas orbital overlaps (as we also saw in polyethylene) are unequal for each type of σ interaction. Therefore, the energy of a particular π orbital is governed *equally* by intraunit and interunit interactions and there is no need to first generate the MOs in a CH dimer unit. Let us consider a $(CH)_{16}$ chain. There is one 2*pz* VAO per carbon, so we will use 16 GOs to describe the 16 MOs for this chain. These orbitals are illustrated in Figure 12.10 and they follow the order of increasing energy: 1*s*, 2*px*, 2*s*, 3*px*, 3*s*, 4*px*, 4*s*, 5*px*, 5*s*, 6*px*, 6*s*, 7*px*, 7*s*, 8*py*, 8*s*, 9*px*. In our designation of GOs to the 16 π orbitals, we have neglected the node in the *xy* plane. All the orbitals contain this node, and it does not contribute to the relative energies of the individual orbitals. The HOMO will be the MO generated by the 5*px* GO, and this MO is weakly bonding. Table 12.5 counts the number of π-bonding and π-antibonding contacts in this fragment of polyacetylene and Generator Table 12.6 summarizes our results. In $(CH)_{16}$, the HOMO has eight bonding and seven antibonding contacts, and the LUMO has seven bonding and eight antibonding contacts. If we consider long uniform chains (i.e., $(CH)_N$), the HOMO has $N/2$ bonding and $N/2 - 1$ antibonding contacts, the LUMO has $N/2$-1 bonding and $N/2$ antibonding contacts. As N goes to infinity, the energy gap between the HOMO and LUMO is eliminated to obtain a half-filled band of orbitals, with no gap (i.e., a metal).

(a)

(b)

(c)

(d)

(e)

(f)

(g)

(h)

(i)

(j)

(k)

(l)

(m)

(n)

(o)

(p)

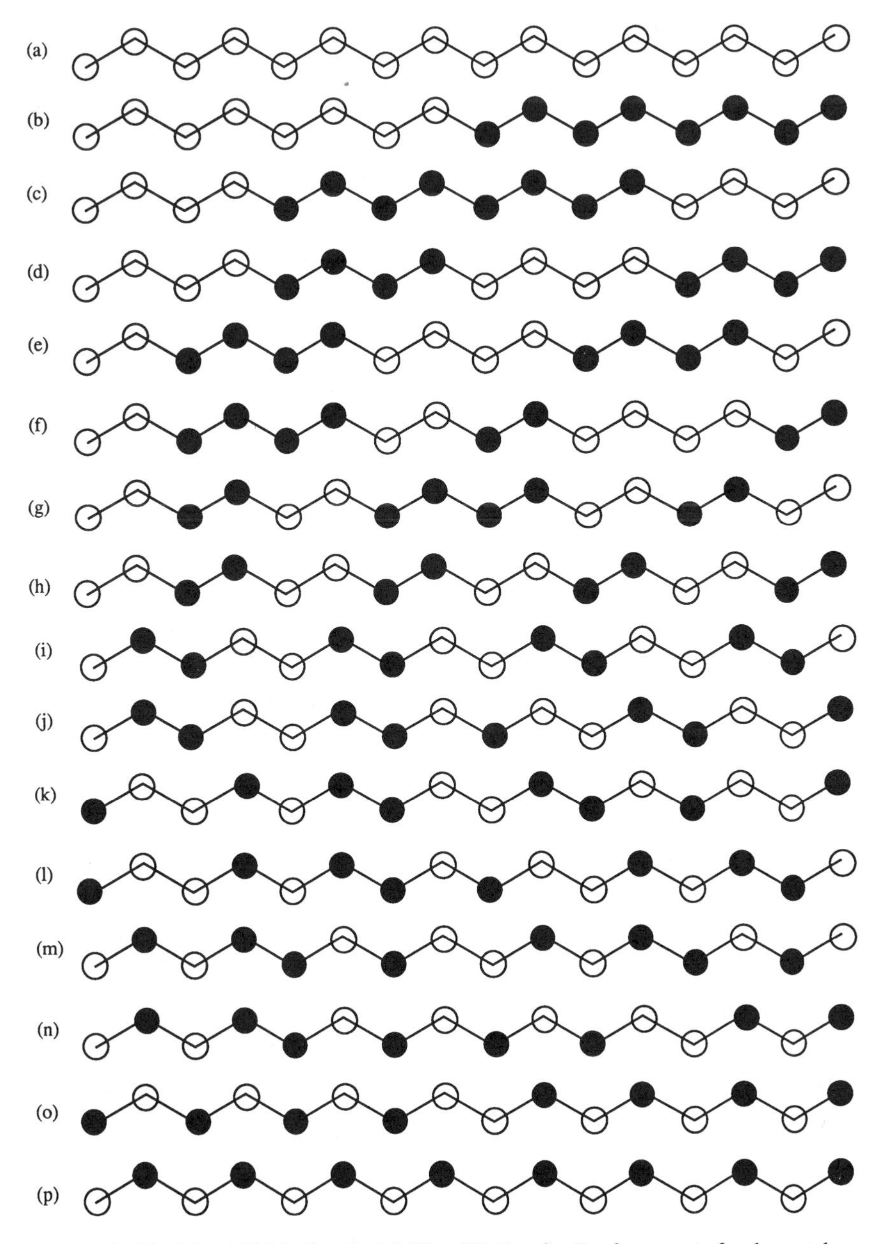

FIGURE 12.10 The 16 MOs for the 16 C2pz VAOs of a C_{16} fragment of polyacetylene generated by the 1s (a), 2px (b), 2s (c), 3px (d), 3s (e), 4px (f), 4s (g), 5px (h), 5s (i), 6px (j), 6s (h), 7px (l), 7s (m), 8px (n), 8s (o), and 9px (p) GOs. Note that the circular and/or planar nodes have not been depicted.

TABLE 12.5 The Number of Bonding and Antibonding Contacts in the π MOs of a C_{16} Fragment of Polyacetylene

GO	Number of Bonding Contacts	Number of Antibonding Contacts
1*s*	15	0
2*px*	14	1
2*s*	13	2
3*px*	12	3
3*s*	11	4
4*px*	10	5
4*s*	9	6
5*px*	8	7
5*s*	7	8
6*px*	6	9
6*s*	5	10
7*px*	4	11
7*s*	3	12
8*px*	2	13
8*s*	1	14
8*px*	0	15

TABLE 12.6 Generator Table for the 16 π MOs in a C_{16} Fragment of Polyethylene

	GOs															
Equivalence Set	1*s*	2*s*	2*px*	3*s*	3*px*	4*s*	4*px*	5*s*	5*px*	6*s*	6*px*	7*s*	7*px*	8*s*	8*px*	9*px*
16 C2*pz*	b	b	b	b	b	b	b	a	b	a	a	a	a	a	a	a

The reason for the semiconducting behavior of undoped polyacetylene arises from a distortion of the uniform chain into alternating short-long C—C distances along the chain. When such an atom displacement is turned on, it can be seen from the GOs in Figure 12.10 that orbitals with more π-bonding interactions in the short distance will drop in energy, while those with more π-antibonding interactions in the short distance will rise in energy, and thus create an energy gap between HOMO and LUMO at the half-filled band electronic configuration. This is called a *Peierls distortion.* When polyacetylene is doped with Br_2, for example, partial oxidation removes some electrons from the HOMOs creating positive "holes," which, like electrons, also conduct electricity. This results in the metallic conductivity observed upon doping this polymer.

If we conceptually remove one set of H atoms on one side of the polyacetylene chain, and then link two such chains together via a C—C contact, we form the planar polymer polyacene, —$[C_2(CH)_2]$— shown in Figure 12.9b. The repeat unit indicated contains four carbon atoms and two terminal hydrogens. We can also imagine polyacene to form by the condensation of benzene rings through opposite edges of the C_6 hexagon with the elimination of H_2 molecules. Therefore, polycyclic molecules like naphthalene and anthracene represent molecular analogues to the quasi-infinite polyacene. Therefore, we will begin with these two model molecules which will demonstrate the utility of using both planar as well as spherical (circular) nodes in these aromatic systems. To start, recall the π orbitals of benzene (Chapter 7). The center of the molecule is equidistant from all six carbon atoms, so that there is only one "shell" of π orbitals. In this case, GOs with spherical nodes are not necessary and the sequence of GOs is $1s$, $2px$, $2py$, $3dx^2 - y^2$, $3dxy$ and $f3'$ when we disregard the node in the xy plane. The other f orbital, $f3''$ is not applicable because its three planar nodes pass through the six C2pz VAOs.

Now, let us consider the π orbitals of naphthalene, $C_{10}H_8$. There are ten C2pz VAOs that can form ten π molecular orbitals. Structurally, $C_{10}H_8$ has three shells of carbon atoms with respect to the molecular center, namely, two atoms approximately 0.7 Å, four atoms approximately 1.86 Å, and four atoms approximately 2.52 Å from the center. The ten GOs are constructed using both planar and spherical (circular) nodes. With three shells of atoms, there can be a maximum of two finite spherical nodes. This is analogous to saying that a $3s$ orbital has three shells of electron density but only two finite spherical nodes. To develop a systematic method for the construction of these GOs, we make the following rules:

1. Only one *vertical planar node* may intersect a bonded pair of atoms, unless a planar node exists that passes through *both* atoms of the pair. In that case, a second planar node is allowed.
2. *Spherical nodes* may pass either through shells of atoms or between adjacent shells of atoms. There are four possibilities to consider for a given GO with m spherical nodes drawn for a system with n shells of atoms:
   a. m even; n even: all m spherical nodes between shells
   b. m odd; n even: all m spherical nodes between shells
   c. m even; n odd: all m spherical nodes between shells
   d. m odd; n odd: $m - 1$ spherical nodes between shells; 1 spherical node at the middle shell

These two rules are presented in order to develop a systematic recipe for constructing GOs in these polycyclic aromatic systems. Without these rules, we have the undesirable situation that there is always an arbitrary choice for the placement of spherical nodes and there can also be several planar nodes intersecting the same bond. We will find that some MOs will be nearly equienergetic (i.e., equal in energy) as determined by counting the number of bonding and antibonding contacts, but these MOs will be fundamentally different in their nodal characteristics.

By keeping these rules in mind, we can now return to naphthalene. Figure 12.11 illustrates the ten MOs generated by the GO sequence $1s$, $2py$, $2px$, $2s$, $3dxy$, $3py$, $3px$, $3s$, $4dxy$, and $4px$. Note that $3dx^2 - y^2$ is missing. This GO is

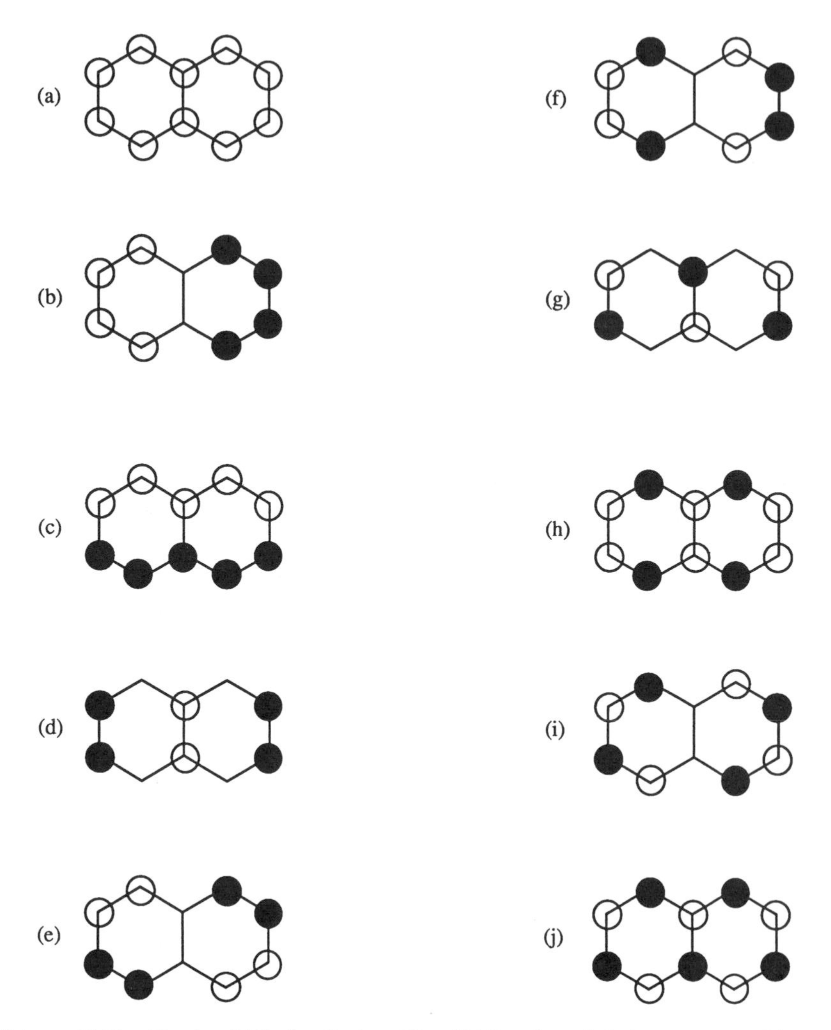

FIGURE 12.11 The ten MOs for the ten C2*pz* VAOs of naphthalene generated by 1*s* (a), 2*py* (b), 2*px* (c), 2*s* (d), 3*dxy* (e), 3*pz* (f), 3*px* (g), 3*s* (h), 4*dxy* (i) and 4*px* (j) GOs. Note that the circular and/or planar nodes have not been depicted.

disallowed by rule 1 above, because two planar nodes would intersect the central C—C bond. Table 12.7 lists the number of bonding and antibonding contacts for each GO-generated MO and the energy calculated for each orbital using a simple Hückel approximation to obtain the spectrum of energy levels. Generator Table 12.8 summarizes our pictorial results. There are five bonding and five antibonding molecular orbitals, which are separated by a gap of $(\sqrt{5} - 1)\beta$ where β is the interaction energy (resonance integral) between C2*pz* VAOs on adjacent sites. In terms of the sequence of GOs (*nlm*), note that for a given "shell" *n*, the energy of a GO goes up as *l*

TABLE 12.7 The Number of Bonding and Antibonding Contacts and Hückel Energies for the π MOs of Naphthalene

GOs	Number of Bonding Contacts	Number of Antibonding Contacts	E (Hückel)
$1s$	11	0	−2.303
$2py$	6	0	−1.618
$2px$	8	3	−1.303
$2s$	3	0	−1.000
$3dxy$	4	2	−0.618
$3py$	2	4	+0.618
$3px$	0	3	+1.000
$3s$	3	8	+1.303
$4dxy$	0	6	+1.618
$4px$	0	11	+2.303

TABLE 12.8 Generator Table for the π MOs in Naphthalene

	GOs									
Equivalence Sets	$1s$	$2s$	$2px$	$2py$	$3s$	$3px$	$3py$	$3dxy$	$4px$	$4dxy$
10 C2pz	b	b	b	b	a	a	a	b	a	a

decreases. This inverse relationship between energies of GOs and l for a specific nlm occurs because the potential energy of the electron density arises from nuclei that are on spherical shells rather than at the spherical center. We shall comment on this effect in the next chapter when we address cage compounds.

The next stage of condensation of benzene rings produces anthracene ($C_{14}H_{10}$). With respect to its molecular center, there are three shells of atoms: the central six-carbons, the outer four carbons and the four carbons lying at an intermediate distance from the center. We conclude, then, that the maximum number of spherical nodes is two. The central six-membered ring constituting the inner shell of atoms further limits the number of planar nodes to three (i.e., $f3'$ and or $f3''$ GOs). Figure 12.12 shows the GOs and MOs for the π system of anthracene and Table 12.9 summarizes the number of bonding, nonbonding, and antibonding orbital overlaps within each MO. This table also lists the number of spherical and planar nodes in each MO. Notice that we have represented 18 MOs, yet the π MOs in $C_{14}H_{10}$ are limited to 14 in number (one for each C2pz VAO). The GOs create redundancy! How can we eliminate this? In Chapter 3, we discussed the inversion operation in a molecule. Functions that are *even* with respect to inversion do not change sign: $f(-x,-y,-z) = +f(x,y,z)$. Functions that are *odd* with respect to inversion do change sign: $f(-x,-y,-z) = -f(x,y,z)$. With respect to inversion, s and d GO wave functions (and hence the orbitals

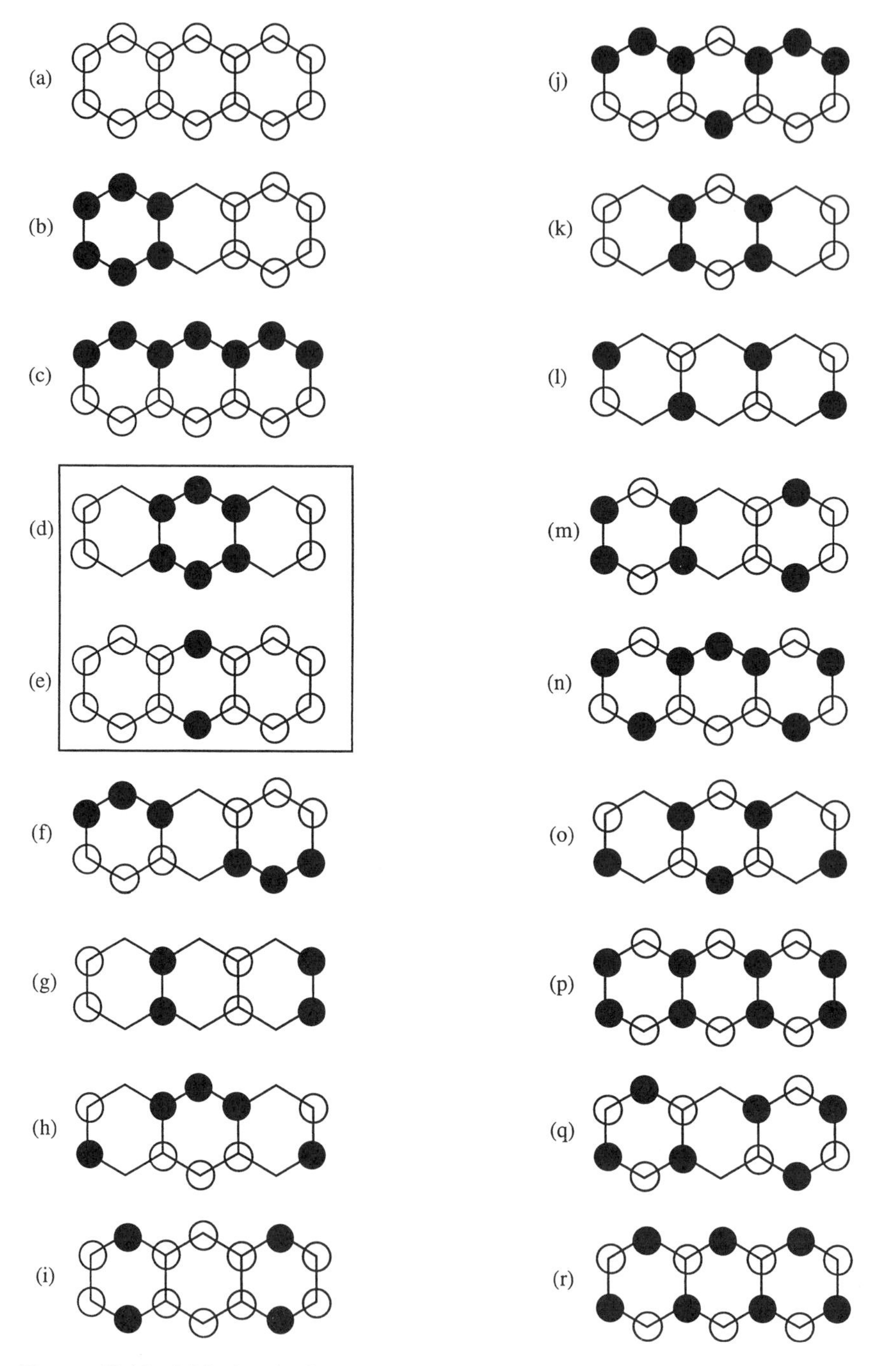

FIGURE 12.12 MOs for the fourteen C2pz VAOs of anthracene generated by 1s (a), 2py (b), 2px (c), 2s (d), 3$dx^2 - y^2$ (e), 3dxy (f), 3py (g), 3px (h), 3s (i), 4f'' (j), 4$dx^2 - y^2$ (k), 4dxy (l), 4py (m), 4px (n), 5f'' (o), 5$dx^2 - y^2$ (p), 5dxy (q), and 6f'' (r) GOs. Note that the circular and/or planar nodes have not been depicted.

TABLE 12.9 Number of Bonding, Nonbonding, and Antibonding Contacts and the Number of Spherical and Planar Nodes Associated with the Eighteen GO-generated MOs for Anthracene

GO	Number of Bonding Contacts	Number of Nonbonding Contacts	Number of Antibonding Contacts	Number of Spherical Nodes	Number of Planar Nodes
$1s$	16	0	0	0	0
$2py$	12	4	0	0	1
$2px$	12	0	4	0	1
$2s$	8	8	0	1	0
$3dx^2 - y^2$	12	0	4	0	2
$3dxy$	8	4	4	0	2
$3py$	4	12	0	1	1
$3px$	4	8	4	1	1
$3s$	8	0	8	2	0
$4f''$	8	0	8	0	3
$4dx^2 - y^2$	4	8	4	1	2
$4dxy$	0	12	4	1	2
$4py$	4	4	8	2	1
$4px$	4	0	12	2	1
$5f''$	0	8	8	1	3
$5dx^2 - y^2$	4	0	12	2	2
$5dxy$	0	4	12	2	2
$6f''$	0	0	16	2	3

themselves) are *even*, *p* and *f* GOs are *odd*. To eliminate redundant GOs (MOs) we take into account the net bonding overlap (i.e., the number of bonding – the number of antibonding contacts). If two MOs with equal bonding overlap are either both even or both odd, we eliminate one MO. If two MOs with equal bonding overlap are even and odd, we do not eliminate either one. Table 12.10 lists the net bonding overlap, and even or odd character of each MO in Figure 12.12. This procedure allows us to eliminate 4 MOs and we are left with 14 MOs that represent the π MOs of anthracene. Table 12.10 also includes the energy of each MO according to a Hückel calculation. Notice that the two MOs with *no* net bonding character do not agree qualitatively with the Hückel calculation of the energies of these MOs. The numerical coefficients of the VAOs in the mathematical expression for these two MOs will differ in magnitude, but these differences are not explicitly derived from the GOs. However, the GOs do correctly reproduce the coefficient signs of the AOs in the MOs generated by the Hückel method. Generator Table 12.11 summarizes the results we obtained pictorially.

Tetracene, $C_{18}H_{12}$ consists of four condensed benzenoid rings. In this molecule, there are five shells of atoms, which limits the maximum number of spherical nodes to four. The center of the molecule lies on a C—C bond, and the maximum number of planar nodes is two (Figure 12.13).

Rather than draw the entire set of MOs for this molecule, we summarize the results in Table 12.12. Here, we list each GO along with the number of

TABLE 12.10 Net Bonding Contacts, Inversion Character and Hückel Energies for the 18 π MOs in Anthracene Generated by the GOs

GO	Net Bonding Contacts	Even or Odd	Hückel E[a]
$1s$	16	even	−2.414
$2py$	12	odd	−2.000
$2px$	8	odd	
$2s$	8	even	−1.414 (2x)
$3dx^2-y^2$	8	even	
$3dxy$	4	even	−1.00 (2x)
$3py$	4	odd	
$3px$	0	odd	−0.414
$3s$	0	even	
$4f''$	0	odd	+0.414
$4dx^2-y^2$	0	even	
$4dxy$	−4	even	+1.000 (2x)
$4py$	−4	odd	
$4px$	−8	odd	
$5f''$	−8	odd	+1.414 (2x)
$5dx^2-y^2$	−8	even	
$5dxy$	−12	even	+2.000
$6f''$	−16	odd	+2.414

[a] 2x means that these energies occur twice in the Hückel MO energy level diagram.

nodal surfaces, and the number of bonding, nonbonding, and antibonding orbital contacts between connected pairs of carbon atoms in the MOs. The geometrical restrictions on the maximum number of nodes provides an upper limit of 25 GOs we can use. Since we can have only 18 π MOs in tetracene, 7 of the GOs and the MOs they generate must be redundant. We have pointed out the redundant GOs in Table 12.12.

The first redundancy occurs when we compare the $3s$ and $4py$ GOs (Figure 12.14). We could arbitrarily eliminate the $4py$ GO because of the two nodal surfaces passing through the middle section of the molecule. However, we note that the energies of these two orbitals are identical: five noninteracting bonding overlaps versus four noninteracting bonding overlaps within the

TABLE 12.11 Generator Table for the 14 π MOs in Anthracene

	GOs													
Equivalence Sets	$1s$	$2s$	$2px$	$2py$	$3s$	$3px$	$3py$	$4px$	$4py$	$4dxy$	$4dx^2-y^2$	$5dxy$	$5dx^2-y^2$	$6f''$
14 C2pz	b	b	b	b	n	n	b	a	a	a	n	a	a	a

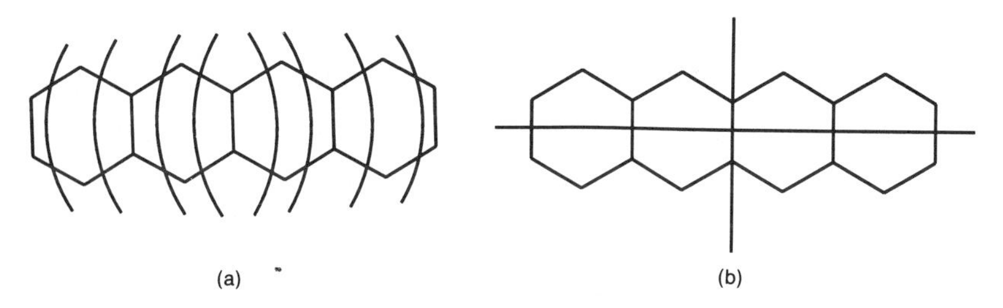

FIGURE 12.13 The five shells of atoms (a) and the two planar nodes (b) in tetracene.

TABLE 12.12 Number of Bonding, Nonbonding, and Antibonding Contacts and the Number of Spherical and Planar Nodes Associated with the 25 GO-generated MOs for Tetracene

GO	Number of Bonding Contacts	Number of Nonbonding Contacts	Number of Antibonding Contacts	Number of Spherical Nodes	Number of Planar Nodes	Remarks
$1s$	21	0	0	0	0	
$2s$	11	10	0	1	0	
$2px$	16	0	5	0	1	
$2py$	16	5	0	0	1	
$3s$	5	16	0	2	0	
$3px$	8	10	3	1	1	
$3py$	6	15	0	1	1	
$3dxy$	12	5	4	0	2	
$3dx^2 - y^2$	13	8	0	0	2	Redundant with $2s$
$4s$	3	10	8	3	0	
$4px$	0	16	5	2	1	
$4py$	4	17	0	2	1	Redundant with $3s$
$4dxy$	4	15	2	1	2	
$4dx^2 - y^2$	7	14	0	1	2	Redundant with $3s$
$5s$	5	0	16	4	0	
$5px$	0	10	11	3	1	
$5py$	2	15	4	3	1	
$5dxy$	0	17	4	2	2	Redundant with $4px$
$5dx^2 - y^2$	5	16	0	2	2	Redundant with $3s$
$6px$	0	0	21	4	1	
$6py$	4	5	21	4	1	
$6dxy$	0	15	6	3	2	
$6dx^2 - y^2$	3	14	4	3	2	Multiple nodes through bonds
$7dxy$	0	15	16	4	2	
$7dx^2 - y^2$	5	8	8	4	2	Multiple nodes through bonds

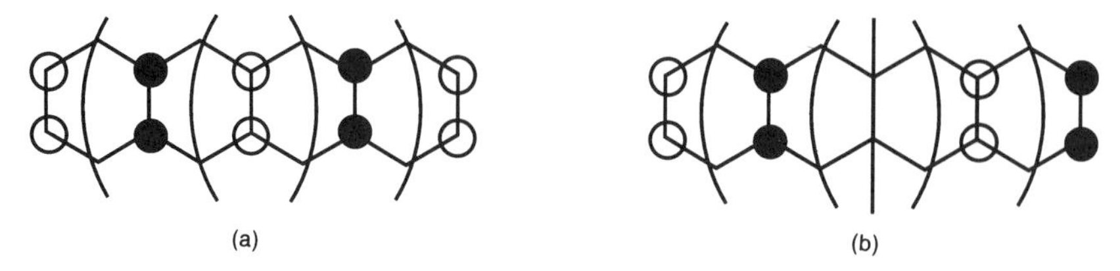

FIGURE 12.14 The 3s-generated (a) and the 4py-generated (b) MOs of tetracene.

same molecule. A similar situation occurs for the 4px and 5dxy-generated MOs. The other type of redundancy involves, for example, the 2s and $3dx^2 - y^2$ GOs (Figure 12.15). In fact, depending upon how we place the nodal surfaces, the two labels 2s and $3dx^2 - y^2$ are interchangeable here. Therefore, we eliminate GOs in our list due to the redundancy of $3dx^2 - y^2$ and $4dx^2 - y^2$, and we can eliminate $5dx^2 - y^2$, $6dx^2 - y^2$ and $7dx^2 - y^2$ GOs due to multiple nodal surfaces cutting the same bond. After these two types of redundancies are recognized, we are left with 18 GOs, one for each π MO in tetracene.

We can continue the process of condensing benzenoid rings beyond tetracene until we reach the polymer polyacene $C_{4n+2}H_{2n+4}$. These oligomers become larger as n increases in. With one electron assigned to each C2pz AO, $2n + 1$ π MOs are occupied and the gap between HOMO and LUMO steadily decreases as n increases. As n increases to infinity in the polymer, the gap disappears and polyacene becomes a metallic conductor.

The condensation of benzene rings can also occur to form larger two-dimensional PAHs that can be expanded by still further condensation to give graphite, a quasi-infinite two-dimensional planar sheet of fused six-membered rings. We shall consider only one example a larger PAH, namely, pyrene ($C_{16}H_{10}$), which consists of five shells of atoms, and whose molecular center lies on the central C—C bond (Figure 12.16). Notice that this molecule is elongated, and so spherical nodes will be distorted to resemble an oblate spheroid. A cross section of such a node will look like an ellipsoid. For a justification of using spherical nodes that must be distorted because of a molecule's lack of high symmetry, see Appendix V. The nodal surfaces are limited to a maximum of four spherical nodes and two planar nodes. Therefore, there is a maximum number of 25 GOs from the 1s through the 5d GO

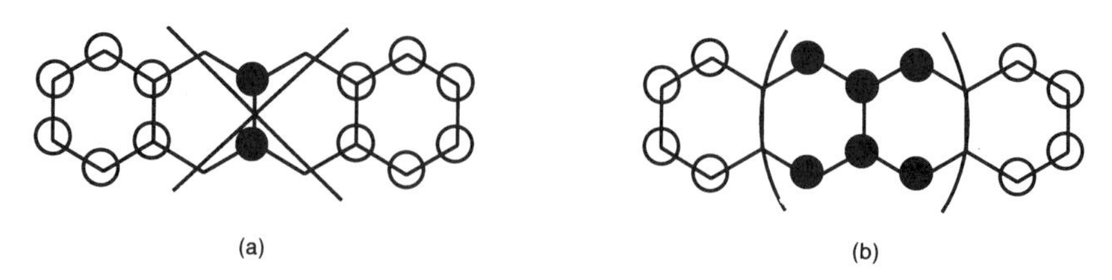

FIGURE 12.15 The 2s-generated (a) and $3dx^2 - y^2$-generated MOs of tetracene.

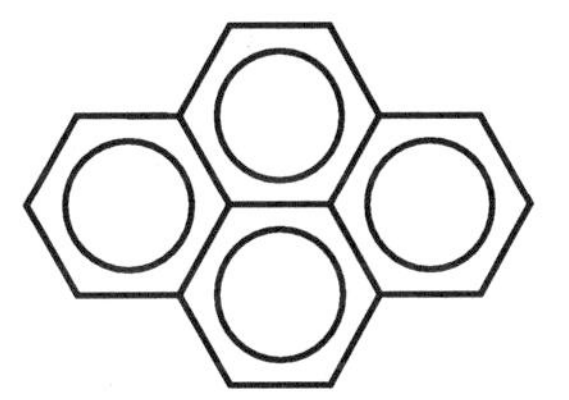

FIGURE 12.16 Pyrene.

sets. The molecule provides only 16 π MOs and so we have 9 redundant orbitals generated in our construction. Rather than diagram all the MOs, we point out that the GOs increase in relative energy along the series $1s < 2py < 2px < 2s < 3dxy < 3dx^2 - y^2 < 3py < 3px$. The first eight GOs account for all the bonding MOs and the $3px$-generated MO is the HOMO (Figure 12.17a). The LUMO is generated by $4dxy$ which is followed by the remaining ABMOs generated by $4dx^2 - y^2$, $4px$, $4py$, $4s$, $5dxy$, $5px$, and $5py$. The $3s$ GO is redundant with $3dx^2 - y^2$. The remaining eight GOs as defined by the above criteria ($5dx^2 - y^2$, $6s$, $6px$, $6py$, $6dx^2 - y^2$, $6dxy$, $7dx^2 - y^2$ and $7dxy$) are also redundant. In this way, we can account for the 9 redundant GOs and generate 16 orbitals that describe the spectrum of π MOs in pyrene.

We shall not carry on condensations of more six-membered rings to pyrene to progress toward the graphite structure, but the description presented for polyacene helps us understand what bonding changes occur in this case. Thus as more six-membered rings are added to pyrene, the HOMO-LUMO energy separation drops and eventually reaches zero in the infinite two-dimensional sheet. The HOMO for pyrene shows 6 bonding and 4 antibonding orbital contacts, and the LUMO reveals 4 bonding and 6 antibonding orbital contacts. For larger 2-dimensional structures, the numbers of bonding and antibonding orbital contacts approach one another and become equal for graphite.

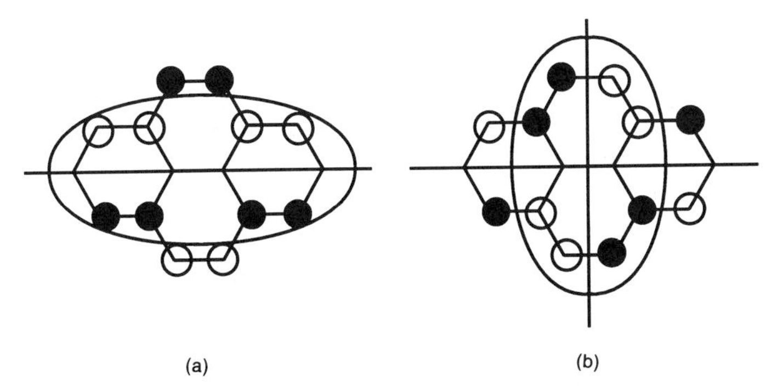

FIGURE 12.17 (a) The $3px$-generated HOMO; and (b) $4dxy$-generated LUMO of tetracene.

Because of the many atoms present in extended arrays such as the polymers discussed in this chapter and the large cages and the solids treated in the next two chapters, the vibrational modes of such systems are complex and are cumbersome to obtain by any method, including the generator orbital approach. We will therefore not derive vibrational modes for extended molecules.

Summary

We have seen in this chapter how the GO approach to MOs can be used to visualize the delocalized MOs in σ-bonded polymers, in π-bonded conjugated polymers, and also in polycyclic aromatic molecular analogues of fragments of graphite. For polycyclic aromatics, we made two rules to avoid arbitrary placement of spherical nodes and the intersection of more than one planar node in a given bond. We also saw how redundancies in our GOs for polycyclic aromatics could be eliminated by considering the behavior of GO-generated sets of MOs upon inversion which have the same net number of bonding contacts. These considerations also need to be taken into account for polymers when using the recipe in Table 11.4.

Exercises

1. The solid compound whose simplest molecular formula is Hg_3AsF_6 is a polymer containing "infinite" chains of mercury atoms. Generate a delocalized view of the bonding in this material, by considering a 16-unit fragment. (Hint: note that a positive charge is present for every three mercury atoms since AsF_6^- is anionic.) Why is this a conducting polymer?
2. The polymer $[Pt(CN)_4]_n$ consists of square planar $Pt(CN)_4$ units connected along their axes by dz^2 VAOs of platinum. Develop a delocalized view for a 16-unit fragment of this compound and comment on why it conducts electricity along its axial sigma framework when it is doped with halogens.
3. Polyacetylene can be doped with sodium metal to cause conduction. How do you suppose conduction occurs under these conditions?
4. Polymeric cumulenes $(=C=)_n$ have not been reported, although short molecular versions of this systems are known. Develop a delocalized view of such a polymer by considering a 16-unit fragment. How might you make such a polymer behave as a metallic conductor?
5. Interest has been developing in making "polycarbon" polymers $(\equiv C-)_n$ for molecular wires. Small analogues of such systems have been synthesized that contain twenty carbons using metal complexes as end groups. Develop a delocalized MO view of a 16-unit fragment of polycarbon. (Disregard the end group.)
6. Polyphenylene consists of repeating units of the type shown below. Develop a delocalized view of a three-unit segment of this polymer by assuming a flat structure.

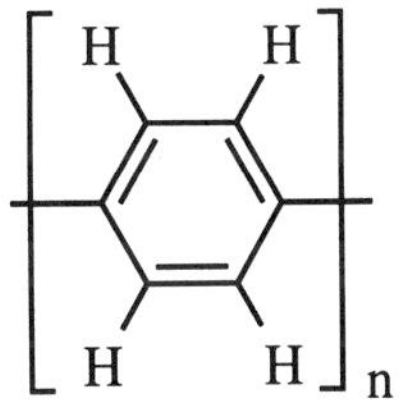

7. Although a polycarbonyl (see below) has not yet been synthesized, such a polymer is predicted to be an insulator. Develop a delocalized view of a 16-unit segment of this polymer and rationalize its predicted insulating nature.

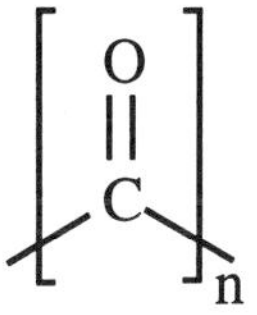

Chapter 13

Cages

Two-dimensional extended structures, such as graphite and convex polyhedral molecules as found in buckminsterfullerene (C_{60}), are topologically similar in that every carbon atom is bonded to three neighbors. There are, however, profound differences in their chemical and physical behavior. Graphite is weakly metallic. In fact, it is called a *zero-gap* semiconductor because the valence and conduction bands just touch. Buckminsterfullerene molecules, on the other hand, condense to form a semiconducting molecular solid.

Geometrically, C_{60} consists of both pentagonal and hexagonal rings of carbon atoms on the surface of a spherical shell. Graphite, an infinite structure, contains only planar six-membered rings. Note that the regular hexagon is a polygon that allows the construction of a planar, infinitely extending two-dimensional network. However, when pentagons occur in the network of three-connected carbon atoms, this surface of carbon atoms and bonds must curve, and in the case of C_{60}, a soccer ball-shaped cage is formed. Actually, there are many variations on the fullerene structural theme, including spheroidal C_{70}, open-ended long tubes, and closed tubes that resemble worms. Fullerenes constitute a sixth allotropic form of carbon; the other five are graphite, diamond, diamond pseudomorphs, glassy carbon, and amorphous carbon.

The electronic structure of graphite has been studied because of its interesting conducting properties. Since the two-dimensional sheets of hexagons are held together by van der Waals attractions (which are weaker than chemical bonds), many of these investigations have examined the effects of these attractions on the conducting properties of graphite. The molecular orbital pattern for C_{60} has also been extensively examined because the material becomes superconducting once it has been doped with alkali metal ions as, for example, in K_3C_{60}. In this chapter we will explore the application of the generator orbital approach to the problem of molecular orbital diagrams for cage molecules. We will limit our attention to cages involving a single shell of atoms. In these cases, the GO method can be used to obtain a semiquantitative energy level diagram.

13.1. A Definition of Cages

Polyhedral structures, such as those found for C_{60}, $B_{12}H_{12}^{2-}$ and $B_{10}C_2H_{12}$, may be considered as either *clusters* or *cages*. What differentiates a cluster from a cage compound? In principle, there is no difference, but we will differentiate them based upon the degree of connectivity of each vertex. We can classify clusters as structures that involve a denser packing of atoms than cages do, that is, polyhedra with triangular and quadrilateral (square or rectangular) faces. The connectivity of one vertex with its neighbors is also greater than or equal to four, as is found in octahedra and icosahedra. Cages use pentagons, hexagons, and higher-order polygons as faces, so the connectivity of vertices is less than four. In the family of cages related to C_{60}, each carbon atom is three-connected to other carbon atoms.

Euler's theorem is useful for calculating geometrical features of these cages. This theorem says that the number of vertices (V) minus the number of edges (E) plus the number of faces (F) is equal to 2:

$$V - E + F = 2 \tag{13.1}$$

or

$$V + F = E + 2 \tag{13.2}$$

If every vertex is three-connected, the number of edges is $3V/2$. Let us consider cages that consist of only five- and six-membered rings as faces. Let f_5 = the number of pentagonal faces, and f_6 = the number of hexagonal faces. Then, the number of edges is expressed as $(5f_5 + 6f_6)/2$ (Appendix VI). Using Euler's relation and the equality $3V = 5f_5 + 6f_6$, we find that $f_5 = 12$. That is, all C_N cages in which each carbon atom is three-connected and the faces are pentagons or hexagons require exactly 12 pentagons. The number of hexagons is exactly $N/2 - 10$. In this chapter, we will briefly discuss the pattern of MOs for C_{20} ($f_6 = 0$), C_{28} ($f_6 = 4$), and C_{60} ($f_6 = 20$).

Since the surface π MOs are the most interesting in these cages, we will focus our attention on them. Because each carbon atom is three-bonded to neighboring atoms, we can construct sp^2 hybrid VAOs at each atom that provide the necessary orbitals for the σ bonding and σ antibonding MOs. This construction leaves a single radially directed $2p$ orbital at each atom. It is most convenient to assign a local z-axis at each carbon atom, which passes through the geometrical center of the cage (i.e., an axis directed radially). Therefore, the surface of the polyhedron is perpendicular to these axes at each carbon atom. The $2pz$ VAOs are also perpendicular to the sp^2 hybrid AOs, and overlap with neighboring $2pz$ VAOs in a quasi π manner (Figure 13.1). In constructing the π MOs for these cage molecules, we will focus on the MOs derived from these $2pz$ VAOs.

As we have seen, the basis for the generator orbital method is the use of members of the series of atomic orbitals placed at the geometrical center of a molecule as a means of constructing qualitative representations of the MOs

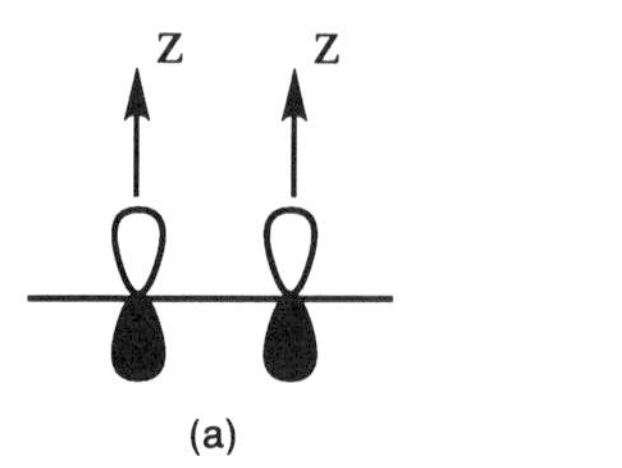

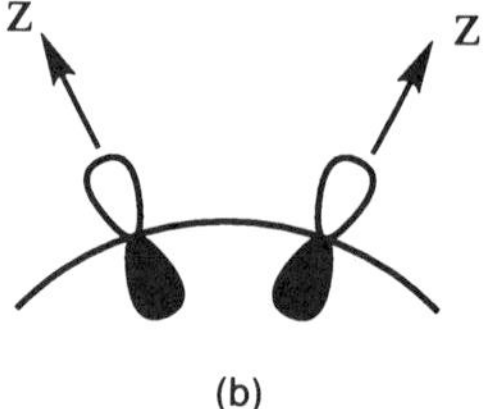

FIGURE 13.1 True π overlap of two $2pz$ AOs (a) and the quasi overlap they experience in a polyhedral molecule (b).

for the molecule in question. In cage molecules, the distribution of electron density is mostly confined to a spherical shell, whose radius is the distance between the molecular center and an atom in the cage. Recall that subshells of atomic orbitals are classified by the quantum numbers n and l. In cage molecules with a single spherical shell of atoms, we can use AOs that have no spherical nodes (i.e., $1s$, $2p$, $3d$, $4f$, etc.). The energies for cages involving N atoms with charge Z on a spherical surface of radius R can be expressed as

$$E_{nl} \approx \frac{-(NZ)^2}{2[n - (2l + 1)/4 - (2NZR)^{1/2}\pi]} \tag{13.3}$$

where $n = 1, 2, 3, \ldots,$ and $l = 0, 1, 2, \ldots, n - 1$ for a given n. Therefore, the energies increase in the order $1s < 2p < 3d < 4f < 5g < 6h < 7i < \cdots$.

13.2. The C_{20} Cage

C_{20}, or dodecahedrene, is a hypothetical molecule, but it is the first member in the series of possible C_N cages that conforms to the restrictions of Euler's theorem. A known related molecule, in which there are no double bonds, is dodecahedrane, $C_{20}R_{20}$. Recently, so-called met-cars have been reported. These molecules, such as Ti_8C_{12}, most likely adopt the pentagonal dodecahedral structure in Figure 13.2.

C_{20} will require 20 GOs to generate the 20 surface π MOs. If we consider a series of 20 GOs of increasing energy, then we will need the following set: $1s$, $2p$, $3d$, $4f$, and $5g$. In fact, we can easily evaluate the highest GO that will incorporate all possible MOs. For the cage C_N, the maximum l value ($l_{\max}$) is determined by the inequality $N \leq \sum_{l=0}^{l_{\max}}(2l + 1)$. In other words, $l_{\max} \geq \sqrt{N} - 1$.

Figure 13.3 shows the MO energies of C_{20} determined from a Hückel calculation. Notice that the three lowest energy levels are labeled as $1s$, $2p$, and $3d$, and show one-, three-, and five-fold degeneracies, respectively. These three orbital sets account for 9 of the 20 MOs needed for C_{20}. The $4f$ and $5g$ GOs that follow give us seven and nine additional GOs, respectively. We can use the seven $4f$ GOs, but only four of the nine $5g$ GOs. It is important at

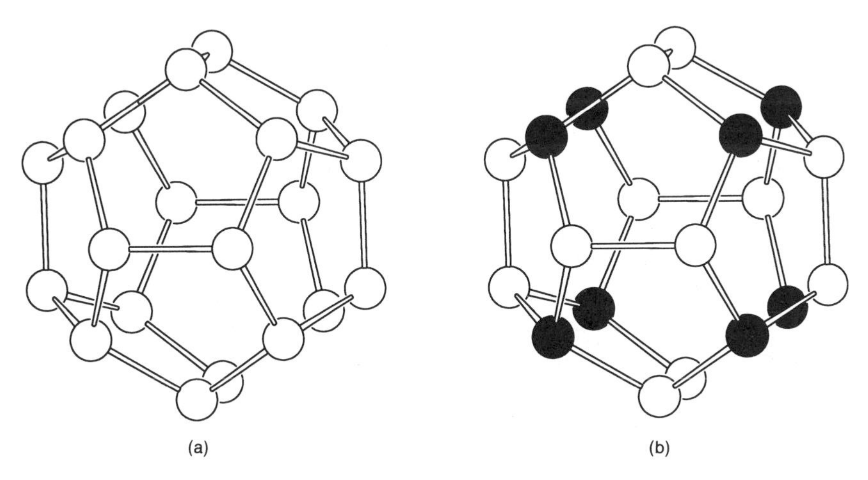

FIGURE 13.2 Pentagonal dodecahedron for C_{20} (a) and the met-car Ti_8C_{12} (b).

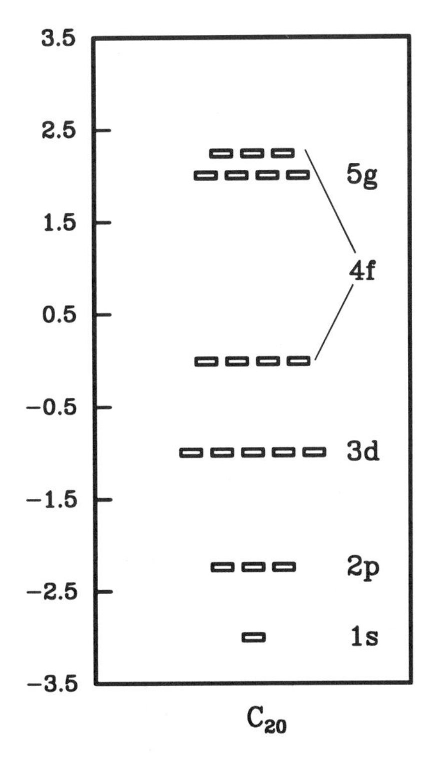

FIGURE 13.3 MO energy level diagram for the surface π orbitals of C_{20} from a Hückel calculation. Labels correspond to the GOs that can be assigned to each MO.

this point to note that the pentagonal dodecahedron has an inversion center [i.e., a carbon atom at the point (x, y, z) in the cluster is matched by another atom at the point $(-x, -y, -z)$]. Recall that with respect to inversion, 1*s*, 3*d*, and 5*g* GOs are even, and 2*p* and 4*f* GOs are odd. Molecules that have an inversion center require their surface π MOs to have an equal number of even and odd MOs with respect to inversion. In the case of C_{20}, then, we must have ten even and ten odd MOs. In this way, we can conclude that the 5*g* GO subshell can contribute only four GOs for generating the correct MOs.

Note also that there are no MOs that are seven-fold or nine-fold degenerate in the MOs for C_{20}. This is a consequence of the symmetry of the molecule. However, we can identify a four-fold degenerate level at $E = 0.000$ and a three-fold degenerate level at $E = +2.236$ as having come from the 4*f* subshell, and a four-fold degenerate level at $E = +2.000$ as having come from the 5*g* subshell. We could use the patterns of these GOs to understand the patterns of the MOs calculated by the Hückel method, but this requires sophisticated mathematical manipulation of the GOs. If you are interested in looking further into this, see Appendix VII.

C_{20} as a neutral molecule contains 20 valence electrons to be assigned to these surface π MOs. Eighteen of these electrons would be assigned to the 1*s*-, 2*p*-, and 3*d*-generated MOs, which leaves two electrons remaining for the 4*f*-generated MOs. Therefore C_{20} would be an "open-shell" molecule (i.e., one that has a partially occupied MO). Even after the degeneracy of the 4*f* GO is broken down into a four- and a three-fold degenerate energy level as shown in Figure 13.3, the open shell character of this molecule remains. However, a loss of two electrons to form $C_{20}{}^{2+}$ results in a closed-shell configuration.

13.3. The C_{28} Cage

This molecule is a polyhedron with 12 pentagonal and 4 hexagonal faces (Figure 13.4). The four hexagons are arranged in a tetrahedral manner about the molecular center. This cage is stabilized *endohedrally* (i.e., internally) by tetrapositive metal atoms such as Ti, Zr, or Hf encapsulated by the cage. Figure 13.4 also depicts the MO energy levels calculated according to the Hückel method. Notice that there is again a shell structure, but that the maximum degeneracy of any MO is four. (In fact, this degree of degeneracy is not allowed for tetrahedral molecules, but it occurs accidentally in this case.) As was the case for C_{20}, some of the MOs of C_{28} are labeled according to the GO that can be easily associated with it. In C_{28}, the 28 MOs are accommodated by the GO sets 1*s*, 2*p*, 3*d*, 4*f*, 5*g*, and three of the 6*h* functions ($l_{max} \geq \sqrt{28} - 1 = 4.3$; $l_{max} = 5$). Since the molecule does not have an inversion center, there are no restrictions on the number of even and odd MOs. As in the case of C_{20}, the MOs assigned to the 4*f*, 5*g*, and 6*h* GOs are difficult to assign. Again, we require a mathematical transformation of functions to

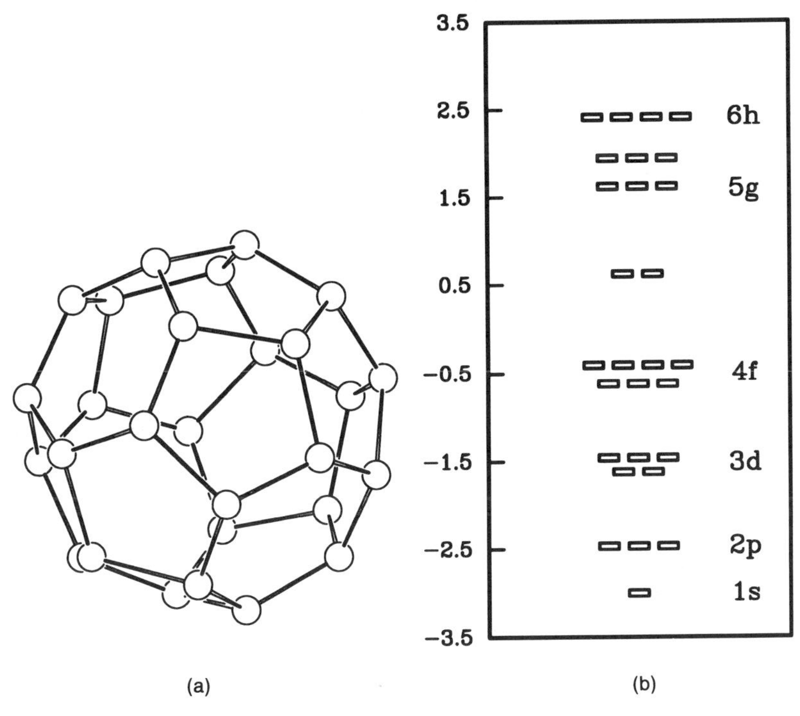

FIGURE 13.4 Polyhedron for C_{28} (a) and the MO energy level diagram for the surface π orbitals of C_{28} from a Hückel calculation. MO levels labeled from 1*s* to 4*f* correspond to the GOs that can be assigned to each MO. Because the upper four MOs are composed of contributions from both the 5*g*- the 6*h*-generated MOs, no assignment of these levels to single GOs can be made.

make a clearer assignment. Nevertheless, the sequence of occupied MOs follows the shell structure quite well, and the GO method can reproduce the occupied orbitals in these cages. Also, a stable electronic configuration occurs when four electrons are added to the neutral molecule (i.e., $C_{28}{}^{4-}$, according to the Hückel calculation), which is consistent with the observation that tetrapositive metal atoms will stabilize this cage.

13.4. The C_{60} Cage

We conclude this chapter with a brief examination of buckminsterfullerene, C_{60}, which has 12 pentagonal and 30 hexagonal faces arranged to give it a soccer-ball shape. Figure 13.5 depicts its structure and the MO energy levels calculated according to the Hückel method. The 60 MOs require GOs up to and including the 8*j* subshell ($l = 7$) of which 11 of the 15 GOs belonging to this subshell are needed. As in the case of C_{20}, the Hückel calculation gives energy levels whose maximum degeneracy is five. Nevertheless, the occupied

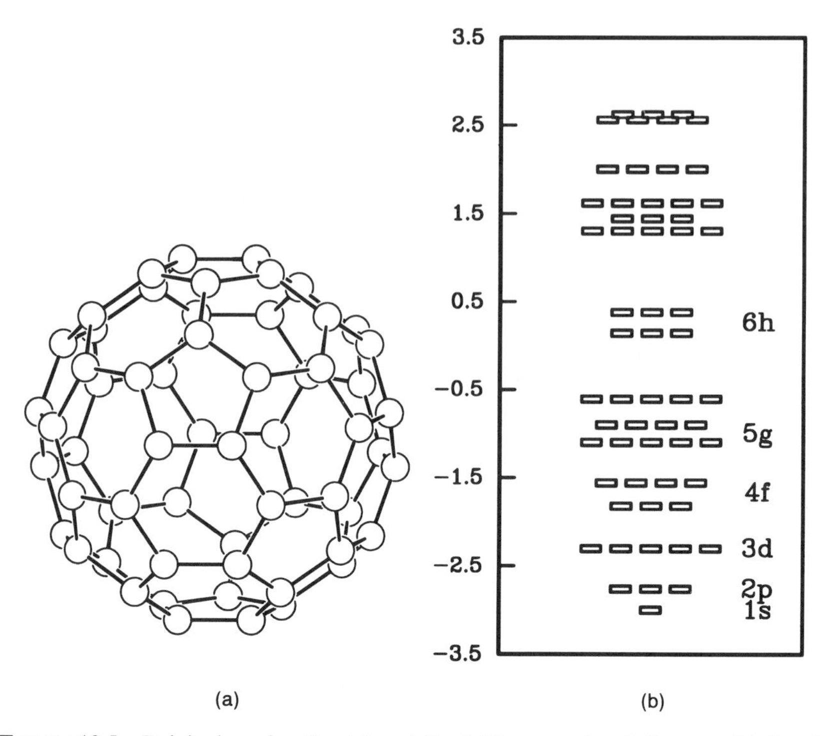

FIGURE 13.5 Polyhedron for C_{60} (a) and the MO energy level diagram (b) for the surface π orbitals of C_{60} from a Hückel calculation. Labels correspond to the GOs that can be assigned to each MO.

orbitals readily show a shell structure using the GO series $1s$, $2p$, $3d$, $4f$, and $5g$. The first problem with assigning each MO to a given GO occurs at the $6h$ subshell. The HOMO in C_{60} is a five-fold degenerate level that is actually derived from the $6h$ GOs. The LUMO, however, which is three-fold degenerate, is derived from the $7i$ subshell. With 60 valence electrons assigned to these surface π MOs, the five-fold degenerate HOMO is completely occupied. Therefore, the degeneracy of the $6h$ subshell is broken by the symmetry of the molecule in such a way that a closed shell electronic configuration is achieved.

Buckminsterfullerene is a semiconductor when the molecules are condensed to form a solid. The structure of the solid is like a face-centered cubic metal (e.g., copper or aluminum).

Summary

Cage molecules of carbon, C_N, in which all the atoms are on a single spherical shell, give π-MO patterns that follow an approximate shell structure $1s < 2p < 3d < \cdots$. The GO approach quickly reproduces this shell structure and provides an assignment of the nodal character of all occupied MOs in these cages. In the antibonding region (unoccupied MOs), the GO approach breaks down in its unambiguous assignment of MOs. Table 11.4 is still a valid summary of the procedures used in this chapter.

Exercises

1. Both C_{60}^{+} and C_{60}^{-} have been reported. Comment on the effect of the formation of these ions on the MO energy level diagram for C_{60} in Figure 13.5 and on the electrical conductivity around the cage.
2. Apply the methods of this chapter to construct an MO energy level diagram of the "radial orbitals" of icosahedral $B_{12}H_{12}^{3-}$. Assign each level as 1*s*, 2*p*, 3*d*, 4*f*....
3. $B_{32}H_{32}$ has been suggested to be a viable gas phase molecular analogue of C_{60}. $B_{32}H_{32}$ arises by placing a BH group at the center of each face of the C_{60} polyhedron. $B_{32}H_{32}$ is one of a series of polyhedra having only triangular faces and having vertices that are both five and six-connected. Derive a set of rules for the cluster B_NH_N analogous to those listed on pages 358 and 359 in Appendix VI for C_N.
4. Construct a MO energy level diagram for $B_{32}H_{32}$ using GOs. Which closed shell configuration would be stable?

CHAPTER 14

Solid State Materials

Solids that have ordered arrays of atoms in three dimensions have a wide range of compositions and properties. Like polymers, they can be insulators, conductors, semiconductors, and superconductors. They can be made up of neutral atoms of a single kind or more than one kind, they can be composed of charged atoms or groups of atoms, and they can have nonstoichiometric as well as stoichiometric ratios of different elements in them. They also come in a large variety of configurations (i.e., geometrical arrangements of their atoms or atom groupings). Our objective in this chapter is to take a brief look at a few examples of very simple solids to appreciate their delocalized bonding patterns and some of the consequences of these patterns.

The example we will choose is a metal that has a simple cubic structure. A metal having such a structure is the alpha form of polonium. To see how an infinitely large cubic lattice of atoms is held together, we need only consider a small fragment. A convenient fragment containing 16 atoms is shown in Figure 14.1, in which E stands for any element having this structure. Let us assume that the valence orbital in which we are interested is of the s type as indicated in Figure 14.1b. Let us further assume that this VAO is doubly occupied. A convenient strategy is first to generate the four LCAO MOs for this set along the left front edge of our fragment. Second, we will take linear contributions of these permitted MOs with an identical set of four MOs on the right edge of the front face. Third, we will take linear combinations of the permitted MOs so generated on the front face with an identical set on the back face.

Using the MO-generation technique we learned in Chapter 12 for polymers, we generate the SOs shown in Figures 14.1c–f. This information is summarized in Generator Table 14.1. Next, we generate an identical set of four MOs on the right front edge of our fragment. Then by taking linear combinations of identical pairs of MOs on both front edges of our fragment (i.e., those MOs generated by the same GO) we generate the set of eight MOs shown in Figure 14.2. An identical set of these eight MOs can be generated for the back face of our fragment and taking linear combinations of each one of these MOs with its twin (generated by the same GO) on the front face produces the 16 MOs drawn in Figure 14.3.

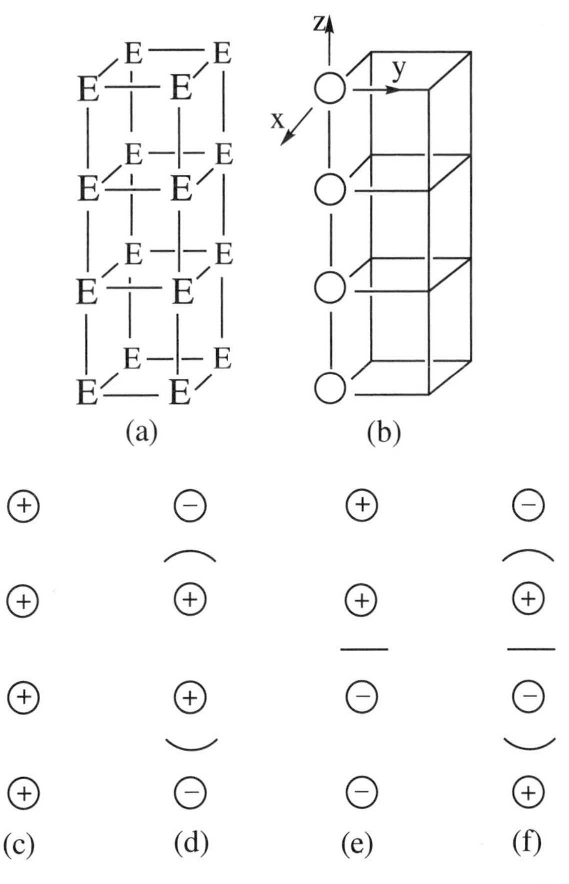

FIGURE 14.1 A simple cubic 16-atom fragment of E atoms (a) and the generation from the valence *s* orbitals of one edge (b) of the symmetry MOs using a 1*s* (c), 2*s* (d), 2*pz* (e), and 3*pz* (f) GOs.

with its twin (generated by the same GO) on the front face produces the 16 MOs drawn in Figure 14.3.

To sketch an MO energy level diagram of the MOs in Figure 14.3, we must first arrange them of their bonding (or antibonding) character. We construct Table 14.2 in which each of these MOs is labeled by its figure number and its number of *adjacent* bonding and antibonding interactions. This leads to the qualitative MO diagram in Figure 14.4 in which the lower eight MOs are net bonding, and would be filled with electrons in our example, while the upper eight are net antibonding and are unoccupied.

Because we chose a cubic fragment with an even number of layers, only bonding and antibonding MOs are created. As we will now see, an odd number of layers produces nonbonding MOs as well as bonding and anti-

TABLE 14.1 Generator Table for E_{16} (32 e^-)

VAO Equivalence Sets	GOs			
	1*s*	2*s*	2*pz*	3*pz*
four E*s*	b	a	b	a

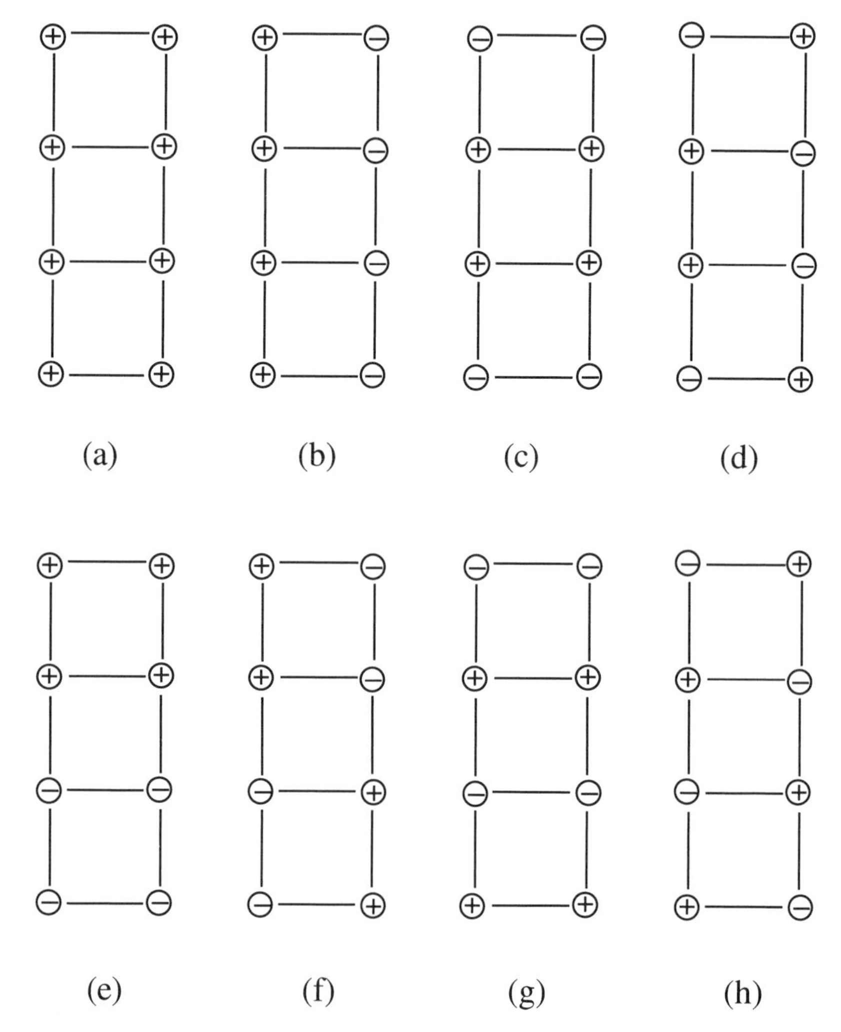

FIGURE 14.2 Linear combinations of MOs on the left and right front edges of simple cubic E_{12} generated by *s* (a–d) and *p* (e–h) GOs.

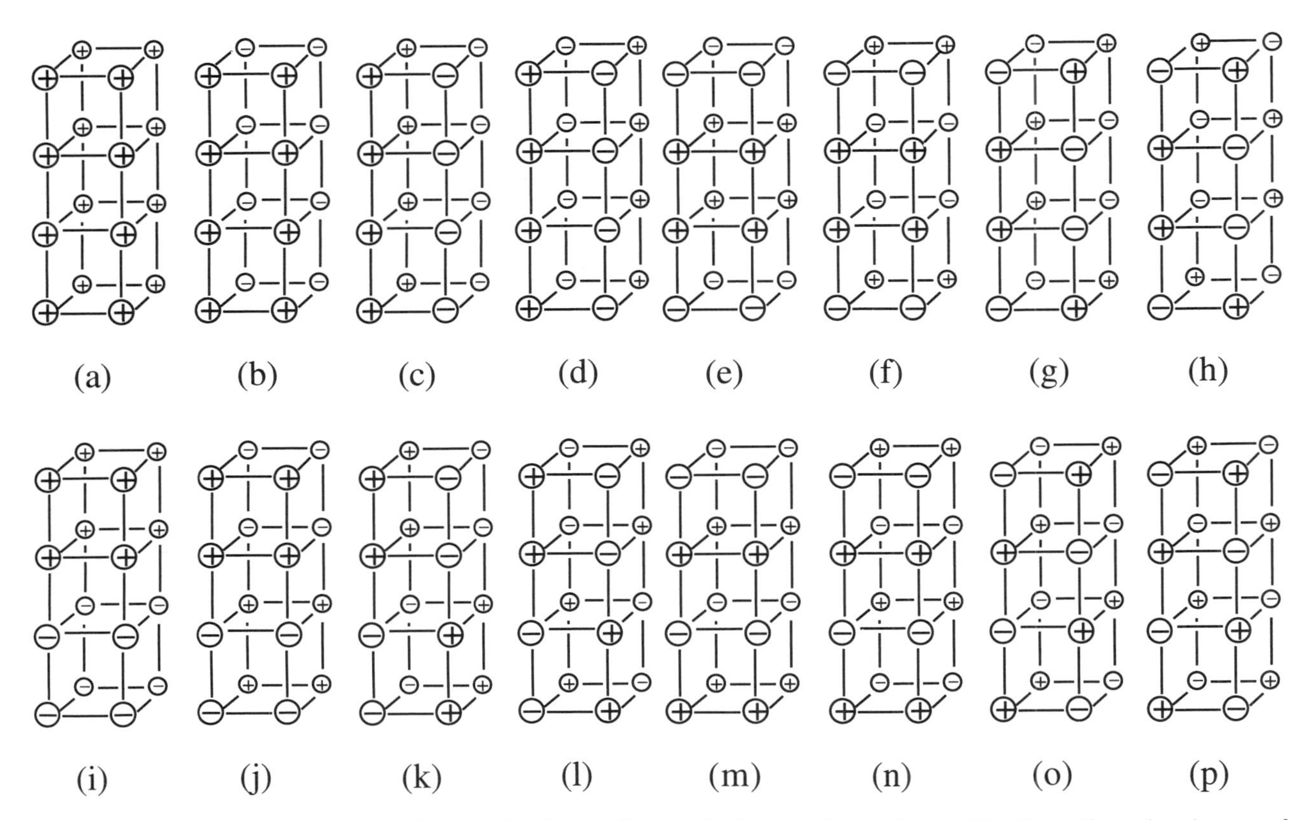

FIGURE 14.3 Linear combinations of identical pairs of MOs on the front and back faces of the three-dimensional array of atoms in our simple cubic E_{16} fragment.

TABLE 14.2 Number of Adjacent Bonding or Antibonding Contacts for the MOs in Figure 14.2.

MO	Number of Bonding Contacts	Number of Antibonding Contacts
Figure 14.3a	28	0
Figure 14.3b	20	8
Figure 14.3c	20	8
Figure 14.3d	12	16
Figure 14.3e	20	8
Figure 14.3f	12	16
Figure 14.3g	12	16
Figure 14.3h	4	24
Figure 14.3i	24	4
Figure 14.3j	16	12
Figure 14.3k	16	12
Figure 14.3l	8	20
Figure 14.3m	16	12
Figure 14.3n	8	20
Figure 14.3o	8	20
Figure 14.3p	0	28

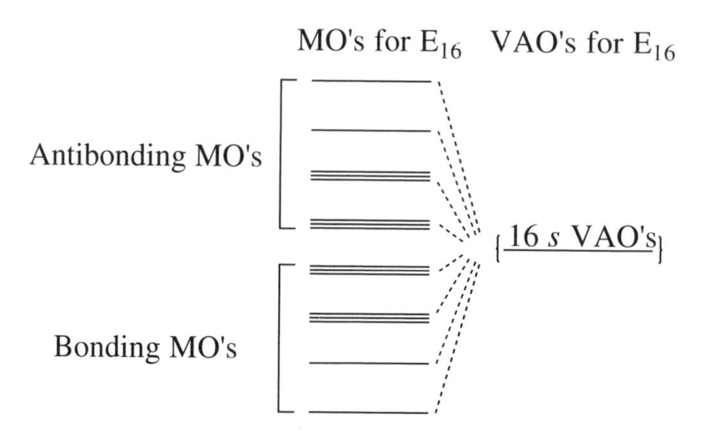

FIGURE 14.4 MO energy levels for an E_{16} simple cubic fragment of atoms.

bonding MOs. Figure 14.5a depicts such an array with three layers of *s* VAOs. The symmetry MOs for the extreme pair are generated as before and are shown in Figures 14.5b–d. Having generated three SOs for the front left string we take linear combinations of an identical set on the right front edge, as shown in Figure 14.6. The linear combinations of the six MOs of the front side (depicted in Figure 14.6) with those of an identical set on the back edge are shown in Figure 14.7, and the bonding and antibonding interactions of the nearest-neighbor orbital pairs are tabulated in Table 14.3. Table 14.4 shows that the MOs in Figures 14.7j and k have equal numbers of bonding

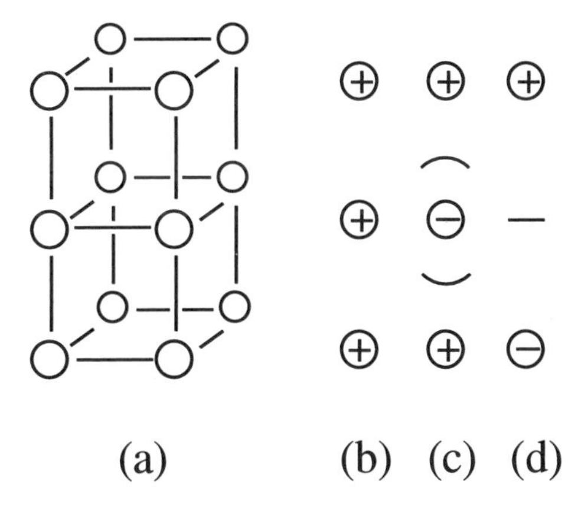

FIGURE 14.5 Array of VAOs in the E_{12} simple cubic fragment (a), and the SOs generated in the VAOs of the left front edge by 1*s* (b), 2*s* (c) and 2*pz* (d).

and antibonding interactions. Thus they are essentially nonbonding. Figure 14.8a shows the relative ordering of the MOs for the cubic E_{12} array, recall that both the E_{16} and E_{12} arrays discussed here are part of an "infinitely" large simple cubic lattice.

To visualize the MOs of a body-centered cubic lattice (e.g., the Group 1 metals, Ba, Cr, Fe, and W) we could insert an *s* VAO into the centers of the three cubes in Figure 14.1a or such a VAO into the centers of the two cubes of Figure 14.5a. We will do the latter as shown in Figure 14.9a and leave the former approach as an exercise. From previous experience we expect an *s*- and a *pz*-generated symmetry MO from the two body-centered VAOs. The former has a symmetry compatible with the MO in Figure 14.7a (which was generated by an *s* GO), while the latter matches the *p*-generated symmetry of the MO in Figure 14.7i. The linear combinations of these MOs are shown in Figure 14.9b–e, and their qualitative effect on the MOs of the E_{12} array is shown in Figure 14.8b. (The relative splittings in Figure 14.8b cannot be gauged accurately without calculations.)

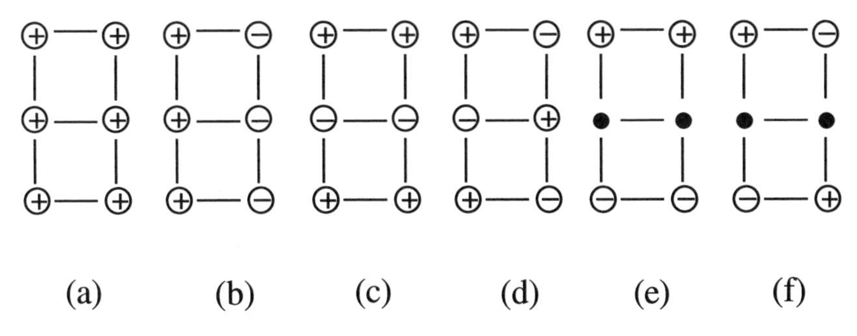

FIGURE 14.6 Linear combinations of identical MOs on the left and right front edges of simple cubic E_{12}.

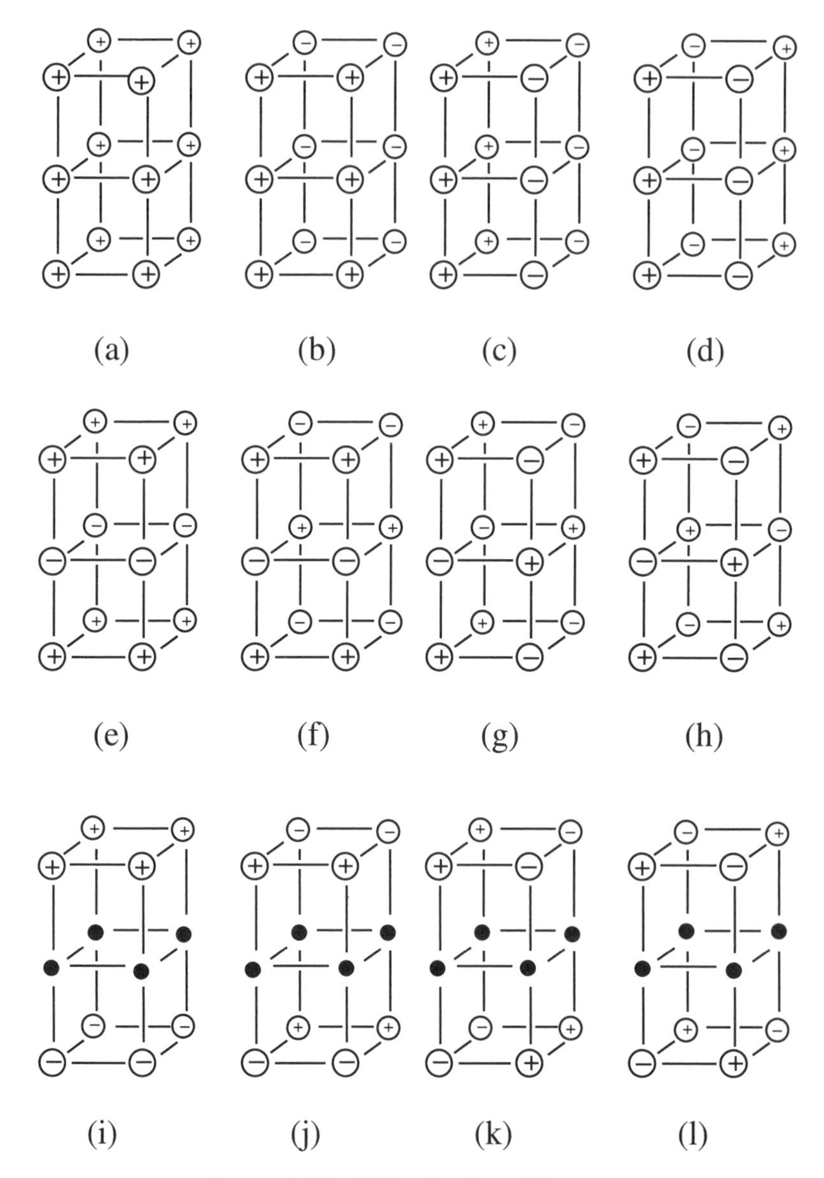

FIGURE 14.7 MOs of an E_{12} cubic array.

Use the Homework Menu in Node Game to repeat the steps taken here for developing delocalized views of a simple and a body-centered cubic lattice.

In addition to the above examples, a face-centered system (e.g., Ag, Au, Ca, Cu, Ni, Pb, or Pt) could have been chosen for illustration. This repeating unit cell is found in cubic close-packed systems that stack their layers ABCABC.... For further discussion of cubic close packing, see virtually any

TABLE 14.3 Number of Adjacent Bonding or Antibonding Contacts for the MOs in Figure 14.7.

MO	Number of Bonding Contacts	Number of Antibonding Contacts
Figure 14.7a	20	0
Figure 14.7b	14	6
Figure 14.7c	14	6
Figure 14.7d	8	12
Figure 14.7e	12	8
Figure 14.7f	6	14
Figure 14.7g	6	14
Figure 14.7h	0	20
Figure 14.7i	8	0
Figure 14.7j	4	4
Figure 14.7k	4	4
Figure 14.7l	0	8

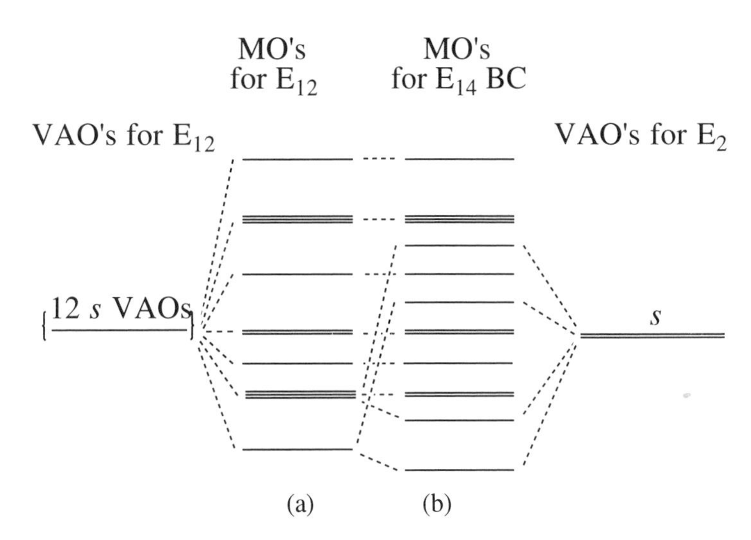

FIGURE 14.8 MO energy levels for an E_{12} cubic array of atoms (a) and for an E_{14} body-centered cubic array (b).

freshman or inorganic text book. Although somewhat more time-consuming, the delocalized MOs of a larger simple cubic chunk (e.g., a cubic array of 27 atoms with 3 atoms along each edge) could be developed. Instead of *s* VAOs, *p* VAOs that link a linear set of a reasonable number of atoms could be explored for their MOs. Such a set would represent infinitely long linear strings of atoms along any of the three Cartesian coordinates in a cubic array.

The last type of metallic unit cell we will mention is that shown in Figure 14.10. The metals Be, Cd, Co, Mg, Ti, and Zn have this repeating

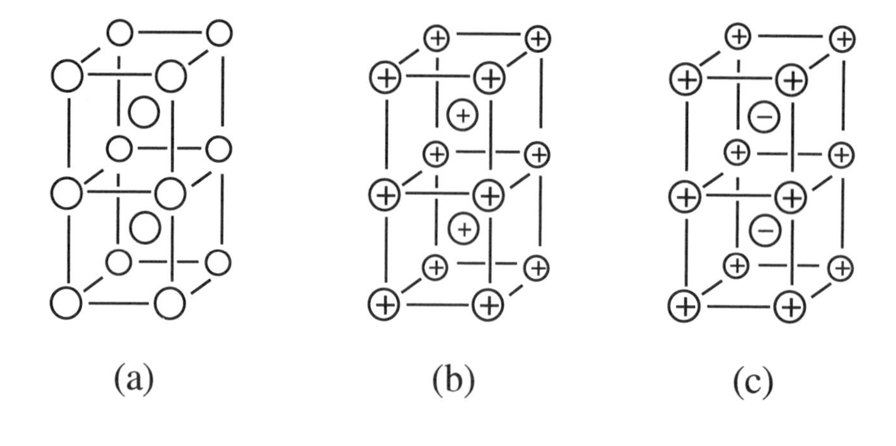

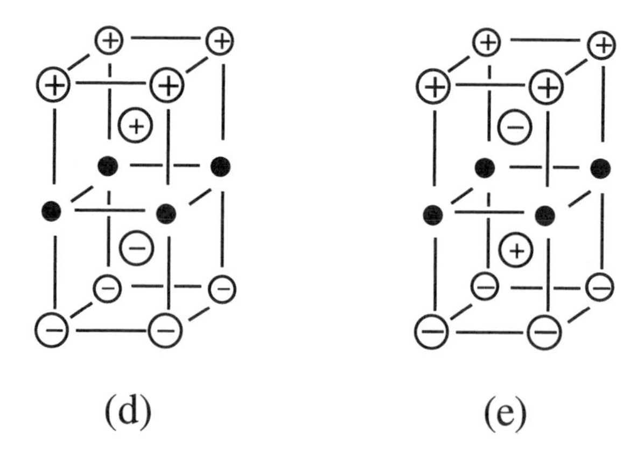

FIGURE 14.9 Body-centered array of E_{14} (a) and the MOs associated with the *s*-generated (b, c, d, e) and the *p*-generated (f, g) MOs of the cubic array and of the body-centered atom pair.

unit cell which is found for hexagonally close packed atoms that stack their layers ABABAB....

It may seem odd that we chose to develop our delocalized view of cubic solids by treating one edge of a chunk at a time, rather than placing a single GO center at the center of the array. We could, of course, have done the latter and this is left as an exercise at the end of the chapter. Such a process is more complicated for cubic chunks, however, in that higher more complex GOs are required. If we had chosen the simplest chunk of our cubic solid, namely, a cube of the simple, body-centered or face-centered type, we would have placed a GO center at the center. (Recall cubane, for example, in Chapter 11.) This brings up the interesting piont that α-polonium is a three-dimensional

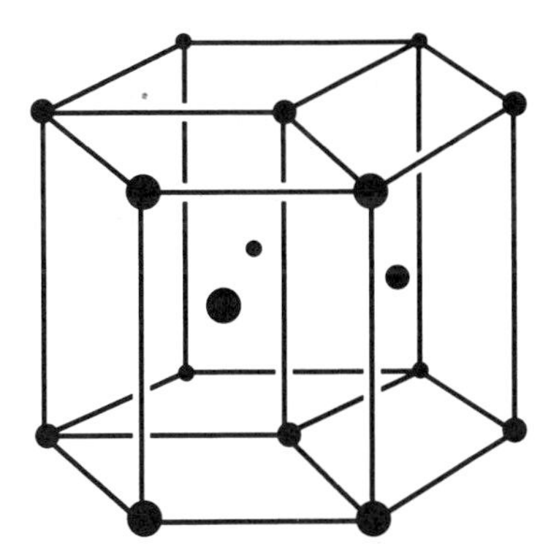

FIGURE 14.10 Unit cell for hexagonally close-packed spheres in ABAB... layers. The three atoms inside the hexagonal prism lie in the middle of their respective compartments.

cubic polymer for which the carbon cluster in cubane structurally represents a monomeric unit.

With the approach we have taken, you can see that because both the most bonding MO and the most antibonding MO (e.g., in Figure 14.4) are not expected to change appreciably in energy as the model E_{16} chunk is infinitely enlarged in three dimensions, the many additional MOs introduced are crowded between these two energy extremes, giving rise to a band. For metals with one *s* valence electron, the *s* band is half-full. This allows easy promotion of electrons at and near the nonbonding level (called the *Fermi* level) to be promoted into empty MOs (i.e., the conduction band) that are close in energy to the Fermi level. The MO energy level diagram in Figure 14.4 also indicates that the density of electronic states for an "infinite" number of MOs in an "infinite" cubic array will rise smoothly and rapidly from the MO energy extremes to the center of the band. By considering bands from a set of *s* and a set of *p* VAOs in a model chunk of solid, you can see why a band gap occurs if the *s* and *p* bands do not overlap in energy. If the bands do overlap, promotion of an electron into the *p* band is quite easy under an electric potential, and electron flow (conduction) occurs in metals such as beryllium, whose *s* band is filled. If the band gap is large enough, an insulator is the result.

Summary

Fragments of two types of solids (cubic and body-centered cubic) were employed to gain an understanding of the delocalized bonding in their corresponding extended arrays. This approach allowed us to appreciate the concepts of MO bands and band gaps, which are important in understanding the properties of conductors and of semiconductors and insulators, respectively.

Because no fundamentally new principles relating to our basic methods have been introduced in this chapter, Table 11.4 is a valid summary of our procedures.

Exercises

1. Develop a delocalized view of the *s*-band of a body-centered Group 1 metal using the 16-atom chunk shown in Figure 14.1a and compare the molecular energy level diagram you obtain with that in Figure 14.9b. Do the same using an eight-atom cubic array.

2. Develop a delocalized view of the p-band in an E_{16} chunk of (a) α-polonium, (b) an E_{14} chunk of a body-centered cubic metal such as potassium.

3. Develop a delocalized view of the *s*-band in a face-centered metal using a chunk composed of sixteen atoms on the corners of the array shown in Figure 14.1a plus an atom in each cubic face.

4. Develop a delocalized view for the *s*-band in an E_{12} simple cubic array assuming a single GO center at the center of the array. Compare your results with the view obtained via the procedure developed in this chapter. (Hint: see Figure 14.7).

5. Repeat problem 4 for a 27-atom cube of α-polonium consisting of 3 atoms on each edge.

6. Draw the permitted vibrations of an E_{14} chunk of a body-centered cubic metals such as barium.

7. Develop a delocalized view for the *s* band of a chunk of hexagonally close packed Be consisting of 27 atoms (i.e., a stack of two of the repeating units shown in Figure 14.10).

Appendix I.A

Wave Amplitude

The amplitude or displacement function for a wave is given by Equation 1; alternatively the sine function may be substituted by the cosine function:

$$f(x,t) = A \sin 2\pi \left(\frac{x}{\lambda} - \frac{t}{\tau} \right). \tag{1}$$

Here x measures the distance along the string, t is the time, A is the maximum amplitude, and λ and τ are constants having the same units as x and t, respectively. The dimension of f is that of A, which in the case of the string would be in units of length.

The amplitude of a standing wave is given by any one of the expressions in Equation 2:

$$\begin{gathered} A\left(\sin 2\pi \frac{t}{\tau}\right)\left(\sin 2\pi \frac{x}{\lambda}\right) \\ A\left(\sin 2\pi \frac{t}{\tau}\right)\left(\cos 2\pi \frac{x}{\lambda}\right) \\ A\left(\cos 2\pi \frac{t}{\tau}\right)\left(\sin 2\pi \frac{x}{\lambda}\right) \\ A\left(\cos 2\pi \frac{t}{\tau}\right)\left(\sin 2\pi \frac{x}{\lambda}\right). \end{gathered} \tag{2}$$

In the first expression the amplitude vanishes for $x = 0$ and at every half-wavelength [i.e., $x = (1/2)\lambda, \lambda, (3/2)\lambda, 2\lambda$, etc.]. The antinodes are seen from this expression to appear halfway between the nodes. In the remaining expressions, nodes and antinodes appear at different x values. In all the expressions, however, the nodes and antinodes are at fixed values of x and are independent of time.

Appendix I.B

Normalization and Orthogonality

The fraction of the electron cloud which is found at the time t in the infinitesimally small volume element $dV = dxdydz$ enclosing the point x, y, z is given by $|\psi(x, y, z, t)|^2 dV$, and this quantity is always positive. (Recall that the letter d in dx etc., simply means an infinitesimally small increment.) This means that at time t the volume element dV makes a contribution of $m|\psi|^2 dV$ to the total electron mass and a contribution $-e|\psi|^2 dV$ to the total electron charge. The quantity $|\psi|^2$ is called the *orbital density*. More rigorously, it describes the *probability* of finding the electron in dV.

Because the time factor in Equation 1.2 in Chapter 1 is of the form $\exp(i\alpha)$ with α being real, one finds $|\psi|^2 = \phi^2|f|^2 = \phi^2(x, y, z)$. This shows that bound orbital densities do not change with time and that they can be calculated.

Since ϕ^2 describes one whole electron in the entire space and $\phi^2 dV$ represents the fraction of this electron in the volume dV, we can add up all the $\phi^2 dV$ in the space and say that it is equal to one electron. The mathematical procedure for such addition is called *integration* and its notation is shown in Equation 3.

$$\int dV \phi^2(x, y, z) = 1 \tag{3}$$

This requirement is called the *normalization* of the orbital ϕ, and it says that the single electron is contained in the space defined by adding up all the infinitesimally small volume elements dV that contain electron density. In the case of a one-dimensional standing electron wave (Figure 1.7 in Chapter 1) $\phi_n^2(x)dx$ is the fraction of an electron in ϕ_n that lies between x and $(x + dx)$. The probability densities $\phi_n^2(x)$ have the forms shown in Figure 1.8 in Chapter 1. The integral in Equation 4 (for which we will use the shorthand

$$\int dV \phi(x, y, z)\psi(x, y, z) = \langle\phi|\psi\rangle \tag{4}$$

notation $\langle\phi|\psi\rangle$) is called the *overlap integral* between ϕ and ψ. This integral is zero for two normalized orbitals on the same atom. Orbitals for which the overlap integral is zero are said to be *orthogonal* to each other. Strictly speaking, this applies only to nondegenerate orbitals (i.e., orbitals of differing energies). Degenerate orbitals such as $2px$, $2py$, and $2pz$ are indeed orthogonal to one another, but as we shall see in Chapter 2, linear combinations of these orbitals need not be. Orthogonality does not imply that the two orbitals are at right angles to each other in space. It means only that their net overlap is zero.

APPENDIX I.C

Using an Orbital Matrix

Suppose we have the three orbitals ϕ_1, ϕ_2, ϕ_3, which are known, and another set ψ_1, ψ_2, ψ_3 which we constructed from ϕ_1, ϕ_2, ϕ_3 by Equation 5, wherein the subscripts in the T_{jk} coefficients are merely labeling numbers:

$$\begin{aligned} \psi_1 &= T_{11}\phi_1 + T_{12}\phi_2 + T_{13}\phi_3 \\ \psi_2 &= T_{21}\phi_1 + T_{22}\phi_2 + T_{23}\phi_3 \\ \psi_3 &= T_{31}\phi_1 + T_{32}\phi_2 + T_{33}\phi_3 \end{aligned} \tag{5}$$

We can write a shorthand version of Equation 5 called a *matrix* (Equation 6):

$$\begin{array}{c|ccc} & \phi_1 & \phi_2 & \phi_3 \\ \hline \psi_1 & T_{11} & T_{12} & T_{13} \\ \psi_2 & T_{21} & T_{22} & T_{23} \\ \psi_3 & T_{31} & T_{32} & T_{33} \end{array} \tag{6}$$

in which the first index on the coefficients T_{jk} denotes the *row* and the *second* denotes the *column* of the matrix. Each of Equations 5 is obtained by multiplying the ϕ_n by the coefficient below them in the row corresponding to the ψ_n and adding. Because orbitals must be normalized (Equation 3) and mutually orthogonal (i.e., zero overlap, Equation 4), the *two sets* of integrals $\langle\phi_j|\phi_k\rangle$ and $\langle\psi_j|\psi_k\rangle$ must both be 1 for $j = k$ and 0 for $j \neq k$. We now demonstrate the these conditions lead to important restrictions on the values the coefficients T_{jk} can take.

Selecting ψ_1 and ψ_2 as an example, we substitute the expansions for these orbitals given in Equation 5 into the overlap integral $\langle\psi_1|\psi_2\rangle$ in Equation 7:

$$\langle\psi_1|\psi_2\rangle = \langle T_{11}\phi_1 + T_{12}\phi_2 + T_{13}\phi_3|T_{21}\phi_1 + T_{22}\phi_2 + T_{23}\phi_3\rangle. \tag{7}$$

Carrying out the multiplication indicated in Equation 7, we obtain Equation 8:

$$\begin{aligned}\langle\psi_1|\psi_2\rangle = T_{11}T_{21}\langle\phi_1|\phi_1\rangle + T_{11}T_{22}\langle\phi_1|\phi_2\rangle + T_{11}T_{23}\langle\phi_1|\phi_3\rangle \\ + T_{12}T_{21}\langle\phi_2|\phi_1\rangle + T_{12}T_{22}\langle\phi_2|\phi_2\rangle + T_{12}T_{23}\langle\phi_2|\phi_3\rangle \\ + T_{13}T_{21}\langle\phi_3|\phi_1\rangle + T_{13}T_{22}\langle\phi_3|\phi_2\rangle + T_{13}T_{23}\langle\phi_3|\phi_3\rangle\end{aligned} \quad (8)$$

which simplifies to Equation 9 when we apply the normalization and orthogonality criteria for ϕ_1, ϕ_2, and ϕ_3:

$$\langle\psi_1|\psi_2\rangle = T_{11}T_{21} + T_{12}T_{22} + T_{13}T_{33}. \quad (9)$$

We also require, of course, that ψ_1 and ψ_2 be orthogonal and normalized, that is:

$$\langle\psi_1|\psi_2\rangle = T_{11}T_{21} + T_{12}T_{22} + T_{13}T_{23} = 0 \quad (10)$$

$$\langle\psi_1|\psi_1\rangle = T_{11}{}^2 + T_{12}{}^2 + T_{13}{}^2 = 1 \quad (11)$$

$$\langle\psi_2|\psi_2\rangle = T_{21}{}^2 + T_{22}{}^2 + T_{23}{}^2 = 1. \quad (12)$$

When functions are both orthogonal *and* normalized, we say they are *orthonormal.* The coefficients T_{jk} must satisfy Equation 10 to 12 in order for ψ_1 and ψ_2 to be orthonormal *and* for ϕ_1, ϕ_2, and ϕ_3 to be orthonormal. An example of such an array of numbers T_{jk} that satisfies Equations 10, 11, and 12 is shown in Equation 13. It is called an *orthogonal matrix*, and Equations 5, which generate the ψ_m from the ϕ_n, are called an *orthogonal transformation.*

$$\begin{bmatrix} T_{11} & T_{12} & T_{13} \\ T_{21} & T_{22} & T_{23} \\ T_{31} & T_{32} & T_{33} \end{bmatrix} = \begin{bmatrix} 1/6 & -5/6 & 10^{1/2}/6 \\ 10^{1/2}/6 & 10^{1/2}/6 & 4/6 \\ -5/6 & 1/6 & 10^{1/2}/6 \end{bmatrix} \quad (13)$$

Note that there was no need to be concerned about the explicit wave functions or the meaning of any of the implied integrations! The array of numbers in Equation 13 is only one example of an orthogonal array. Other orthogonal 3×3 arrays can be devised.

The two sets of orbitals ψ_n and ϕ_n ($n = 1, 2, 3$), which we have been discussing are equivalent in that all observable electronic properties are independent of our choice of orbital sets. Suppose that ψ_1, ψ_2, and ψ_3 each contain the same number of electrons (say, two). The total density ρ is then given by Equation 14, in which the factor of 2 represents the presence of two electrons:

$$\rho = 2(\psi_1{}^2 + \psi_2{}^2 + \psi_3{}^2). \quad (14)$$

If we now use the relations of Equations 5 and the orthonormality conditions for ψ_1, ψ_2, ψ_3, Equation 15 can be obtained:

$$\rho = 2(\phi_1{}^2 + \phi_2{}^2 + \phi_3{}^2). \quad (15)$$

Appendix II.A

Integrating an Orbital Wave Function

Let us consider the simplest of all atomic orbitals, the 1*s*. Whenever we discuss the *spatial amplitude* of an orbital we parenthesize it [i.e., (1*s*) in this case]. The spatial amplitude of the (1*s*) is described by

$$(1s) = \psi_{1s}(x, y, z) = (\pi\alpha^3)^{-1/2} e^{-r/\alpha} \tag{1}$$

where α is the ratio of the Bohr radius $a_0 = 0.529$ Å to the nuclear charge Z, and the factor $(\pi\alpha^3)^{-1/2}$ normalizes the wave function so that

$$\int dV(\psi_{1s})^2 = 4\pi \int_0^\infty dr\, r^2 (\pi\alpha^3)^{-1} e^{-2(r/\alpha)}$$

$$= (4/\alpha^3) \int_0^\infty dr\, r^2 e^{-2(r/\alpha)} = 1. \tag{2}$$

APPENDIX II.B

Calculating the Angles for Conal Nodes

Table 1 lists the dependences of hydrogen-like AOs on the angle θ (see Figures 2.3 and 2.6). If we wish the calculate this angle for, say, $(f1')$ or $(f1'')$, we set the expression equal to zero because we want the location where the wave function vanishes:

$$(f1')_\theta = (f1'')_\theta = \frac{\sqrt{42}}{8} \sin\theta(5\cos^2\theta - 1) = 0. \qquad (3)$$

Under these circumstances, the normalization coefficient $\sqrt{42}/8$ drops out. One of the roots is $\sin\theta = 0$, which reflects the fact that there is a trivial root for which $\theta = 0$. The remainder of Equation 3 reduces to $\cos\theta = \pm\sqrt{1/5}$, from which we find that there are two angular nodes, one at 63°26′ and the other at 180°–63°26′.

TABLE 1 Hydrogen-Like AO Functions having Angular Nodes

AO	Wave Function
(p)	$(p0)_\theta = \frac{\sqrt{6}}{2}\cos\theta$
(d)	$(d0)_\theta = \frac{\sqrt{10}}{4}(3\cos^2\theta - 1)$
	$(d1')_\theta = (d1'')_\theta = \frac{\sqrt{15}}{2}\sin\theta\cos\theta$
(f)	$(f0)_\theta = \frac{3\sqrt{14}}{4}\cos\theta(\cos^2\theta - 1)$
	$(f1')_\theta = (f1'')_\theta = \frac{\sqrt{42}}{8}\sin\theta(5\cos^2\theta - 1)$
	$(f2')_\theta = (f2'')_\theta = \frac{\sqrt{105}}{4}\sin^2\theta\cos\theta$
(g)	$(g0)_\theta = \frac{9\sqrt{2}}{16}\left(\frac{35}{3}\cos^4\theta - 10\cos^2\theta + 1\right)$
	$(g1')_\theta = (g1'')_\theta = \frac{9\sqrt{10}}{8}\sin\theta\cos\theta\left(\frac{7}{3}\cos^2\theta - 1\right)$
	$(g2')_\theta = (g2'')_\theta = \frac{3\sqrt{5}}{8}\sin^2\theta(7\cos^2\theta - 1)$
	$(g3')_\theta = (g3'')_\theta = \frac{3\sqrt{70}}{8}\sin^3\theta\cos\theta$
(h)	$(h0)_\theta = \frac{15\sqrt{22}}{16}\cos\theta\left(\cos^4\theta - \frac{14}{3}\cos^2\theta + 1\right)$
	$(h1')_\theta = (h1'')_\theta = \frac{\sqrt{165}}{16}\sin\theta(21\cos^4\theta - 14\cos^2\theta + 1)$
	$(h2')_\theta = (h2'')_\theta = \frac{\sqrt{1155}}{8}\sin^2\theta\cos\theta(3\cos^2\theta - 1)$
	$(h3')_\theta = (h3'')_\theta = \frac{\sqrt{770}}{32}\sin^3\theta(9\cos^2\theta - 1)$
	$(h4')_\theta = (h4'')_\theta = \frac{3\sqrt{385}}{16}\sin^4\theta\cos\theta$

Appendix II.C

The Mathematical Meaning of R

Here $R = \int dVr\phi^2$ the summation of the product of the distance of all points from the origin and the weighting factor $\phi^2 dV$ representing the fraction of the orbital charge in dV.

APPENDIX II.D

The Mathematical Meaning of R^*

The volume containing 0.9 electron (i.e., 90% of it) is given by:

$$\int_{V(R^*)}\int\int dV[\psi(x, y, z)]^2 = 0.9. \tag{4}$$

Here $V(R^*)$ denotes the volume inside the sphere with radius R^*.

APPENDIX II.E

The Mathematical Meaning of $v(nlm)$

Since $dV\psi^2(x, y, z)$ is proportional to the electronic charge density around the point x, y, z in the volume dV, its contribution to the potential energy is given by $(\psi^2 dV)(-Ze^2/r)$. The potential energy $v(nlm)$ of the entire orbital $\psi = (nlm)$ is therefore

$$v(nlm) = \iiint dV[\psi^2(x, y, z)]^2(-Ze^2/r) = -Ze^2/R \tag{5}$$

where R [defined by $R^{-1} = \iiint dVr^{-1}\psi^2(x, y, z)$] can be considered as another measure of atom size. This radius, which corresponds to that spherical shell which contains the largest fraction of the electron, also turns out to be the radius of the classical Bohr model for the $(1s)$ orbital.

APPENDIX II.F

Kinetic and Total Electronic Energy

Although $t(nlm)$ can be derived rigorously from quantum mechanics, it can also be obtained by considering the Bohr model of an electron as a particle circling the nucleus at a distance R. Quantum mechanics requires that the circumference of the path of such a particle in a circular box must contain an integral number of wavelengths (i.e., $\lambda = 2\pi R/n$). A free particle has only kinetic energy, which is given by $t(nlm) = h^2/2m\lambda^2$. We can therefore write Equation 6 wherein $\hbar = h/2\pi$:

$$t(nlm) = \left(\frac{\hbar^2}{2mR^2}\right)n^2. \tag{6}$$

Although we do not do so here, it can be shown that for any stable orbit in a coulombic potential the potential energy = −2(kinetic energy). This equality, known as the virial theorem, is also true for quantum mechanical energies. Therefore we can say $(1/2)v(nlm) = -t(nlm)$ and from Equation 7 we see that $\varepsilon(nlm) = (1/2)v(nlm) = -t(nlm)$. Substituting into this equation the expressions for $v(nlm)$ from Equation 5 and $t(nlm)$ from Equation 6 we have

$$\varepsilon(nlm) = \frac{-Ze^2}{2R} = -\left(\frac{\hbar^2}{2mR^2}\right)n^2 \tag{7}$$

so that $R = (\hbar^2/me^2)(n^2/Z) = a(n^2/Z)$ where a is the Bohr radius (i.e., $a = R$ when n and $Z = 1$). This means that

$$\varepsilon(nlm) = \frac{-Ze^2}{2a(n^2/Z)} = -\frac{e^2}{2a}\left(\frac{Z}{n}\right)^2 \tag{8}$$

where $-e^2/2a$ is the total energy of the $(1s)$ orbital electron in hydrogen.

Appendix II.G

Reorienting a Set of (p) Orbitals

Let us see what happens when we take linear combinations of a set of three (p) orbitals. To calculate values of the $(2px)$, $(2py)$, and $(2pz)$ orbital amplitudes at various locations (x, y, z) in space, we consider Equations 9

$$
\begin{aligned}
(2px) &= [\pi(2\alpha)^3]^{-1/2}(x/2\alpha)e^{-(r/2\alpha)} = Kxe^{-kr} \\
(2py) &= [\pi(2\alpha)^3]^{-1/2}(y/2\alpha)e^{-(r/2\alpha)} = Kye^{-kr} \\
(2pz) &= [\pi(2\alpha)^3]^{-1/2}(z/2\alpha)e^{-(r/2\alpha)} = Kze^{-kr}
\end{aligned} \tag{9}
$$

wherein the constants K and k are the same for x, y, and z and α has the same meaning as discussed earlier. So that we can conveniently take linear combinations of the $(2p)$ orbitals in Equations 9, we will examine some simple properties of vectors. Let $\mathbf{r}$ be the vector which points from the origin to the point (x, y, z) as shown in Figure 1a and let $\mathbf{e}$, with components (ξ, η, ζ), be a *unit vector* as depicted in Figure 1b. A unit vector $\mathbf{e}$ is one unit long as measured in whatever units we are working with. In Figure 1b the unit vector has been placed along the y axis. Then the *projection of* $\mathbf{r}$ *onto* $\mathbf{e}$, denoted by $P_e(\mathbf{r})$ and shown in Figure 1c is given by the so-called "dot product" $(\mathbf{e}\cdot\mathbf{r})$ which is defined in Equation 10:

$$P_e(\mathbf{r}) = (\mathbf{e}\cdot\mathbf{r}) = \xi x + \eta y + \zeta z. \tag{10}$$

Since the three orbitals in Equations 9 are each a function of a single coordinate, Equation 10 can be recast into Equations 11 which are algebraic expressions for the projections of r onto the three coordinate axes:

$$
\begin{aligned}
&P_{e_x}(\mathbf{r}) = (\mathbf{e}_x\cdot\mathbf{r}) = \xi x,\ P_{e_y}(\mathbf{r}) = (\mathbf{e}_y\cdot\mathbf{r}) = \eta y,\ P_{e_z}(\mathbf{r}) = (\mathbf{e}_z\cdot\mathbf{r}) = \zeta z \\
&\text{or } P_{e_j}(\mathbf{r}) = (\mathbf{e}_j\cdot\mathbf{r}).
\end{aligned} \tag{11}
$$

Here $\mathbf{e}_x = (1,0,0)$, $\mathbf{e}_y = (0,1,0)$, and $\mathbf{e}_z = (0,0,1)$ are the unit vectors in the x, y, and z directions. Equations 11 then permit us to rewrite Equations 9 as Equation 12:

$$(2pj) = KP_{e_j}(\mathbf{r})e^{-kr} \qquad \text{where } j = x,\ y,\ z. \tag{12}$$

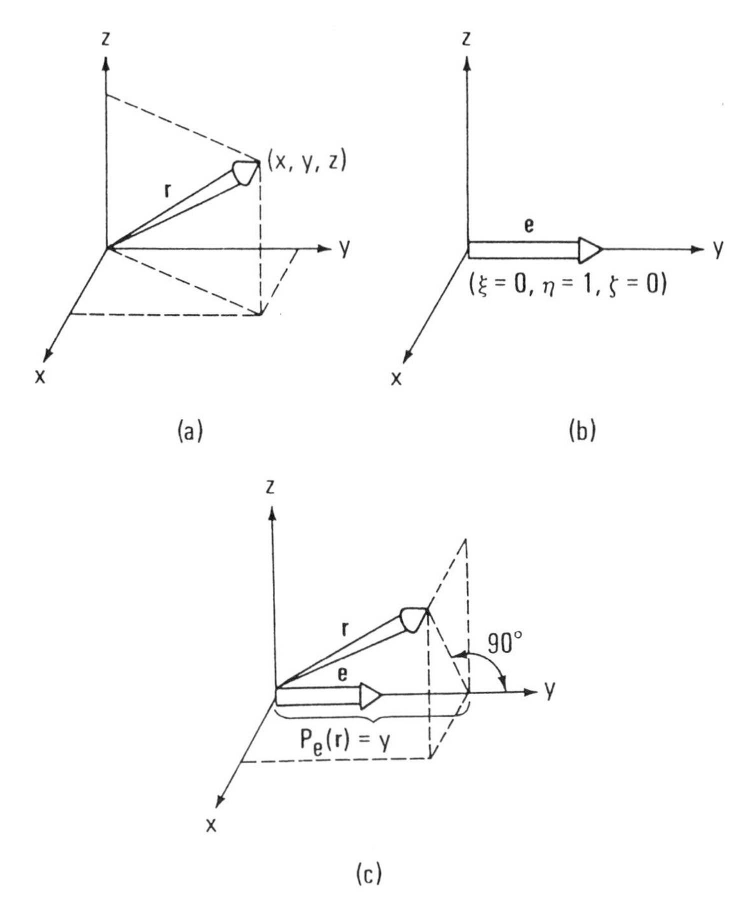

FIGURE 1 Drawings depicting a vector **r** (a), a unit vector **e** (b), and the projection of **r** onto **e** (c).

We already know, of course, that (2*px*), (2*py*), and (2*pz*) possess identical shapes and differ only in the directions along which they are oriented. Thus, for example, the (2*px*) orbital is concentric around x and, by convention, we think of it as pointed along the $+x$ direction when the positive amplitude is in the $+x$ region and the minus amplitude is in the $-x$ region. Similar arguments apply to (2*py*) and (2*pz*).

Equation 12 actually allows us to think of a (2*p*) orbital as pointed along any direction we choose, since we can formulate this equation for an arbitrary direction along **e** as in Equation 13 where $P_e(\mathbf{r})$ is the projection of $\mathbf{r} = (x, y, z)$ onto $\mathbf{e} = (\xi, \eta, \zeta)$:

$$(2p\mathbf{e}) = KP_e(\mathbf{r})e^{-kr}. \tag{13}$$

If we now insert the right-hand side of Equation 10 into Equation 13, we obtain

$$(2p\mathbf{e}) = K(\xi x + \eta y + \zeta z)e^{-kr}$$
$$= \xi K x e^{-kr} + \eta K y e^{-kr} + \zeta K z e^{-kr} \tag{14}$$

which upon substitution with Equations 9 gives

$$(2p\mathbf{e}) = \xi(2px) + \eta(2py) + \zeta(2pz). \tag{15}$$

Equation 15 says that any linear combination of the orbitals $(2px)$, $(2py)$, and $(2pz)$ (such that $\xi^2 + \eta^2 + \zeta^2 = 1$) gives again an orbital of the $(2p)$ type but pointing in the direction of the unit vector $\mathbf{e} = (\xi, \eta, \zeta)$.

With this result, we can rotate *the set of all three p orbitals*, while keeping them orthogonal. Suppose that we have three *mutually orthogonal* unit vectors $\mathbf{e}_1, \mathbf{e}_2, \mathbf{e}_3$, which are related to $\mathbf{e}_x$, $\mathbf{e}_y$, $\mathbf{e}_z$ by the orthogonal transformation matrix in Equation 16:

	$\mathbf{e}_x$	$\mathbf{e}_y$	$\mathbf{e}_z$
$\mathbf{e}_1$	T_{11}	T_{12}	T_{13}
$\mathbf{e}_2$	T_{21}	T_{22}	T_{23}
$\mathbf{e}_3$	T_{31}	T_{32}	T_{33}

(16)

wherein the T_{ik} satisfy the relations $T_{i1}T_{k1} + T_{i2}T_{k2} + T_{i3}T_{k3} = 1$ if $i = k$ and $=0$ if $i \neq k$ (Appendix I.C.). Suppose now that we transform the orbitals $(2px)$, $(2py)$, $(2pz)$ into three $(2p)$ orbitals $(2p\mathbf{e}_1)$, $(2p\mathbf{e}_2)$, $(2p\mathbf{e}_3)$ pointing in the directions $\mathbf{e}_1$, $\mathbf{e}_2$, and $\mathbf{e}_3$, respectively. We can then express this transformation by the transformation matrix in Equation 17:

	$(2px)$	$(2py)$	$(2pz)$
$(2p\mathbf{e}_1)$	T_{11}	T_{12}	T_{13}
$(2p\mathbf{e}_2)$	T_{21}	T_{22}	T_{23}
$(2p\mathbf{e}_3)$	T_{31}	T_{32}	T_{33}.

(17)

APPENDIX II.H

Linearly Combining (2*s*), (2*px*), (2*py*), and (2*pz*)

Linear combinations of a set of carbon VAOs, for example can be obtained by means of the transformations implied in the matrix:

$$\begin{array}{c|cccc} & (2s) & (2px) & (2py) & (2pz) \\ \hline (h_1) & T_{10} & T_{11} & T_{12} & T_{13} \\ (h_2) & T_{20} & T_{21} & T_{22} & T_{23} \\ (h_3) & T_{30} & T_{31} & T_{32} & T_{33} \\ (h_4) & T_{40} & T_{41} & T_{42} & T_{43}. \end{array} \qquad (18)$$

(See also Appendix I.C.) We know that all the canonical (*nlm*) orbitals are mutually orthonormal (i.e., $\langle nlm|n'l'm'\rangle = 1$ if $n = n'$, $l = l'$, $m = m'$ and $=0$ otherwise). For the (*h*) set to be orthonormal, the transformation in Equation 18 must be an orthogonal one, as discussed previously. The (*h*) set is equivalent to the canonical AO set (2*s*), (2*px*), (2*py*), (2*pz*) in that any linear combination of the latter set can be expressed as a linear combination of the four members of the (*h*) set and vice versa.

APPENDIX II.I.

The Angle between Orthogonal Digonal *sp* Hybrids

Let us first examine two (p) orbitals pointing along two arbitrary unit vector directions $\mathbf{e}_1 = (\xi_1, \eta_1, \zeta_1)$ and $\mathbf{e}_2 = (\xi_2, \eta_2, \zeta_2)$ so that

$$\begin{aligned}(2p\mathbf{e}_1) &= \xi_1(2px) + \eta_1(2py) + \zeta_1(2pz) \\ (2p\mathbf{e}_2) &= \xi_2(2px) + \eta_2(2py) + \zeta_2(2pz).\end{aligned} \tag{19}$$

For the overlap integral between these two orbitals, we find that

$$\begin{aligned}\langle(2p\mathbf{e}_1)|(2p\mathbf{e}_2)\rangle = {} & \xi_1\xi_2\langle(2px)|(2px)\rangle + \xi_1\eta_2\langle(2px)|(2py)\rangle \\ & + \xi_1\zeta_2\langle(2px)|(2pz)\rangle + \eta_1\xi_2\langle(2py)|(2px)\rangle \\ & + \eta_1\eta_2\langle(2py)|(2py)\rangle + \eta_1\zeta_2\langle(2py)|(2pz)\rangle \\ & + \zeta_1\xi_2\langle(2pz)|(2px)\rangle + \zeta_1\eta_2\langle(2pz)|(2py)\rangle \\ & + \zeta_1\zeta_2\langle(2pz)|(2pz)\rangle\end{aligned} \tag{20}$$

which upon imposition of the condition that these two AOs must be orthogonal, simplifies to Equation 21:

$$\langle(2p\mathbf{e}_1)|(2p\mathbf{e}_2)\rangle = \xi_1\xi_2 + \eta_1\eta_2 + \zeta_1\zeta_2. \tag{21}$$

The right-hand side of Equation 21 is actually the so-called dot product between two directions $\mathbf{e}_1$ and $\mathbf{e}_2$, which we can write as Equation 22:

$$\langle(2p\mathbf{e}_1)|(2p\mathbf{e}_2)\rangle = (\mathbf{e}_1 \cdot \mathbf{e}_2) = \cos\gamma_{12} \tag{22}$$

in which γ is the angle between the two directions. It follows then that if these two p AOs are to be orthogonal (i.e., for Equation 22 to be equal to zero), γ must be 90° (i.e., $\mathbf{e}_1$ and $\mathbf{e}_2$ must be orthogonal).

Now moving on to our hybrid quadruple, consider any pair [say, (h_1) and (h_2)] pointing in the directions $\mathbf{e}_1$ and $\mathbf{e}_2$. We can write normalized expressions for these hybrids as Equation 23 (wherein $0 < \alpha < 90°$, $0 < \beta < 90°$):

$$\begin{aligned}(h_1) &= \cos\alpha(2s) + \sin\alpha(2p\mathbf{e}_1) \\ (h_2) &= \cos\beta(2s) + \sin\beta(2p\mathbf{e}_2)\end{aligned} \tag{23}$$

so that all four trigonometric coefficients are positive. Consequently, the positive lobe of each hybrid AO (h_k) points in the direction $\mathbf{e}_k$. The parameters α and β determine the $(s\text{–}p)$ mixtures of (h_1) and (h_2), respectively. Proceeding now in a manner analogous to the one above for two orthogonal (p) orbitals we write the overlap integral in Equation 24:

$$\begin{aligned} \langle (h_1)|(h_2)\rangle &= \cos\alpha\cos\beta + \sin\alpha\sin\beta(\mathbf{e}_1\cdot\mathbf{e}_2) \\ &= \cos\alpha\cos\beta + \sin\alpha\sin\beta\cos\gamma_{12}. \end{aligned} \tag{24}$$

For (h_1) and (h_2) to be orthogonal, Equation 24 must be equal to zero, from which Equation 25 follows:

$$\cos\gamma_{12} = -\cot\alpha\cdot\cot\beta. \tag{25}$$

This equation says that when two $(s\text{–}p)$ hybrids are orthogonal to each other, their degrees of hybridization uniquely determine the angle γ_{12} between their directions $\mathbf{e}_1$ and $\mathbf{e}_2$. Moreover, according to Equation 25, this angle is always greater than 90° unless at least one of the hybrids has 100% (p) character ($\cos\alpha = 0$ or $\cos\beta = 0$), in which case this (p) orbital forms 90° angles with *all* the other hybrids.

How are the energies of hybrid AOs related to those of their constituent canonical AOs? Since the hybrid AOs are constructed by means of an orthogonal transformation, it can be shown that

$$\varepsilon(h_k) = T_{k0}{}^2\varepsilon(2s) + (T_{k1}{}^2 + T_{k2}{}^2 + T_{k3}{}^2)\varepsilon(2p) \tag{26}$$

where $\varepsilon(2p) = \varepsilon(2px) = \varepsilon(2py) = \varepsilon(2pz)$. Since $T_{k0}{}^2 + T_{k1}{}^2 + T_{k2}{}^2 + T_{k3}{}^2 = 1$, we have Equation 27:

$$\varepsilon(h_k) = T_{k0}{}^2\varepsilon(2s) + (1 - T_{k0}{}^2)\varepsilon(2p) \tag{27}$$

which for an orbital of the type described in Equation 2.12 in Chapter 2 can be written

$$\varepsilon(h) = A^2\varepsilon(2s) + B^2\varepsilon(2p). \tag{28}$$

Appendix II.J

Transformation Matrix for *sp* Digonal Hybrids

The expressions in Equations 2.15 require the following transformation:

$$
\begin{array}{c|cccc}
 & (2s) & (2px) & (2py) & (2pz) \\
\hline
(h_1) & 2^{-1/2} & 2^{-1/2} & 0 & 0 \\
(h_2) & 2^{-1/2} & -2^{-1/2} & 0 & 0 \\
(h_3) & 0 & 0 & 1 & 0 \\
(h_4) & 0 & 0 & 0 & 1.
\end{array}
\qquad (29)
$$

Appendix II.K

Reorienting *sp* Digonal Hybrids

The hybrid quadruple set associated with our equivalent s^1p^1 hybrids involves reorienting the quadruple set along the unit vectors $\mathbf{d}_1$, $\mathbf{d}_2$, $\mathbf{d}_3$, $\mathbf{d}_4$, which are generated pictorially in Figure 2 by linearly combining (px), (py), and (pz) AOs. From Figures 2a and b we conclude that (d_+) and (d_-) point along $\mathbf{d}_1$ and $\mathbf{d}_2$, respectively. For the pure (p) hybrids $(2p\mathbf{d}_3)$ and $(2p\mathbf{d}_4)$ to be at right angles to the new digonal hybrids (d_+) and (d_-), they must be formed as shown in Figure 2c and d. The unit vectors $\mathbf{d}_1$, $\mathbf{d}_2$, $\mathbf{d}_3$, $\mathbf{d}_4$ are seen from Figures 2a–d to have the directions $\mathbf{d}_1 = (-2^{-1/2}, 2^{-1/2}, 0)$, $\mathbf{d}_2 = (2^{-1/2}, -2^{-1/2}, 0) = -\mathbf{d}_1$, $\mathbf{d}_3 = (1/2, 1/2, 2^{-1/2})$, $\mathbf{d}_4 = (-1/2, -1/2, 2^{-1/2})$. These directions are the coefficients of the $(2p)$ orbitals, which allow our reoriented p hybrids $(2p\mathbf{d}_1)$, $(2p\mathbf{d}_2)$, $(2p\mathbf{d}_3)$, and $(2p\mathbf{d}_4)$ in Figure 2 to be normalized. Thus the sum of the squares of the coefficients is 1 in all cases. That these coefficients are consistent with the geometrical constraints of our reorientation can be seen by considering Figure 2a as an example. Here we are taking the linear combination of two vectors, each of which is one unit in length from the origin to either end. Taking the positive lobe as our reference, this unit vector has the coordinates $(-2^{-1/2}, 2^{-1/2}, 0)$ as shown in Figure 3. In Figure 4 is depicted a composite of the relative orientations of the unit vectors $\mathbf{d}_1$, $\mathbf{d}_2$, $\mathbf{d}_3$, $\mathbf{d}_4$ along which lie the reoriented (s^1p^1) hybrids (h_1), (h_2), (h_3), (h_4) shown in Equation 30:

$$\begin{aligned}(h_1) &= (d_+) = 2^{-1/2}[(2s) + (2p\mathbf{d}_1)]\\(h_2) &= (d_-) = 2^{-1/2}[(2s) + (2p\mathbf{d}_2)]\\(h_3) &= (2p\mathbf{d}_3)\\(h_4) &= (2p\mathbf{d}_4).\end{aligned} \tag{30}$$

Here the positive lobes of the (s^1p^1) digonal hybrids point away from each other in the $\mathbf{d}_1$ and $\mathbf{d}_2$ directions, and the two pure (p) orbitals lie along $\mathbf{d}_3$ and $\mathbf{d}_4$.

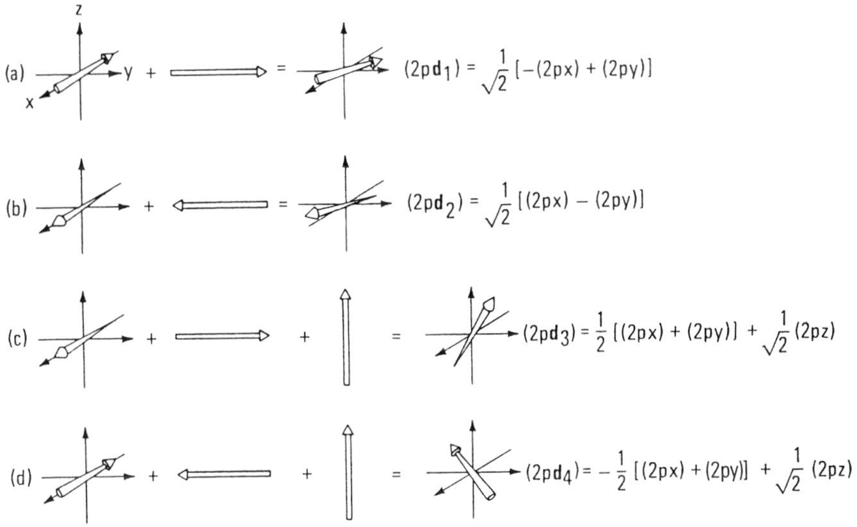

z
(a)
x
y
$(2pd_1) = \frac{1}{\sqrt{2}}[-(2px) + (2py)]$
(b)
$(2pd_2) = \frac{1}{\sqrt{2}}[(2px) - (2py)]$
(c)
$(2pd_3) = \frac{1}{2}[(2px) + (2py)] + \frac{1}{\sqrt{2}}(2pz)$
(d)
$(2pd_4) = -\frac{1}{2}[(2px) + (2py)] + \frac{1}{\sqrt{2}}(2pz)$

FIGURE 2 The orientation of the full set of $2p$ hybrids along the new unit vectors $\mathbf{d}_1$, $\mathbf{d}_2$, $\mathbf{d}_3$, $\mathbf{d}_4$.

$-\frac{1}{\sqrt{2}}$
1
y
$\frac{1}{\sqrt{2}}$
x

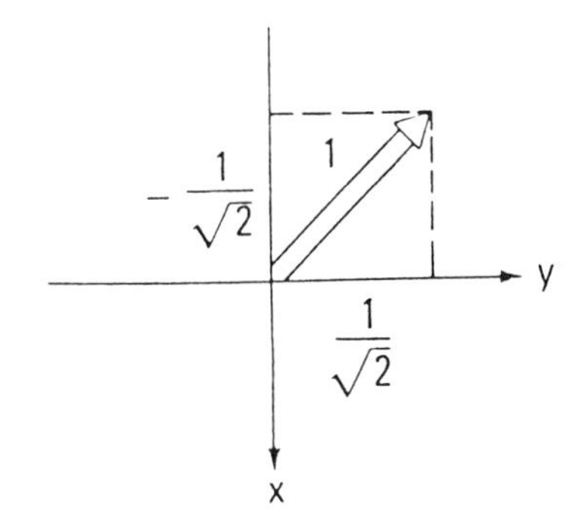

FIGURE 3 The orientation of the unit vectors $\mathbf{d}_1$ in the xy plane.

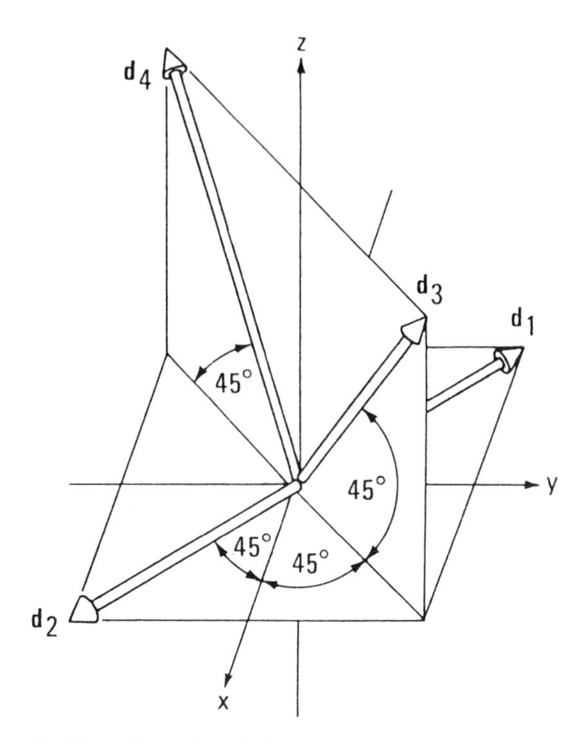

FIGURE 4 The orientation of the unit vectors $\mathbf{d}_1$, $\mathbf{d}_2$, $\mathbf{d}_3$, $\mathbf{d}_4$ in space.

APPENDIX II.L

Transformation Matrix for sp^2 Trigonal Hybrids

The orthogonal matrix for the hybrid quadruple set associated with equivalent sp^2 orbitals is:

	(2s)	(2px)	(2py)	(2pz)	Explicit Expressions
(h_1)	$\frac{1}{\sqrt{3}}$	$\sqrt{\frac{2}{3}}$	0	0	$\frac{1}{\sqrt{3}}(2s) + \sqrt{\frac{2}{3}}(2px) = (tr_1)$
(h_2)	$\frac{1}{\sqrt{3}}$	$\frac{-\sqrt{2}}{2\sqrt{3}}$	$\frac{1}{\sqrt{2}}$	0	$\frac{1}{\sqrt{3}}(2s) - \frac{1}{\sqrt{6}}(2px) + \frac{1}{\sqrt{2}}(2py) = (tr_2)$
(h_3)	$\frac{1}{\sqrt{3}}$	$\frac{-\sqrt{2}}{2\sqrt{3}}$	$\frac{1}{\sqrt{2}}$	0	$\frac{1}{\sqrt{3}}(2s) - \frac{1}{\sqrt{6}}(2px) - \frac{1}{\sqrt{2}}(2py) = (tr_3)$
(h_4)	0	0	0	1	$(2pz)$.

(31)

These hybrid AOs can also be written in more general form as in Equation 32:

$$\begin{aligned}
(h_1) &= \frac{1}{\sqrt{3}}(2s) + \sqrt{\frac{2}{3}}(2p\mathbf{f}_1) = (tr_1) \\
(h_2) &= \frac{1}{\sqrt{3}}(2s) + \sqrt{\frac{2}{3}}(2p\mathbf{f}_2) = (tr_2) \\
(h_3) &= \frac{1}{\sqrt{3}}(2s) + \sqrt{\frac{2}{3}}(2p\mathbf{f}_3) = (tr_3) \\
(h_4) &= (2pz)
\end{aligned} \tag{32}$$

wherein the $(2p\mathbf{f}_k)$ orbitals are defined in the case at hand by Equation 33:

$$\begin{aligned}
(2p\mathbf{f}_1) &= (2px) \\
(2p\mathbf{f}_2) &= -\frac{1}{2}(2px) + \frac{1}{2}\sqrt{3}(2py) \\
(2p\mathbf{f}_3) &= -\frac{1}{2}(2px) - \frac{1}{2}\sqrt{3}(2py).
\end{aligned} \tag{33}$$

The $(2p\mathbf{f}_k)$ orbitals point in the directions $\mathbf{f}_1 = (1,0,0)$, $\mathbf{f}_2 = (-1/2, 3^{1/2}/2, 0)$, $\mathbf{f}_3 = (-1/2, -3^{1/2}/2, 0)$. These vectors are of unit length so that the $(2p\mathbf{f}_k)$ are normalized. As shown in Figure 5, the unit vector directions are at 120° to one another as are the (sp^2) hybrids that point along these directions.

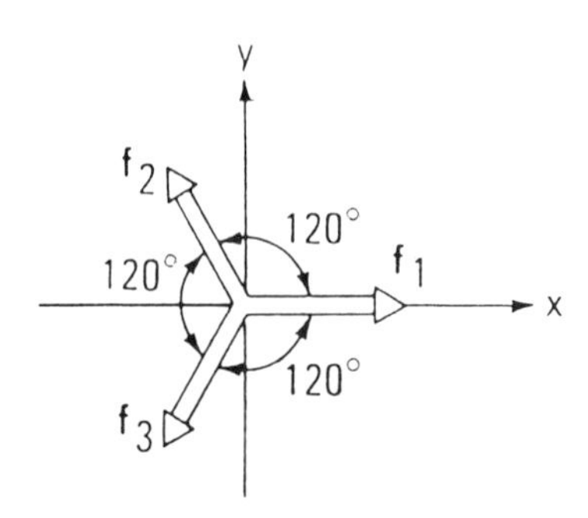

FIGURE 5 Orientation of the (sp^2) trigonal hybrid set of orbitals.

APPENDIX II.M

Transformation Matrix for sp^3 Tetrahedral Hybrids

The orthogonal matrix for the hybrid quadruple set associated with the sp^3 orbitals is:

	$(2s)$	$(2px)$	$(2py)$	$(2pz)$	Explicit Expressions
(h_1)	$\frac{1}{2}$	$\frac{1}{2}$	$\frac{1}{2}$	$\frac{1}{2}$	$\frac{1}{2}(2s) + \frac{1}{2}(2px) + \frac{1}{2}(2py) + \frac{1}{2}(2pz) = (th_1)$
(h_2)	$\frac{1}{2}$	$-\frac{1}{2}$	$-\frac{1}{2}$	$\frac{1}{2}$	$\frac{1}{2}(2s) - \frac{1}{2}(2px) - \frac{1}{2}(2py) + \frac{1}{2}(2pz) = (th_2)$
(h_3)	$\frac{1}{2}$	$\frac{1}{2}$	$-\frac{1}{2}$	$-\frac{1}{2}$	$\frac{1}{2}(2s) + \frac{1}{2}(2px) - \frac{1}{2}(2py) - \frac{1}{2}(2pz) = (th_3)$
(h_4)	$\frac{1}{2}$	$-\frac{1}{2}$	$\frac{1}{2}$	$-\frac{1}{2}$	$\frac{1}{2}(2s) - \frac{1}{2}(2px) + \frac{1}{2}(2py) - \frac{1}{2}(2pz) = (th_4)$.

(34)

These hybrids can also be expressed as Equation 35:

$$(th_k) = \tfrac{1}{2}(2s) + \tfrac{1}{2}\sqrt{3}(2p\mathbf{g}_k) \tag{35}$$

using the four $(2p)$ orbitals in Equation 36 oriented along the four unit vectors $\mathbf{g}_k = (\xi_k, \eta_k, \zeta_k)$

$$(2p\mathbf{g}_k) = \xi_k(2px) + \eta_k(2py) + \zeta_k(2pz), \tag{36}$$

which point to the corners of a regular tetrahedron as given by their coordinates in Figure 6.

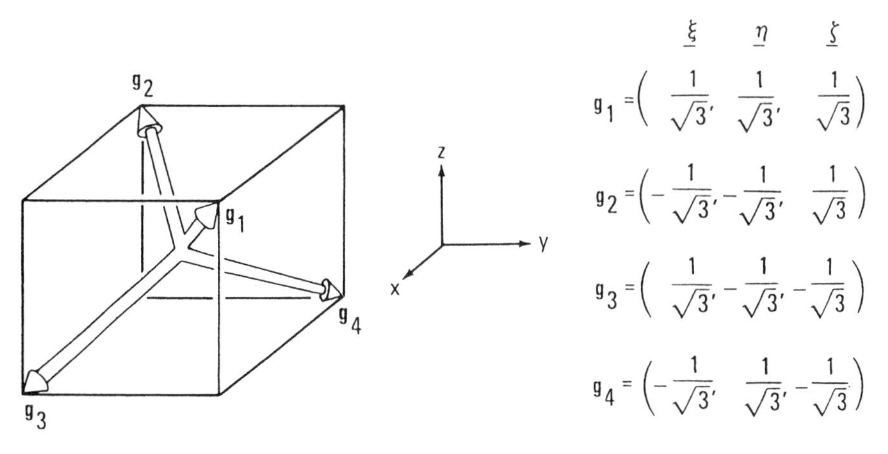

FIGURE 6 Orientation and coordinates of the four unit vectors along which lie the (sp^3) hybrid AOs (th_k).

Appendix II.N

Gradual Changes in Character between Two Orbitals

Let us consider the transition from (2*s*) plus (2*px*) to (*sp*). We can generalize the transformation of (2*s*) and (2*px*) to the digonal hybrids (h_+) and (h_-) given earlier in Equations 2.15 and the transformation matrix in Appendix II.J by substituting A and B as shown in Equations 37:

	(2*s*)	(2*px*)	(2*py*)	(2*pz*)
$(h_1) = (h_+)$	A	B	0	0
$(h_2) = (h_-)$	B	$-A$	0	0
(h_3)	0	0	1	0
(h_4)	0	0	0	1

(37)

As long as $A^2 + B^2 = 1$, this transformation is orthogonal for any values of A and B. Assuming $A > 0$ and $B > 0$, (h_+) points in the $+x$ direction and (h_-) points in the $-x$ direction, and the two hybrids always remain 180° apart. Furthermore, the (*s*) and (*p*) characters of (h_+) are the converse of those of (h_-). Thus if (h_+) has 80% (*s*) character and 20% (*p*) character, then (h_-) possesses 20% (*s*) character and 80% (*p*) character, and the shapes of these orbitals as well as their energies reflect these differences. The transformation in Equation 37 affords the possibility of a continuous transition from (2*s*) plus (2*px*) to the digonal (d_+) and $(d_-)(s^1p^1)$ set. For $A = 0$, $B = 1$ we have $(h_-) = (2s)$ and $(h_+) = (2px)$. As A increases to positive values (and B decreases concomitantly to preserve orthogonality) (h^-) gains (*p*) character and (h_+) accumulates (*s*) admixture until for $A = B = 2^{-1/2}$ both hybrids assume that same shape and energy and they become (d_+) and (d_-).

Appendix II.O

Relationship between Trigonal and Tetrahedral Hybrids

Before we consider the relationships among the various hybrid quadruple sets, we shall first briefly discuss an alternative set of hybrids $(\hat{th}_k)$ using the definition in Equation 38 in order to visualize the relationships between the tetrahedral and trigonal hybrids. In these equations the new directional unit vectors $\hat{\mathbf{g}}_k = (\xi_k, \eta_k, \zeta_k)$ are given by Equation 39:

$$
\begin{aligned}
(\hat{th}_k) &= \tfrac{1}{2}(2s) + \tfrac{1}{2}\sqrt{3}(2p\hat{\mathbf{g}}_k) \\
(2pg\hat{\mathbf{g}}_k) &= \xi(2px) + \eta(2py) + \zeta(2pz)
\end{aligned}
\tag{38}
$$

$$
\begin{aligned}
\hat{\mathbf{g}}_1 &= \tfrac{1}{3}\sqrt{8}\mathbf{f}_1 + \tfrac{1}{3}\mathbf{e}_z \\
\hat{\mathbf{g}}_2 &= \tfrac{1}{3}\sqrt{8}\mathbf{f}_2 + \tfrac{1}{3}\mathbf{e}_z \\
\hat{\mathbf{g}}_3 &= \tfrac{1}{3}\sqrt{8}\mathbf{f}_3 + \tfrac{1}{3}\mathbf{e}_z \\
\hat{\mathbf{g}}_4 &= -\mathbf{e}_z.
\end{aligned}
\tag{39}
$$

Here, the new vectors $\hat{\mathbf{g}}_k$ are expressed in terms of the three previously discussed trigonal unit vectors $\mathbf{f}_k$ in the xy plane, and the z axis unit vector $\mathbf{e}_z$. Thus the $\hat{\mathbf{g}}_k$ point in the directions shown in Figure 7 as is seen from the sole contribution of $-\mathbf{e}_z$ to $\hat{\mathbf{g}}_4$ and the positive contributions of $\mathbf{e}_z$ to the trigonal unit vectors $\mathbf{f}_k$. By inserting into Equation 39 the values of $\mathbf{f}_k$ given in Appendix II.M, we obtain the coordinates shown in Figure 7 for the $\hat{\mathbf{g}}_k$. The angle between any pair of $\hat{\mathbf{g}}_k$ is the tetrahedral angle of 109°28′ as can be demonstrated by taking the dot product between any two unit vectors:

$$\cos\gamma_{jk} = (\hat{\mathbf{g}}_j \cdot \hat{\mathbf{g}}_k) = -\tfrac{1}{3} \qquad \text{for} \quad jk = 12, 13, 14, 23, 24, 34. \tag{40}$$

For example, $(\hat{\mathbf{g}}_1, \hat{\mathbf{g}}_2) = -4/9 + 0 + 1/9 = -3/9 = -1/3$. Furthermore, it is clear from Equation 35 in Appendix II.M that these sp^3 hybrids have 25% (s) and 75% (p) character. The only difference between the $(\hat{th}_k)$ and the (th_k) hybrids examined earlier is that the two hybrid sets are oriented differently with respect to the Cartesian axes.

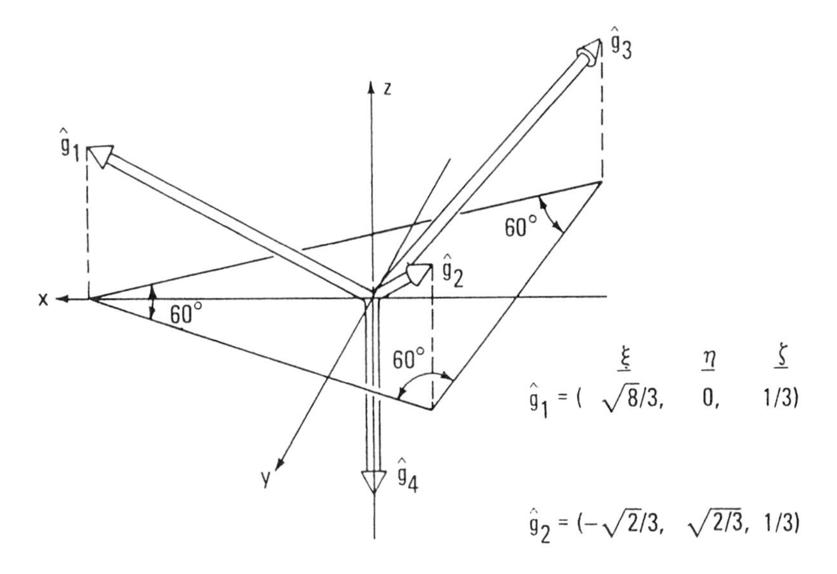

$$\hat{g}_3 = (-\sqrt{2}/3,\ -\sqrt{2/3},\ 1/3)$$

$$\hat{g}_4 = (0,\quad 0,\quad -1)$$

FIGURE 7 Orientation and coordinates of the four unit vectors along which the (sp^3) hybrid AOs $(\hat{t}\hat{h}_k)$ lie.

APPENDIX III.A

Symmetry, Molecular Orbitals, and Generator Orbitals

Molecules, such as H_2, intuitively appear to be less symmetrical than their constituent atoms because a hydrogen atom has a spherical nucleus and a spherical electron density. On the other hand a hydrogen molecule possesses two nuclei and its electron density resembles a sausage. The peculiar shapes of AOs are a consequence of the spherical symmetry of atoms. The physical scientist defines symmetry in terms of symmetry elements. These elements apply to any object, and they represent physical manipulations of, or "operations" on the object that do not allow an observer to detect that any operation occurred if he/she had been temporarily blindfolded. For example, consider the equilateral triangle (Figure 1) which has three identical vertices. One of its symmetry elements is a three-fold axis Rot_3 perpendicular to its plane. If the triangle is rotated around this axis once, twice, or three times by 120° (in either a clockwise or counterclockwise direction), a temporarily blindfolded observer would not realize what had occurred because all the points of the triangle are identical. We say that the operation associated with this symmetry element leaves the triangle "invariant". There are also three two-fold rotation axes (Figure 1b) about which rotations by 180° leave the triangle invariant. Similarly there is a reflection plane (Figure 1c) through which any point in the top half of our triangular object reflected directly across this plane to a point equidistant below the plane leaves the triangle invariant. There are also three reflection planes perpendicular to the plane of the triangle (Figure 1d). In addition, there is another symmetry element called the *identity* element whose operation leaves the object untouched and hence invariant. (There is one more symmetry element, namely an "improper axis" of rotation, which need not concern us here.)

A spherical object (such as an atom) has all of the symmetry elements of a triangle plus many more. Although a sphere (like any other object) has only one identity operation, it possesses an infinite number of n-fold rotation axes where n can be any number from one to infinity. It also contains an infinite number of reflection planes. Hence a sphere is the most symmetrical geometrical form we know of. A triangle, or any object which is not a sphere, must therefore possess lower symmetry, and this conclusion is reflected by the fact

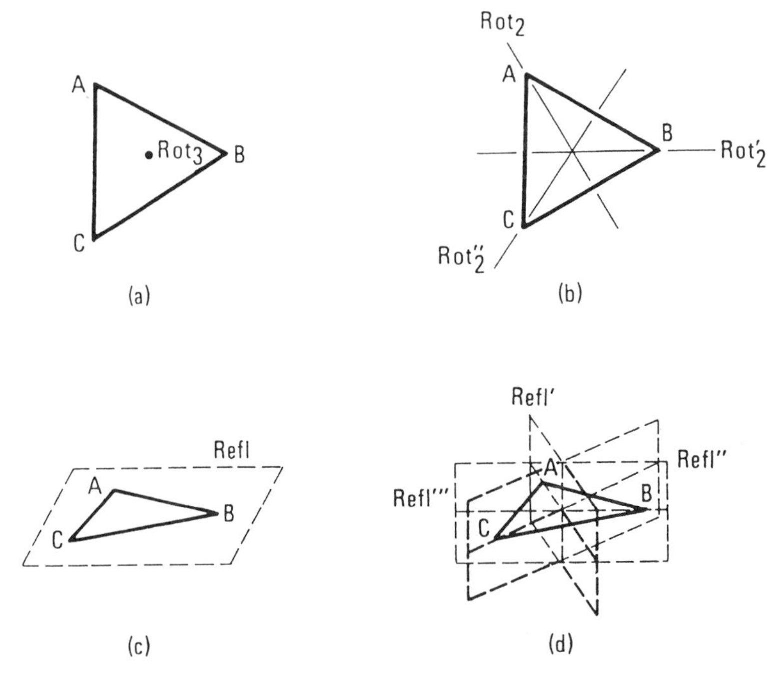

FIGURE 1 The three-fold rotation axis (a), two-fold rotation axes (b), and planes of reflection (c), (d) of an equilateral triangle having three equivalent vertices.

that the symmetry elements of nonspherical objects in every case constitute a subset of the symmetry elements of a sphere. Because of this fact, the shapes of canonical AOs (i.e., their nodal and lobal patterns) can be related to the symmetry of any molecule and hence they also can be related to the shapes of the molecule's orbitals. In other words, molecular orbitals have nodal and lobal patterns that are analogous to those of AOs.

The question is, how do we construct or generate MOs for a given molecule? We must find an appropriate method which compares molecular orbitals with AOs while taking into account the symmetry of the molecule we are considering. One such method is embodied in group theory, which is generally taught in a graduate-level course. A simpler (but nonsimplistic) method is the one developed in this book, which uses atomic orbitals as pictorial generators of molecular orbitals.

APPENDIX III.B

The Overlap Integral

In the case of $(\sigma_{1s,1s})$, for example:

$$\int dV(\sigma_{1s,1s})^2 = N_1{}^2\left[\left[\int dV(1sA)^2 + \int dV(1sB)^2 + 2\int dV(1sA)(1sB)\right]\right] \quad (1)$$

which simplifies to Equation 2 when we recall that the integral of the square of an orbital equals 1:

$$N_1{}^2\left[1 + 1 + 2\int dV(1sA)(1sB)\right] = 1. \quad (2)$$

The integral $\int dV(1sA)(1sB)$ is a measure of the overlap between (1sA) and (1sB) and is called the *overlap* integral S. Thus we have $N_1 = [2(1 + S)]^{-1/2}$ [i.e., $S = \langle\phi|\psi\rangle = \int dV\phi(x, y, z)\psi(x, y, z)$].

The overlap integral is the most widely used *quantitative* measure of the degree to which two orbitals overlap. Clearly, all nonzero contributions to S come from those volume elements in which both orbitals have substantial nonzero values, i.e., where they overlap. The largest value of S (which because of normalization equals 1) occurs when the orbitals are identical and superposed. S equals zero if ϕ and ψ are orthogonal. The overlap integral in molecules can have positive or negative values between -1 and 1, depending on whether the overlapping orbital segments are of the same or opposite sign, respectively. Overlap integrals between wave functions ϕ, ψ centered on different nuclei in molecules never achieve the maximum value, i.e., $|1|$. To do so requires superposition of atoms which in turn requires coalescence (nuclear fusion) of nuclei and electron clouds. Repulsive forces, of course, prevent nuclei from approaching each other very closely (except in nuclear fusion reactions). Why then does the equilibrium distance in molecules occur where it does? Some insight into this question can be gained by considering what happens to the energy splitting between two orbitals as the overlap changes (Figure 2). As the two orbitals approach one another, both their interaction energy and overlap integral increase but only up to a point. For reasons that need not concern us here, the energy splitting decreases after $|S|$ reaches 0.5 to 0.65.

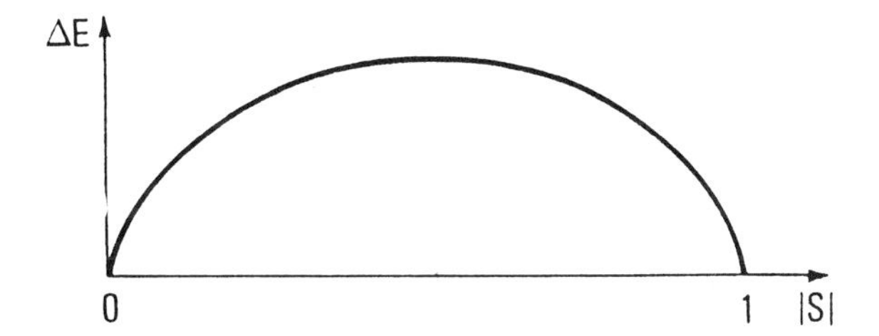

FIGURE 2 A plot of the energy splitting (ΔE) between a BMO and its partner ABMO versus the absolute value of the overlap integral ($|S|$).

APPENDIX III.C

A Kit for Visualizing MOs in Three Dimensions

Atomic orbitals and the spatial orientations they assume in symmetry orbitals in various molecular geometries can be visualized using a set of simple three-dimensional models. A kit[a] for this purpose is easily constructed from Styrofoam orbitals and a wooden support as illustrated in Figures 3 and 4, respectively. In Figure 5 is an example of how the apparatus can be used to view the px- and py-generated SOs of the (Cl3pz) VAOs in PCl_3. The kit can be made in a range of sizes. A convenient size for the base is about 6″ × 6″ × 3/4″ with orbitals constructed from 1″ and 1 1/4″ Styrofoam balls and 9″ × 3/16″ dowels. Tempera paints can be used to denote the different signs of the orbital wave functions.

[a] J. G. Verkade, K. Ruedenberg "Mechanical System for Visually Determining Permitted Electron Distributions within Molecules," U.S. Patent 3,967,388, July 6, 1976.

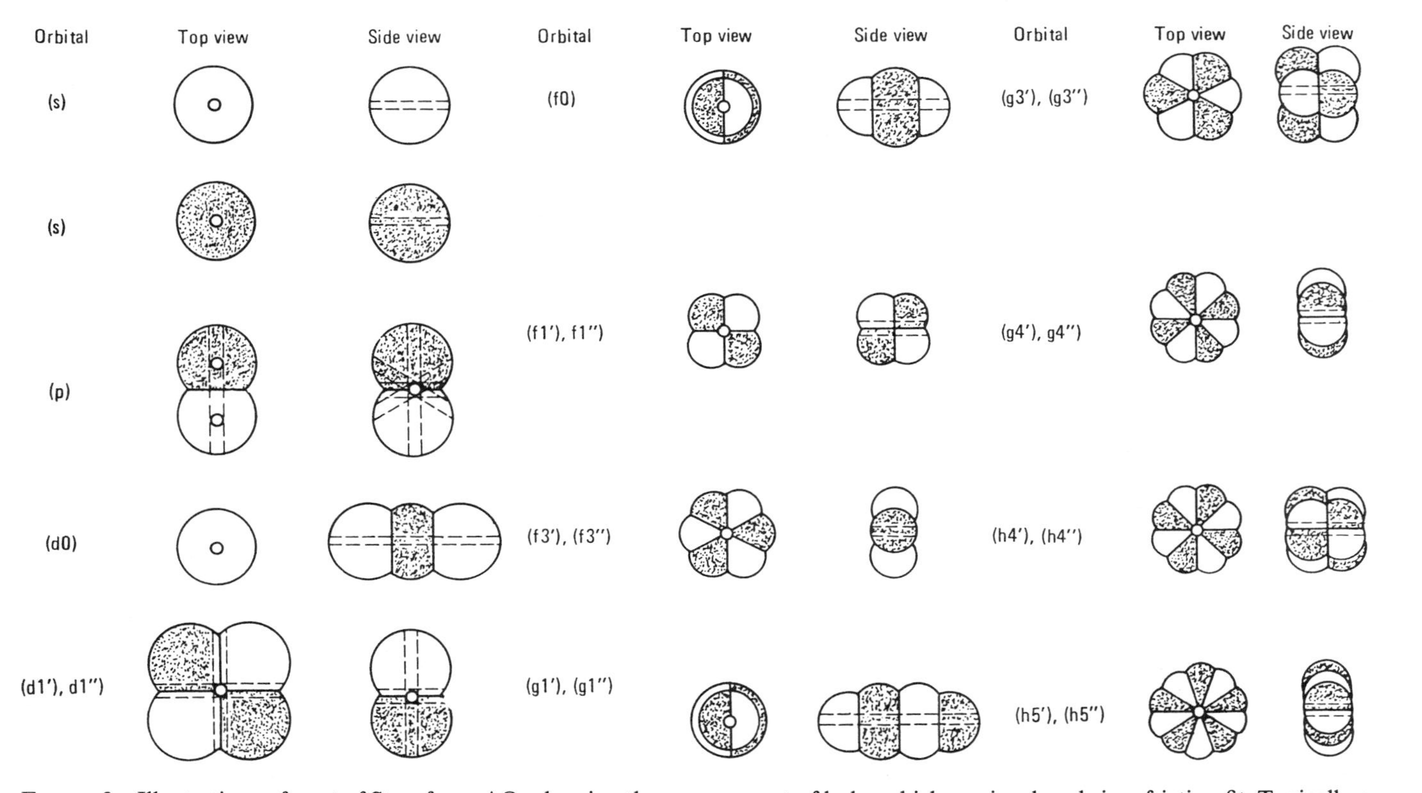

FIGURE 3 Illustrations of a set of Styrofoam AOs showing the arrangement of holes which receive dowels in a friction fit. Typically a set of AOs consisting of 30 (*s*) (20 of one sign and 10 of the other), 20 (*p*), 7 (*d*1′), 7 (*d*1″), 1 (*f*2′), 1 (*f*2″) (not shown), and 1 of each of the remaining AOs shown is useful.

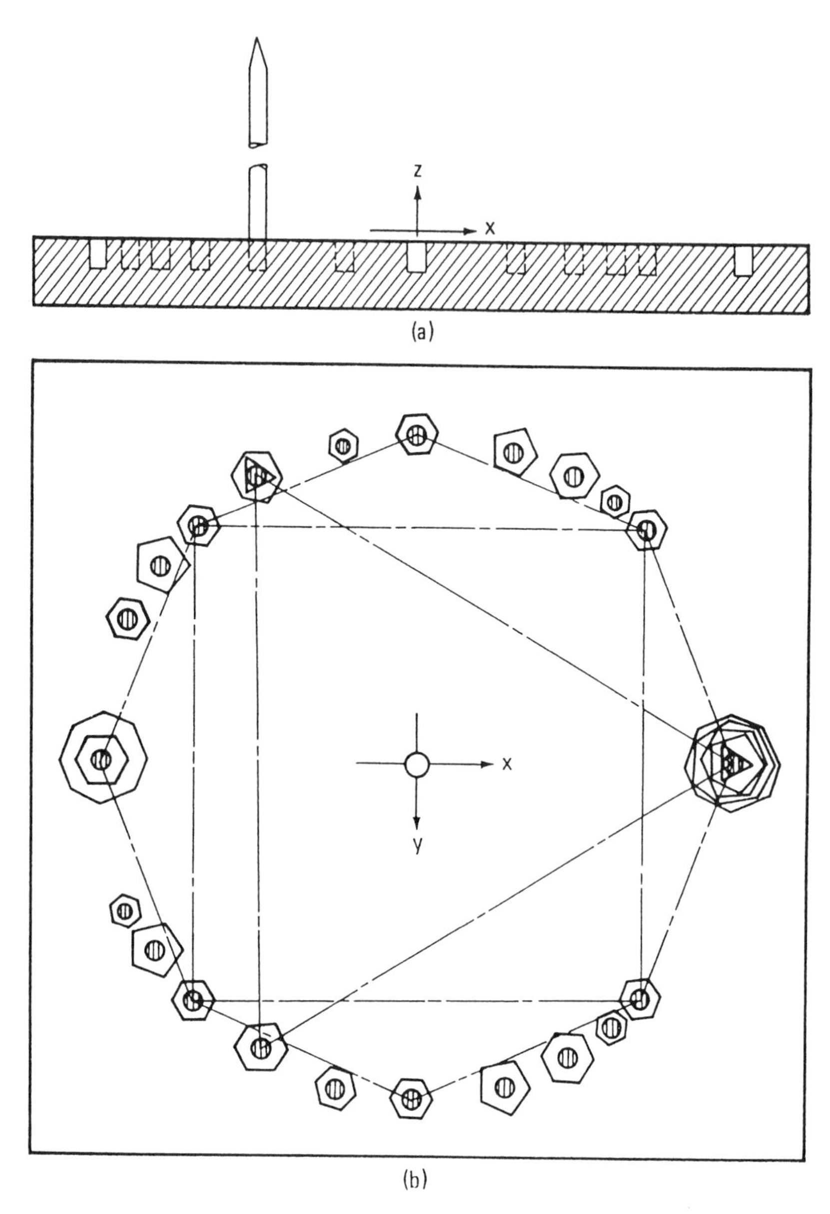

FIGURE 4 Side (a) and top (b) view of the base plate. Holes in the plate (in the top view) are located at the vertices of regular polygons and receive friction-fitting dowels, which may be pointed for easier insertion into the orbitals.

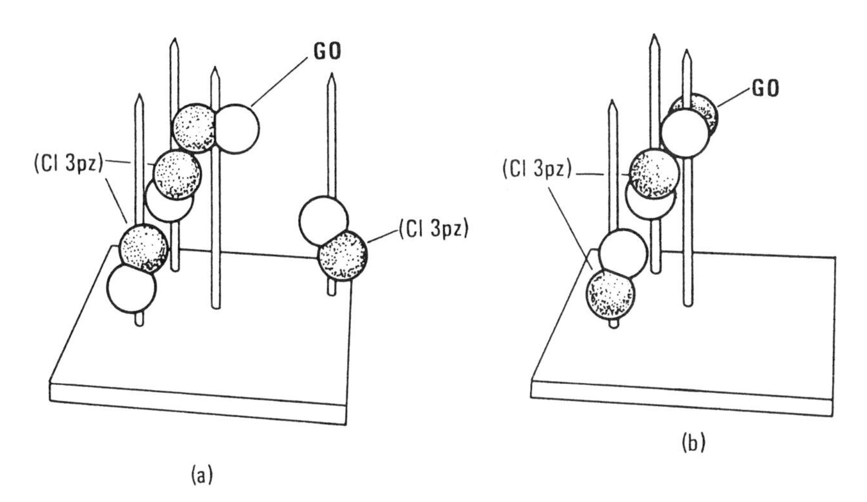

FIGURE 5 Arrangement of *px*- (a) and *py*-generated (b) SOs of the chlorine (3*pz*) VAOs in PCl_3.

APPENDIX IV

Hybrid GOs and Noncooperating Central Atoms

Here we show that there must be a small antibonding contribution to the localized MOs of FHF^- shown in Figure 4.6b. Thus we will see that in the following expressions for these MOs, the coefficient α is smaller than β and of opposite sign:

$$\hat{\sigma}_A = \alpha(F_B 2pz) + \beta(F_A 2pz) + \gamma(H1s) \tag{1}$$

$$\hat{\sigma}_B = \beta(F_B 2pz) + \alpha(F_A 2pz) + \gamma(H1s). \tag{2}$$

For normalization, we require that

$$\begin{aligned} 1 = \langle \hat{\sigma}_A | \hat{\sigma}_A \rangle = \alpha^2 + \beta^2 &+ 2\alpha\beta\langle (F_A 2pz)|(F_B 2pz)\rangle \\ &+ 2\alpha\gamma\langle (F_B 2pz)|(H1s)\rangle + \gamma^2 \\ &+ 2\beta\gamma\langle (F_A 2pz)|(H1s)\rangle, \end{aligned} \tag{3}$$

which can be simplified to

$$1 = \alpha^2 + \beta^2 + \gamma^2 + 2\gamma(\alpha + \beta)\langle (F2pz)|(H1s)\rangle, \tag{4}$$

if we assume that the overlap between $(F_A 2pz)$ and $(F_B 2pz)$ is zero and that the values of the wave functions $(F_A 2pz)$ and $(F_B 2pz)$ are the same. The expression in Equation 4 is identical for $\langle \hat{\sigma}_B | \hat{\sigma}_B \rangle$. For orthogonality, we require that

$$\begin{aligned} 0 = \langle \hat{\sigma}_A | \hat{\sigma}_B \rangle = (\alpha^2 + \beta^2)&\langle (F_A 2pz)|(F_B 2pz)\rangle + \gamma^2 + 2\alpha\beta \\ &+ \gamma(\alpha + \beta)\langle (F_A 2pz)|(H1s)\rangle \\ &+ \gamma(\alpha + \beta)\langle (F_B 2pz)|(H1s)\rangle \end{aligned} \tag{5}$$

which can be simplified to

$$0 = \gamma^2 + 2\alpha\beta + 2\gamma(\alpha + \beta)\langle (F2pz)|(H1s)\rangle, \tag{6}$$

if we use the same assumptions used in writing Equation 4. To simplify our task still further, we make use of the so-called zero neighbor overlap approximation, which says that the overlap between neighboring atoms is in fact usually much less than unity because the overlap integral is nonzero only in regions of space where *both* VAOs are appreciably nonzero. (See also Appendix III.B.) This means Equations 4 and 6 can be reduced to Equations 7 and 8, respectively, for qualitative purposes. It is reasonable to assume also that

$$1 = \alpha^2 + \beta^2 + \gamma^2, \tag{7}$$

$$0 = \gamma^2 + 2\alpha\beta, \tag{8}$$

($F_B 2pz$) and ($H1s$) in Equation 1 are fully called in, with the consequence that the coefficients of these functions in Equation 1 are the same (i.e., both VAOs are fully engaged in this MO). By substituting $\alpha = \gamma$ into Equation 8, we obtain the result that $0 = \alpha(\alpha + 2\beta)$ for which the only reasonable solution is that $-(1/2)\alpha = \beta$. By substituting $-(1/2)\alpha = \beta\ [= -(1/2)\gamma]$ into Equation 7, we find that the normalized values of these coefficients are $\alpha = 2/3^{1/2} = \gamma$ and $\beta = -3^{-1/2}$. We therefore conclude that there are indeed negative contributions of one of the ($F2pz$) VAOs in each of the MOs $\hat{\sigma}_A$ and $\hat{\sigma}_B$, which are only half as large as the contributions of the other ($F2pz$) VAO. Moreover, these different ($F2pz$) contributions are necessary to preserve orthogonality between $\hat{\sigma}_A$ and $\hat{\sigma}_B$. Thus in Equation 8, β could not possibly be zero, for then γ would have to be zero, too, and there would be *no* hydrogen ($1s$) contribution.

In FXeF (Chapter 4.2) there is a situation similar to that described above for FHF^- except that there is a xenon (p) VAO which is the only central atom VAO called in by the sp_1 and sp_r GOs. By applying analogous reasoning to expressions for the MOs depicted in Figure 4.11, it can be shown that there are again smaller contributions from one of the ($F2pz$) VAOs, but this time, these contributions are not opposite in sign to that of the larger ($F2pz$) coefficient.

Appendix V

Circular Nodes in Rectangularly Shaped Molecules

Pyrene is an example of a molecule that has a rectangular shape (i.e., its extensions in two mutually perpendicular directions are different). In such molecules, energy levels and orbitals may also be derived from the problem of a particle in a two-dimensional box.

A particle in a one-dimensional box of length L has energies and orbitals given by

$$E_n = \left(\frac{h^2}{8m}\right)\frac{n^2}{L^2} \tag{1}$$

$$\phi_n(x) = A \sin\frac{2\pi x}{\lambda_n} \tag{2}$$

In two dimensions, a box with lengths L_1 and L_2 generates energies and wave functions with two quantum numbers n_1 and n_2:

$$E_{n_1 n_2} = \left(\frac{h^2}{8m}\right)\left[\frac{{n_1}^2}{{L_1}^2} + \frac{{n_2}^2}{{L_2}^2}\right] \tag{3}$$

$$\phi_{n_1 n_2}(x, y) = A \sin\frac{2\pi x}{\lambda n_1} \sin\frac{2\pi y}{\lambda n_2} \tag{4}$$

In these two boxes, n and $n_1 + n_2$ are the node counters. Consider $n_1 = 2$ and $n_2 = 1$. On the left side of Figure 1, we show the box with two nodes perpendicular to L_1 and one node perpendicular to L_2. The signs in the box correspond to the sign of $\phi_{n_1 n_2}(x, y)$. On the right, we illustrate the correspondence between the particle-in-a-box picture and the use of planar and "circular" nodes in the rectangular box. The circular nodes are better represented as ellipses in these cases. As further examples, consider Figures 2 and 3.

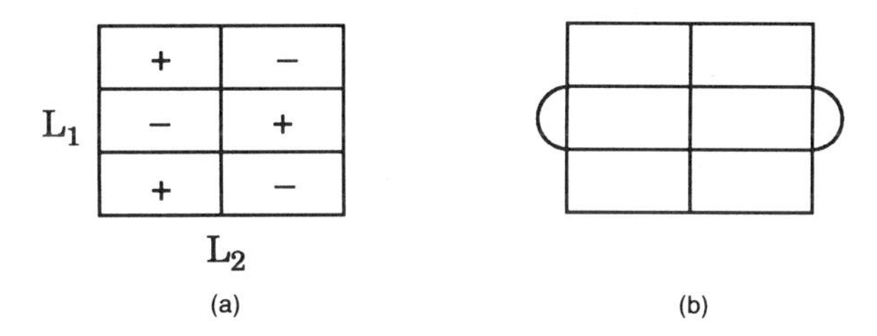

FIGURE 1 A box in which $n_1 = 2$ and $n_2 = 1$ (a) and its representation by a "3*py*" GO (having a planar node, i.e., the vertical line) and a "circular" node (the "ellipse") (b).

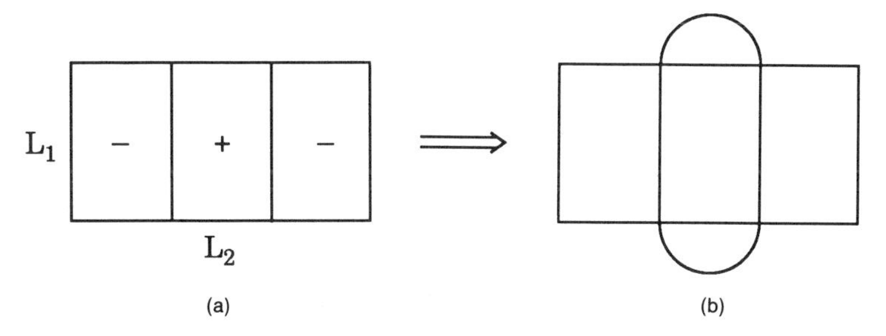

FIGURE 2 A box in which $n_1 = 0$ and $n_2 = 2$ (a) and its representation by a "2*s*" GO (having no planar nodes and one "circular" node) (b).

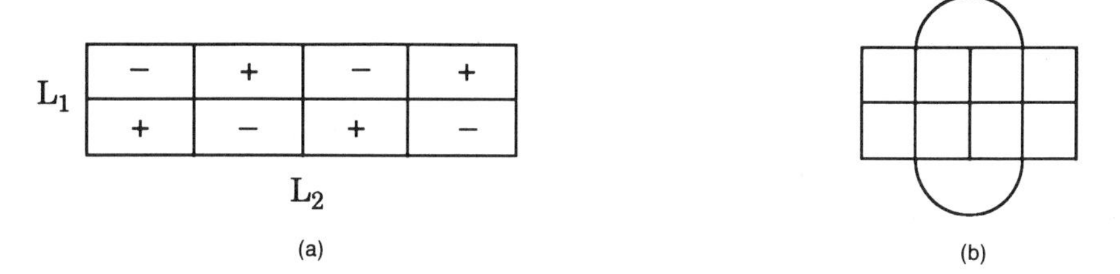

FIGURE 3 A box in which $n_1 = 1$ and $n_2 = 3$ (a) and its representation by a "4*dxy*" GO (having 2 planar nodes and one "circular" node) (b).

Appendix VI

Euler's Theorem and Cages

Euler's relation between the number of vertices (V), edges (E), and faces (F) in Equation 13.1, is the basis for the conclusions that in polyhedra in which each vertex is three-connected and whose faces are made up of only pentagons and hexagons:

1. The number of pentagons is exactly 12.
2. The number of hexagons is exactly $V/2 - 10$.
3. The number of vertices must be greater than or equal to 20.

We can derive these conclusions by determining the number of edges, E. With respect to vertices, three edges meet at each vertex, and each edge links two vertices. Therefore,

$$E = \frac{3V}{2} \qquad (1)$$

With respect to faces, let f_5 = the number of pentagons, and let f_6 = the number of hexagons. The total number of faces is

$$F = f_5 + f_6 \qquad (2)$$

Now, five edges surround each pentagon, six edges surround each hexagon, and each edge "bridges" two faces. Therefore,

$$E = \frac{5f_5 + 6f_6}{2} \qquad (3)$$

Equations 1 and 3 give the equality

$$V = \frac{5f_5 + 6f_6}{3} \qquad (4)$$

Therefore, we have obtained V, E, and F in terms of only f_5 and f_6. Inserting these three expressions into Equation 1 gives

$$(5f_5 + 6f_6)/3 - (5f_5 + 6f_6)/2 + (f_5 + f_6) = 2 \qquad (5)$$

or

$$f_f = 12 \tag{6}$$

In a C_N cage, $V = N$. Using Equations 4 and 6

$$N = \frac{60 + 6f_6}{3} \tag{7}$$

or

$$f_6 = \frac{N}{2} - 10 \tag{8}$$

Since $f_6 \geq 0$, then $N \geq 20$. Therefore, we have accounted for the rules stated in Chapter 13 for cages of this type.

APPENDIX VII

Hückel Calculations

In a Hückel calculation, we use an orthonormal AO basis set, $\{\chi_i\}$ in which $\langle\chi_i|\chi_j\rangle = \int \chi_i\chi_j \mathrm{dV} = \delta_{ij}$ ($\delta_{ij} = 1, i = j$ and $\delta_{ij} = 0$, $i \neq j$), to construct the molecular orbitals via a linear combination. Thus $\psi_n = \sum_i \mathrm{C}_{ni}\chi_i$. A Hamiltonian matrix is constructed with matrix elements $H_{ij} = \langle\chi_i|H|\chi_j\rangle$ in which $H_{ii} = \alpha = \phi$ (homonuclear structures), $H_{ij} = \beta$ for nearest neighbor contacts, and $H_{ij} = 0$ for all other (longer) contacts. The Hamiltonian matrix is diagonalized according to the equation

$$\det|H_{ij} - E_n\delta_{ij}| = 0 \tag{1}$$

which gives the set of energies $\{E_n\}$, and coefficients in the LCAO expansion $\{\mathrm{C}_{ni}\}$. The MOs $\{\psi_n\}$ have two important orthogonality characteristics:

1. $\langle\psi_m|\psi_n\rangle = \delta_{nm}$: $\{\psi_n\}$ is an orthonormal set of functions.
2. $\langle\psi_m|H|\psi_n\rangle = \mathrm{E}_n\delta_{mn}$: the Hamiltonian matrix is diagonal with these matrix elements being the MO energies $\{E_n\}$.

In the GO construction, the object is to create a set of orbitals that reproduces the nodal characteristics of the MOs. For cage molecules, the geometry is nearly as symmetrical as a sphere. Therefore, we can use the set of spherical harmonic functions, i.e., 1*s*, 2*p*, 3*d*, 4*f*, 5*g*, etc., with respect to a coordinate system whose origin is at the center of the molecule. Since each spherical harmonic function is a function of the Cartesian coordinates [i.e., $Y_l^m(x, y, z)$] we assign the coefficients C_{ni} as $Y_l^m(x_i, y_i, z_i)$ where n is the GO label lm (1*s*: $l = 0, m = 0$; 2*p*: $l - 1, m = -1, 0$, and $+1$; 3*d*: $l = 2, m = -2, -1, 0, +1, +2$; etc.) and (x_i, y_i, z_i) are the coordinates of atom i in the cage. Then, each GO is represented as

$$\mathrm{GO}(n) = \sum_i \mathrm{C}_{ni}\chi_i \tag{2}$$

Although the construction is useful, there are no restrictions on the expressions $\langle\mathrm{GO}(m)|\mathrm{GO}(n)\rangle$ and $\langle\mathrm{GO}(m)|H|\mathrm{GO}(n)\rangle$. There are techniques that allow us to orthogonalize these GOs and to assign the Hückel energy levels to each GO. These methods will be reported in a paper to be published by the authors.

Index

C

D

E

T

U

V

W

X